Add

W. K. KELLOGG
BIOLOGICAL STATION

D0881219

W. K. KELLOGG
BIOLOGICAL STATION

studies in PALEOBOTANY

studies in

John Wiley & Sons, Inc., New York and London

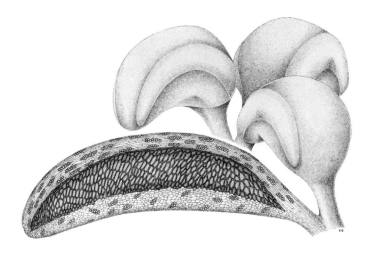

PALEOBOTANY

Henry N. Andrews, Jr.

Professor of Botany,
Washington University, St. Louis

Paleobotanist, Missouri Botanical Garden

with a chapter on palynology by
Charles J. Felix, *Sun Oil Company*

W. K. KELLOGG
BIOLOGICAL STATION

COPYRIGHT © 1961 by JOHN WILEY & SONS, Inc.

*All rights reserved. This book or any part
thereof must not be reproduced in any form
without the written permission of the publisher.*

Library of Congress Catalog Card Number: 61-6768

Printed in the United States of America

B 25698
8-9-63
Gull Lake

In appreciation of the counsel
of three great botanists

HUGH HAMSHAW THOMAS

RAY ETHAN TORREY

ROBERT E. WOODSON, Jr.

PREFACE

In preparing this account of the vegetation of the past, advice of several colleagues was sought as to the most suitable content for an introductory textbook. I was advised to write about the things that interest me and to present a story of fossil plants that would be suitable for advanced botany students as well as for geology classes in which the students have little botanical background. These last two demands are not easy to reconcile, at least in a comparatively small book. A certain amount of material has been incorporated concerning living plants which may be helpful for the nonbotanist, but it must be regarded as introductory. The serious student, whether of geology or botany, will find it necessary to refer occasionally to textbooks dealing more fully with corresponding living plant groups.

Real progress in any field of learning depends on a clear understanding of what is known as well as what is not known; there are several pertinent remarks that may be made in this connection. Perhaps the greatest contribution that paleobotany has made is not in filling gaps in our knowledge of the evolution of the plant kingdom but in showing us how many gaps exist. The fossil record reveals the former existence of major groups, such as the pteridosperms, that thrived for a time in the past and then became extinct; fragmentary remains of many strange plants that are not satisfactorily decipherable suggest unique groups about which we may one day know more. There are many fossil plants in this category of the problematical, and they are important in elucidating the complexity of plant evolution in a way that the extant flora cannot do. This leads to the matter of classification, both in general and in dealing with the problematical fossils. My procedure has been based on the belief that a system of classification should serve to create some semblance of order which we must have in dealing with any accumulation of facts and that it should demonstrate natural relationships where they are evident or probable. However, the introduction of an elaborate classification in an introductory text-book has the habit of establishing itself as fact in the students' minds when actually it may be little more than a working hypothesis. There is a considerable ferment under way at present concerning the classification of vascular (woody) plants, and this is only the beginning. Much of this has resulted from the discovery of new types of plants—strange primitive ones that were among the first inhabitants of the land, many distinctive fossils that are early representatives of pteridophytic lines, unique seed plants, and so on. It is my feeling that the cause of plant evolutionary studies will be more encumbered than aided by immediately confining such plants in a rigid taxonomic system. Thus in several chapters I have reserved a section for especially problematical but highly interesting plants; the entire Chapter 12 is so devoted. Finally, in the matter of classification, the outlook here is a polyphyletic one; the reasons for this are explained in the chapters dealing with the various pteridophytic and gymnospermous groups.

A few comments may be offered to justify the order in which the subject material is presented, although additional notations will be found at appropriate points in the text. There is now some evidence for tentatively accepting an evolutionary sequence beginning with the psilophytes and following through the coenopterid ferns, pteridosperms, and angiosperms. There is good reason to indicate that certain pteridophytic lines, such as the lycopods and articulates, do not fit into this general trend, and some gymnospermous groups quite certainly could not have led to the flowering plants. Thus I have exploited this knowledge in Chapters 2 to 7. It must be remembered, however, that we are dealing here with a very broad stream of evolution, and every effort has been made to avoid a dogmatic presentation. If the student or teacher prefers to deal first with all of the pteridophytic groups, and then the gymnospermous ones, in a classical approach, I believe that it can be done with little inconvenience.

Some aspects of paleobotany are most effectively handled by a floristic or geographical approach. For example, there is a special fascination about polar floras; here are violent contrasts that make a strong appeal to human nature, and although fossil plants from the Arctic zone are brought into many chapters it seemed to me of sufficient interest to compile a separate account on the topic. In

Chapter 16 I have tried to review some of the more important late Paleozoic and early Mesozoic floras of the Asiatic and southern regions. Research in this general paleobotanical province is going ahead very rapidly, and I am sure that textbooks of the future will devote increasingly greater attention to it. Having had but little opportunity to study these floras at first hand, I am aware of the inadequacy of their treatment here.

There is some divergence of opinion among paleobotanists as to whether a general textbook of this sort should attempt to encompass the rapidly expanding field of fossil pollen and spore studies. It is clearly a branch of paleontology in its own right, and the chapter devoted to it is intended only as a brief introduction. Since so many paleobotanists are now engaged in palynological studies exclusively or incidentally, and the field will attract many students in the future, it seemed essential to include this separate chapter, although numerous references to pollens and spores will be found throughout the book. I am grateful to Dr. Charles J. Felix for his willingness to prepare this account.

Finally, after some hesitation, I decided to include an outline of certain of the more useful paleobotanical techniques. Great advances have been made in recent decades in paleobotany as a result of improved methods of extracting information from fossils. It is my hope that this last chapter may serve to encourage younger workers to add to our still scanty repertoire of study methods.

It is not possible to acknowledge adequately the many sources of aid that have, over a period of about two decades, contributed to the development of this book. Information has been drawn from many places, most important of which is the literature created by hundreds of paleobotanists over the last century and a half. Thus reference citations given at the end of each chapter are intended as both a source of more detailed information and an acknowledgment to the original investigator. Thanks are due especially to the many individuals, institutions, and societies for permission to reproduce illustrations and quotations from their publications.

Certain individuals have been most generous in the matter of supplying illustrations and assisting me with the study of plant collections under their charge: Roland W. Brown, Sergius H. Mamay, and James M. Schopf of the United States Geological Survey; Robert W. Baxter, University of Kansas; Ralph W. Chaney, University of California; Rudolf Florin, Hortus Bergianus, Stockholm; Ove Arbo Höeg, The University, Oslo; Suzanne Leclercq, University of Liège; Maria F. Neuburg, Geological Institute, Moscow; Olof H. Selling, Swedish Natural History Museum; Wilson N. Stewart, University of Illinois; François Stockmans, Royal Natural History Museum of Belgium; F. M. Wonnacott, British Museum of Natural History. I am also grateful to Mrs. Ellen Kern Lissant for her assistance with a considerable portion of the drawings and to my student Mr. Tom Phillips for his aid in many ways.

My thanks are also due the John Simon Guggenheim Memorial Foundation and the National Science Foundation. Although grants-in-aid that I have received from these institutions were not made for the preparation of this book, it has been possible in the course of research projects to gather much useful information that could not have been obtained otherwise.

HENRY N. ANDREWS, JR.

St. Louis, Missouri
December 1960

CONTENTS

1

INTRODUCTION
some basic principles

Paleobotany is the story of the preserved vestiges of the plant life of the past and of the men and women who have interpreted the remains that we call fossil plants. If the students of these fossil plants are given a somewhat more prominent status than is usual in a textbook, it is because some acquaintance with them seems particularly essential to an understanding of the subject material. When studying living plants it is often possible, when an obstacle is encountered, to repeat an experiment, to gather more specimens, or to refer to an herbarium collection for additional information. In paleobotanical studies the approach is rather different; there have never been large numbers of investigators, the preserved plants are usually fragmentary, collections are scattered, and all too often they have been badly curated, lost, or destroyed. One is, therefore, frequently forced to depend on published reports and to place considerable weight on the general calibre of the worker. It is not implied that the problems of the paleobotanist are greater than in the various categories of living-plant botany, but they are certainly unusual.

The paleobotanist deals with plants that have lived on the earth during the past half billion years. This is a rather arbitrary figure since we know that plant life has existed for a much longer period, but the record prior to about that point in time is extremely meager. The approach to a study of these plants, as presented here, is primarily a botanical one; that is, we shall be concerned with the kinds of plants that have existed in the past for their own sake and for what they can tell us about the origins of modern floras. However, like any other branch of science the worker in paleobotany cannot function as an isolationist. A considerable knowledge of living plants is necessary to interpret correctly those of past ages and an

1

understanding of the geological surroundings in which the fossils occur is essential. Since many of the students using this book will not be equally well grounded in botany and geology, certain elementary principles will occasionally be brought in which some readers may skip over.

If this chapter takes on a somewhat miscellaneous aspect, it is due to an effort to answer certain basic questions that inevitably arise in the initial phases of any study of plants and animals of the past. Among these are the measurement of geologic time, the conditions under which plants may be fossilized, and the ways in which we find them preserved today; or, in other words, just what is a fossil and how is its age determined. More purely botanical questions center around a correct interpretation of the structures that we can observe, the classification of living and fossil plants, and apparent evolutionary trends. The introductory topics that have been selected are basic to all the chapters that follow and will be enlarged upon as occasion demands.

It is not possible in a study of this sort to acknowledge adequately the many sources from which information has been drawn without constantly inserting reference citations. At the end of each chapter there is a list of some of the more important sources, and the bibliographies which in turn may be found in these articles and books will lead the interested student to as detailed a study as is possible from the literature alone.

The Geologic Time Table

Rather early in the present century Bertram Boltwood, an American radiochemist working in Lord Rutherford's laboratory at Cambridge, suggested that the newly acquired knowledge that the radioactive elements uranium and thorium decay into helium and lead might be used to measure the ages of minerals containing such elements. Since then tremendous strides have been made in increasing the accuracy of our methods for measuring geologic time. A concise and useful summary of the whole field may be found in Bowen's recent book *The Exploration of Time*. This discussion is necessarily limited to aspects of time that apply most directly to the study of fossil plants.

Radioisotopes which have proven to be particularly useful in geologic dating include uranium-238, uranium-235, thorium-232, potassium-40, and rubidium-87. We may confine our attention for the moment to uranium; the various isotopes of this element have correspondingly different half-lives, only the longer ones being useful in geologic dating. In a rock specimen containing uranium-238 and uranium-235 the isotopes of lead (206 and 207) are found which are the respective end products of decay. The disintegration rates of these isotopes of uranium are known; for example, the half-life of uranium-235 is 710 million years.

According to the known rate of decay a uranium mineral will contain specific quantities of the decay products, lead and helium. Thus, in a uranium sample one billion years old, for every 1000 atoms of uranium-238 there should be 166 atoms of lead-206 which is the end product of uranium-238, 11.9 atoms of lead-207 which is the end product of uranium-235, and 1410 atoms of helium. There are, then, four ways in which the age of such a rock can be calculated: the ratio of uranium-235 to lead-207, the ratio of uranium-238 to lead-206, the ratio of the two leads, and the ratio of helium to uranium. Of these, the ratio of the two leads is subject to the least error. There are complicating factors in the technique such as the possible presence of ordinary lead, that is, lead that was present as such from the start, and the fact that helium tends to leak away.

It may be of interest to cite several specific dates and compare them with the ages given on the geologic time chart shown on page 6. A Carboniferous magnetite from the Urals has been dated as 260 million years, a Cretaceous magnetite from British Columbia as 100 m.y. and a Miocene specimen from Utah as 30 m.y. A sufficient number of such datings have now been made so that the ages shown on the chart are reasonably reliable, but taking the world as a whole we have only a brief framework of such determinations. The method is unfortunately very precarious with sedimentary rocks where one cannot be certain of the origin of the radioactive material, if indeed it is found at all.

Developments in radioactive dating techniques have been so rapid in recent decades as to leave little doubt that they will continue to become more accurate and that they can be applied ultimately to a greater number of geological horizons than is presently possible. One result of these studies in very recent years suggests an appreciably greater age for the earth than was supposed a few decades ago. Based in part on analyses of ancient lead from the Rosetta Mine in South Africa and lead found in meteorites a figure of 4.5 billion years is now suggested for the age of the earth. There is evidence, as we shall see later, that plant life has existed for close to half this period.

In view of its widespread use with plant materials it seems appropriate to include a few notes on the time-measuring technique employing carbon-14 that has been developed by W. F. Libby. This is based on our knowledge that cosmic ray impacts with nitrogen atoms (most abundant at the 40,000 foot level in the atmosphere) result in the transmutation of nitrogen atoms into a radioactive isotope of carbon with a mass of 14, or as it is commonly referred to, "carbon-14." When such carbon atoms are combined with oxygen to form CO_2 and reach the earth, they may, through photosynthesis, be incorporated into plant materials. This radioactive carbon-14 has a half-life of 5568 ± 30 years, which means that the quantity present in any particular plant material is reduced to half in that number of years. At the time of the death of the plant the acquisition of the radioactive carbon ceases; thus with a given sample of ancient wood or other plant materials the amount of carbon-14 remaining can be determined and its age calculated. In view of the rather short half-life the use of this method is limited to specimens of not more than 50,000 years old. It is, therefore, of little significance to us here since we shall be dealing chiefly with scores or hundreds of millions of years.

It may be pertinent at this point to consider just what we do mean by a *fossil* plant in terms of time, that is, how old do plant remains have to be to fall in this category. One can of course set an arbitrary point in time, but since time and plant evolution are continuous phenomena it would have no real meaning. Looking to the Latin origin of the word *fossil* we find that it refers to anything dug out of the earth and may be applied to organic as well as inorganic remains, although modern usage has come to eliminate the latter. Perhaps the most reasonable basis for a definition, if one feels more comfortable in establishing one, is the use to which we put vegetable debris that is found buried in the earth. For example, plants of very recent origin, not in excess of five or six thousand years old, have been studied largely by archaeologists and ethnologists and the center of interest here has focused around the use of these plants by ancient societies of men. As another example, the study of plant remains, particularly pollen but not exclusively so, from Pleistocene deposits has been followed for many years by certain botanists and geologists with interests in glacial geology and former plant distribution patterns. They deal with plant species that are, for the most part, still living and the ages involved range from several thousand to perhaps several hundred thousand years. Such plant remains may or may not be referred to as "fossils," but there is obviously little to be gained from arguing the matter. Prior to the Pleistocene, plant remains are without question called fossils, perhaps because of the increased time span, although it is somewhat arbitrary and not based on any precise number of years.

A few comments are next in order on the geologic time chart shown on page 6. Fossil plants and animals now play so important a role in many areas of natural science that almost any brand of biologist must have in mind the correct sequence of the geologic eras and periods. For the beginning student it would seem unnecessary to commit these to memory immediately, but rather refer to the chart as the names appear in the text until the sequence is mastered.

A few of the problems encountered in fixing any given fossil plant or animal in this time chart may be considered. If one notes for example that the Triassic period covered a span of some 30 million years it might be suspected that this would include many rock formations varying in thickness and physical qualities, and scattered very irregularly over the surface of the earth. A great many names have been applied to the numerous rock units which make up a great system like the Triassic and it is one of the important tasks of the geologist to establish the correct stratigraphic sequence of these units in different parts of the world, or, in other words, to *correlate* Triassic rocks of, let us say, North America and Africa. As occasion arises it will be necessary to elaborate on this time chart and consider categories below that of the Period. It may be worth noting that the Geological Society of America has, over a period of some few years, published in its Bulletin numerous special studies dealing with the correlation of rock formations in North America. Many of these are available in reprint form.

Since we shall devote a great deal of time to plants of the Carboniferous system it may be well to note that the Pennsylvanian period is regarded as synonymous with Upper Carboniferous, and Mississippian as synonymous with Lower Carboniferous. The *Carboniferous* terms are generally used by European geologists, whereas *Pennsylvanian* and *Mississippian* are used in this country. Stratigraphically the two do not coincide exactly, but for much of the discussion here they will be considered as essentially identical.

The figures given in millions of years must of course be taken with some reservation. Since it is the primary task of the paleobotanist to aid in bettering our understanding of evolutionary patterns in the

Major Stratigraphic and Time Divisions in Use by the U. S. Geological Survey

Era	System or Period		Series or Epoch	Estimated Ages of Time Boundaries in Millions of Years
Cenozoic	Quaternary		Recent	
			Pleistocene	1 ———
	Tertiary		Pliocene	10 ———
			Miocene	25 ———
			Oligocene	40 ———
			Eocene	60 ———
			Paleocene	
Mesozoic	Cretaceous		Upper (Late) Lower (Early)	125 ———
	Jurassic		Upper (Late) Middle (Middle) Lower (Early)	150 ———
	Triassic		Upper (Late) Middle (Middle) Lower (Early)	180 ———
Paleozoic	Permian			205 ———
	Carboniferous systems	Pennsylvanian	Upper (Late) Middle (Middle) Lower (Early)	
		Mississippian	Upper (Late) Lower (Early)	255 ———
	Devonian		Upper (Late) Middle (Middle) Lower (Early)	315 ———
	Silurian		Upper (Late) Middle (Middle) Lower (Early)	350 ———
	Ordovician		Upper (Late) Middle (Middle) Lower (Early)	430 ———
	Cambrian		Upper (Late) Middle (Middle) Lower (Early)	510 ———
	Precambrian		Informal subdivisions such as upper, middle, and lower, or upper and lower, or younger and older may be used locally.	→ 3,000 ———

plant kingdom, it is more essential that the precise stratigraphic position of a fossil, or collection of fossils, is known, rather than the exact age in years.

A botanist cannot help noting the similarity of his problems with those of the geologist. The latter is confronted with thousands of rock units composing the crust of the earth and deals with problems such as the limitations of these units in a restricted area and their correlation with sequences of presumed similar age in distant regions. The botanist attempts to understand the boundaries of his major biological unit, the species, and interpret the complex evolutionary patterns that he finds over the surface of the earth.

The Preservation of Plants as Fossils

As might be expected from the very wide range of plant structures and the diverse conditions under which they have been preserved the modes of preservation are numerous although, as a matter of convenience, they may be listed under a few fairly distinct categories. In this section we shall consider some of the more significant modes of preservation and the environmental conditions that were involved. Some examples have also been selected to reveal the remarkably fine preservation that is occasionally encountered. Paleontologists are sometimes given to making excuses for the inadequacies of the fossil record; it is evident of course that only fragments of the floras of the past have been preserved and the entire picture can never be restored. Yet if one takes a thoroughly pessimistic viewpoint, which can result only from a complete ignorance of the remarkable discoveries of the past century, there is no point in considering the fossil record. An optimistic outlook is adopted here. We are constantly being surprised at the way plants were preserved to form fossils and it is safe to assume that many surprises are in store, and with the discovery of new sources of material, and improved techniques for dealing with them, the usefulness of the fossil record will continue to command increased respect.

Many factors are involved between the tree that lived in a Triassic landscape some 160 million years ago and the specimens that lie on the laboratory bench awaiting study, or perhaps it might be better to end with the published account that arrived in this morning's mail. Some of these factors are: the nature of the plant material, whether thin delicate leaves or tough leathery ones; the kind of entombing matrix and the rapidity with which the plants were covered and thus prevented from exposure to decay organisms; the condition of the plant materials at the time of deposition; and alterations that the rocks have been subjected to during the thousands or millions of years that have elapsed since their formation. Time does not necessarily play a significant role; that is, the better preserved plants are by no means correlated with recent horizons. The important thing is that the plants were originally preserved quickly and protected against the forces of decay.

In attempting to arrive at an attractive approach to this discussion of the origin of our raw material it was decided to select a few fossil plant deposits, scattered through time and space, which have been a source of both inspiration and fine specimens to the writer. The account is thus somewhat biased by the limitations of my own experience, but I believe the fossils considered are representative.

In 1916 E. W. Berry described a clay deposit near the little town of Puryear in western Tennessee as "the most remarkable leaf-bearing clay that I have ever seen at any geologic horizon." As a result of several visits to this locality I fully share this appraisal. The fossil deposit here is a plastic brown clay about 4 feet thick which is overlain by 10 feet of a lighter colored clay, both being mined for brick manufacture. Distributed through the brown clay are innumerable leaves and occasionally seeds and fruits; it is a deposit that is abundant in specimens and numbers of species, over two hundred having been recorded. This belongs to a series of beds known as the Wilcox group and is of lower Eocene age. Some 50 million years ago the trees and shrubs of the surrounding forest shed their leaves which fell into a nearby lagoon and were quickly covered by a fine sediment. The latter, with the entombed plant debris, became consolidated and the area was elevated sufficiently to form dry land. The clay can be dug out with a stiff shovel and then split along the bedding plane with a knife to reveal beautiful fossil leaves. These vary in the amount of leaf material that remains; in some there is essentially no organic matter preserved at all, thus only the veins and outline of the leaf remain, the fossil being a typical *impression* (Fig. 1-1). With others there is some organic matter present but little more than a trace. It is interesting to note that abundant pollen grains are also found in the clay intermingled with the leaves.

If we next visit a portion of the Yorkshire (England) coast a few miles south of Scarborough a light gray, readily worked shale is encountered that was probably deposited under rather similar circumstances. However, the fossils here appear coal black against the light color of the shale, and it too is a prolific flora, including liver-

Fig. 1-1. A. The clay pit at Puryear, Tennessee; a source of abundant leaf impressions of early Eocene age. B. A leaf *impression;* virtually no organic matter remains.

worts, ferns, ginkgo leaves, cycadophytes, conifers, and other groups; we shall have occasion to refer several times to this classic Jurassic locality. Of particular interest at the moment is the fact that much of the organic matter is preserved so that the fossils, known as *compressions,* can in many instances be removed intact from the rock; this is readily accomplished by simply soaking a specimen in hydrofluoric acid, thus dissolving the sediment and liberating the plant parts. Unlike an impression fossil in which little or no cell

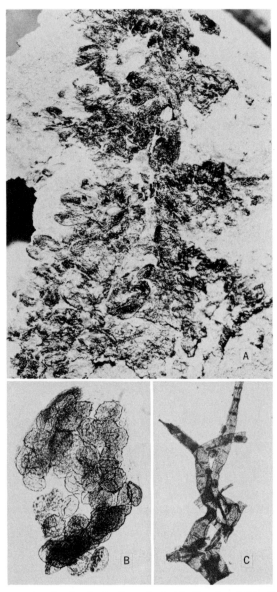

Fig. 1-2. A compression fossil. A. Portion of a fertile fern frond (*Senftenbergia*) from the Upper Carboniferous of Illinois. B. One sporangium (enlarged) has been treated to partially dissolve the wall, revealing the enclosed spores. C. Epidermal hairs (enlarged) found on the rachis of the frond.

structure remains, we may find in compression fossils the cuticular-ized epidermal cells of various plant organs, spores, epidermal hairs, and occasionally other cellular details; thus very valuable diagnostic data may be gleaned from such remains. The Carboniferous compression shown in Fig. 1-2 is close to three times as old as the Puryear fossils and is yet much better preserved.

Numerous examples of compression fossils will be dealt with in the following chapters, excellent specimens having been found in the Lower Carboniferous and even the Devonian. This should not be construed to imply that plant fossils improve with time! It does mean that time is of less importance than certain other factors, such as the kinds of plant structure concerned and the geological conditions responsible for the initial phases in fossilization.

Any discussion of plant preservation would hardly be complete without some mention of the fossil forests of Yellowstone National Park, and they will be considered as an example of a *petrifaction.* The extensive exposures of petrified stumps and logs in the north central part of the park and along the Montana boundary in the northwest corner present a spectacular paleobotanical display that is probably not excelled in any other part of the world. As one ascends Big Horn Mountain in the northwest corner of the park the dense coniferous woods give way to an open slope which is precipitous in spots and littered with fragments of silicified wood and stumps standing in the original position in which they grew. In ascending further it becomes evident that the mountain at this point consists of a series of fossil forests, one above the other (Fig. 1-4).

Going back in time some 50 million years it is not difficult to visualize the sequence of events that resulted in this unique geological formation. A living forest carpeted the lowlands and it was a more diverse one than is found in the Yellowstone today; there were many more broad-leaf trees as well as giant redwoods and others that are not present in that region today. Destruction in the form of ash and breccias from nearby volcanoes entombed portions of the forest, stripping the trunks and branches of their bark and covering the gaunt remains to a depth of many feet. In the ensuing centuries another forest developed over the grave of the first one and in turn met the same catastrophic end. These cycles of forest growth and volcanic destruction continued for many thousands of years and at one point 27 successive forests can be counted. As siliceous waters filtered down through the deposits the wood became infiltrated and replaced to a greater or less degree by silica. The stumps and logs thus ultimately became somewhat harder than the enclosing ash-breccia matrix and, being more resistant to erosion, they stand out on the steep hillside as spectral but magnificent reminders of the distant past (Fig. 1-3).

Plant deposits referred to as fossil forests have been recorded from numerous geographical localities scattered through past ages. At Florissant, Colorado, there is a particularly fine example where great silicified stumps stand in the position in which they grew and the enclosing volcanic ash is rich in leaf impressions and animal remains, particularly insects. In the journal of his *Voyage of the Beagle* Charles Darwin mentions a similar forest near Mendoza in Argentina. In Mesozoic rocks of South Dakota silicified cycads were discovered about a century ago and several hundred specimens, gathered through the efforts of G. R. Wieland and others, are now in the collections of Yale University.

In contrast to what may be termed a true fossil forest, in which the stumps stand in their place of growth, there are many deposits, great and small, of logs that were carried by rivers and ultimately dropped along the stream course, in a delta or elsewhere, to be covered by sediments and later infiltrated with mineral matter. The

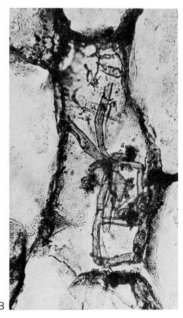

A

B

Fig. 1-3. Petrifaction fossils. A. A silicified tree trunk standing in the original position in which it grew; Yellowstone National Park. B. Fungus filaments in the cortical cells of an Upper Carboniferous fern (highly enlarged).

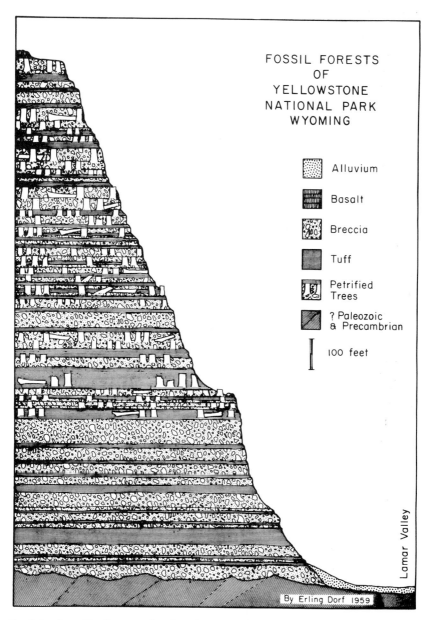

Fig. 1-4. A profile diagram showing the succession of fossil forests in Yellowstone National Park. (Prepared by Erling Dorf.)

fossil forests of Arizona, as well as numerous others in the western states, originated in this manner. A comparable deposit of cordaitean logs has been described from the Permian of southwest Africa.

Among the greatest sources of fossil plant materials found thus far are aggregations of petrified plant materials known as coal balls which are found in coal seams of Pennsylvanian age. Much has been learned about the forests of this age from European and North American coal-ball deposits. Many of the included plants will be considered in later chapters; at this point a notation is inserted simply on the origin of this important source of information.

As vast quantities of plant debris accumulated in the Pennsylvanian swamps small portions were infiltrated, chiefly with calcium and magnesium carbonates, or with iron sulfide, and were thus prevented from being altered into coal. The resultant petrifactions are frequently more or less spherical, varying in size from a centimeter or two in diameter to specimens weighing several hundred pounds. The forms assumed, however, are variable in the extreme, so that the word "ball" is not wholly appropriate. Coal balls have been found in England, on continental Europe from Belgium to the Donetz basin, in Spain, and in the central states of this country from Kentucky and Indiana west to eastern Kansas. At certain localities and in some coal seams they are very abundant and elsewhere they are scarce or never occur; the quality of preservation, although highly variable, may be exquisite. (As examples see Figs. 3-5, 5-1, 8-6.)

Although some cellular structure (occasionally a great deal) is preserved in a compression type fossil, it is generally much better in a petrifaction. Resistant structures such as wood are especially common as petrifactions, yet delicate plant parts, even nuclei and chloroplasts, are sometimes encountered. Indeed, thin-walled cellular structures such as fleshy seed coats and root hairs are sometimes better preserved than tougher ones apparently because they were more readily impregnated with mineral-bearing waters.

Very little is actually known about the petrifying process itself. It was once explained on the basis of molecular interchange, but in view of the known differences in molecular structure between the organic materials and the petrifying minerals (usually carbonates, silica, and iron sulfide) this seems unlikely. The amount and kinds of organic matter present in a petrifaction vary greatly. In the case of some fossil woods the mineral matter may be dissolved out, using hydrofluoric acid with silicified specimens, leaving the original wood intact.

A few studies have been made in recent years which shed a little light on the durability of certain organic materials. Investigations of the shell of the clam (*Mercenaria mercenaria*) have yielded the following information: specimens approximately 1000 years old contained protein essentially unchanged; specimens known to be not less than 100,000 years old showed the protein reduced to a tar-like substance with peptide chains consisting of two or more amino acids; specimens 25 million years old contained only individual amino acids. As a far more ancient example, the armor of the Devonian fish *Dinichthys terelli*, from the Black Shale of Ohio, was found to contain several amino acids including alanine, glycine, proline, and valine.

Illustrating another type of preservation, we may look next to the hills of central Vermont, near the town of Brandon, where there is a small deposit of low rank brown coal that has attracted considerable attention since its discovery about a century ago. Although of low fuel value and too limited in extent to have been of much commercial importance, it has interested geologists because of its occurrence in the New England hills area, and paleobotanists because of the variety of fossil plants it contains. Under the leadership of E. S. Barghoorn the deposit was reopened in 1947 and extensive collections of seeds, fruits, wood, spores, and pollen were obtained. The fossil remains are not petrified and must be stored in water or glycerin to prevent them from drying and shattering. Although the Brandon flora has not yet been studied in full, it promises to give us an especially complete and accurate picture of the flora of Vermont in mid-Tertiary times. Since it includes wood, seeds and fruits, and pollen, a three-way check on identifications is possible; the spores and pollen give one facet of facts, the lignitic wood a second set, and the seeds and fruits still a third.

It is well to emphasize that mineralization is not necessarily a prerequisite to good cellular preservation. A striking instance of this is the occurrence of abundant wood in the Cretaceous clays of northern New Jersey that contains no petrifying mineral matter yet many logs and branches are well enough preserved to suggest that they were buried several decades ago rather than 100 million years. This is an excellent example of the fact that plant materials may be preserved for very long periods of time if they are adequately protected from the action of biological and chemical decay.

In areas where waters are supersaturated with mineral matter, plant materials may become encrusted to form a sort of armor coating. This may be observed at Mamouth Hot Springs in Yellowstone

Park; the hot waters are supersaturated with carbonates which precipitate on cooling, and twigs, leaves, insects, and the like that fall into them quickly become coated to form an *encrustation* fossil.

Amber, which is fossil resin, probably derived from coniferous trees, presents another distinctive mode of fossilization. Amber collected along the shores of the Baltic is famous for its flowers and insects. A remarkable instance of preservation in amber has been recorded by Petrunkevitch; he has described a spider that was entrapped in resin when it was actively making threads and it was possible to trace these to their respective spinning tubes. Although the fossils contained in this medium may be objects of great beauty and well preserved, its occurrence is so rare as to render it of little importance in paleobotany.

A few outstanding instances of preservation will be described next, not in an attempt to exaggerate the potentialities of the fossil record but to indicate what is possible.

A Tertiary *Azolla*. From an early Tertiary chert in India the late Birbal Sahni and H. S. Rao have described the petrified massulae of a species of the water-fern *Azolla*. This is one of several genera of heterosporous plants, certainly not all closely related, known as water-ferns. In *Azolla* several extraordinary structures develop within the microsporangium from plasmodial substance which encloses several microspores. These are known as massulae, and they finally develop numerous peripheral appendages called glochidia which in turn terminate in a harpoonlike tip. In the Tertiary *Azolla intertrappea* (Fig. 1-5) beautifully preserved massulae are known and have been found anchored to the megaspore.

Microorganisms in Coprolites. With the exception of the diatoms, studies of microscopic plant remains were for the most part initiated only a few decades ago and it is my impression that we are just beginning to realize the potentials of the fossil record. An interesting example is the discovery of certain algae and bacteria in an upper Eocene coprolite (fossil feces) from the Bridger formation of Wyoming. Thousands of bacteria, assigned to the genus *Escherichia,* have been isolated from the coprolites as well as numerous cells of the desmid *Staurastrum*. The latter is representative of a distinctive group of unicellular algae; the wall as well as central bodies that are interpreted as shrunken chloroplasts are preserved. Other structures were found which are interpreted as conjugating cells and zygospores of desmids. These and other aquatic organisms contained in the coprolites were probably taken in by chance when the animal drank water from a pool at low stage that contained a strong

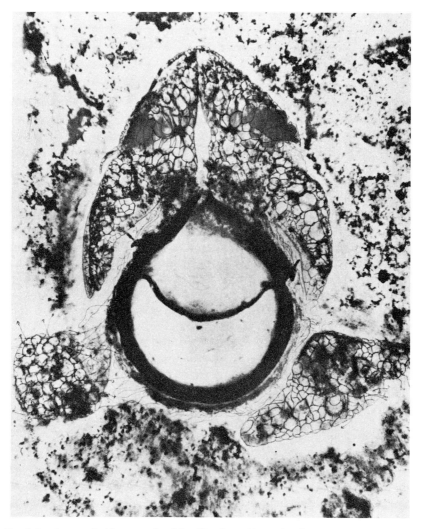

Fig. 1-5. A remarkable example of fossilization. Reproductive structures of *Azolla intertrappea,* an early Tertiary silicified aquatic fern from Chhindwara, India. The massulae are shown anchored to fibrils of the megaspore. Certain details have been slightly intensified. (From Sahni, 1941.)

concentration of the microscopic plants. It is assumed that it was not an herbivore since the desmids, with their cellulose walls, passed unscathed through the digestive tract. It is also suggested by W. H. Bradley in his report on these fossils that the feces, after being dropped by the animal, were exposed for several hours to sunlight with sufficient heat to harden or "fix" the cell contents of the desmids. Shortly thereafter the feces were covered by sediment, probably volcanic ash, and from this the silica was derived which resulted in silicification of the microorganisms.

Cretaceous Flagellates. Another instance of the preservation of delicate organic structures is found in Wetzel's account of unicellular flagellates (*Ophiobolus utinensis*) from a European Cretaceous flint. Flagella may be clearly distinguished in one of his photographs and it is suggested that the organisms were imbedded in a siliceous colloid at the bottom of a Cretaceous sea, the colloid having originated from the dissolution of the shells and skeletons of diatoms, radiolarians, and silicisponges.

Fossil Bananas. The discovery of a fossil banana fruit from a mid- or lower Cretaceous horizon in Colombia offers a complex problem. It is a cast 16 cm long and a little over 3 cm in diameter and although no cellular structure is preserved it presents in its external form a striking resemblance to a small modern banana. Several competent paleobotanists have examined the specimen and seem inclined to accept the identification as valid. It is remarkable in that the banana genus (*Musa*) is considered by many botanists to be a rather recent, highly evolved one and native only to Asia and Africa. However, there is some evidence, from early historical records, that the banana was cultivated in Central and South America before the advent of Europeans. Additional evidence which suggests that the banana is not a newcomer to the New World comes from Tertiary (possibly Oligocene) seeds, also found in Colombia, which compare quite closely with those of the extant *Musa ensete,* one of the seed-bearing, inedible bananas of Africa.

The Geisel Valley Fossils. Certain middle Eocene lignite deposits in the Geisel Valley of Germany stand out as one of the most remarkable fossil discoveries in recent years; although many of the more extraordinary remains are of animals it may not be amiss to mention them here for the exceptional types of preservation encountered. In certain parts of the deposit bones were found to have been destroyed by humic acid, whereas others are described as being as soft as butter and were treated with paraffin or special glue before removing. The brains and spinal cords of frogs were found preserved as calcium compounds of fatty acids. Blue-bottle grubs were found in skulls and could be studied in much the same way as modern larvae. Muscular tissues of fishes, salamanders, lizards, and mammals were preserved and nuclei were observed in the epithelium of frog's skin. Stomach contents of fishes, birds, beetles, and mammals were recognized, and chlorophyll was identified, as well as the coloring of many insects.

In concluding this section a few notations may be appropriate concerning *pseudofossils*. Pitfalls that the unwary paleontologist may fall into are legion and there are few who have escaped entirely unscathed. A wide variety of wholly inorganic structures may resemble plant organs, whereas outright fakes or artifacts of one sort or

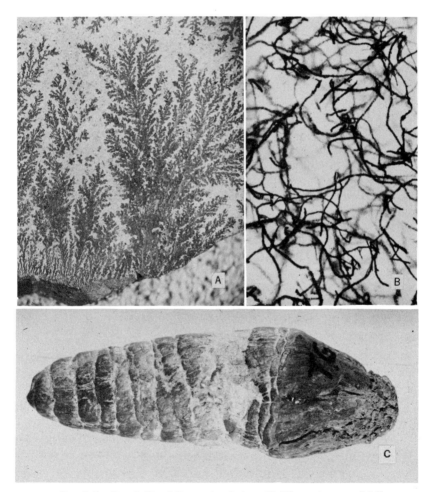

Fig. 1-6. Pseudofossils. A. Dendrites, natural size. B. Filaments, superficially resembling those of an alga, found in a "thunder egg," about 20X. C. Ichthyosaur coprolite from the Upper Cretaceous of Selma Creek, Alabama, about 2X; this is a "pseudofossil" in that it bears a false resemblance to an evergreen cone. (From specimens in the U. S. National Museum.)

another may, without critical examination, appear to represent plant life of the past.

Dendrites (Fig. 1-6A) are rather common pseudofossils, being mineral formations formed in rock crevices; although presenting a superficial resemblance to fern leaves a careful examination reveals an absence of the characteristic regularity of form of such a plant organ.

Figure 1-6C shows an *Ichthyosaurus* coprolite from the Upper Cretaceous of Alabama that superficially resembles the cone of an evergreen tree. It would perhaps be more appropriate to call this a pseudo-plant fossil.

R. W. Brown has recently written an instructive account of the plantlike structures that are found in "thunder-eggs"; these are mineral formations formed in pockets of volcanic rocks and in some of them filamentous structures have been developed (possibly comparable to the silicate "chemical gardens") which resemble certain algae, (Fig. 1-6B). Precise study of such structures reveals details of form and branching pattern which are quite unlike an alga.

Some years ago what appeared to be a remarkable fossil ear of corn (*Zea*) was found in South America and was described as such by a competent paleobotanist. The specimen created considerable interest until another paleobotanist, whose suspicions were aroused, took the trouble to section it and found, indeed, that it was a cleverly made piece of ceramic artistry.

A specimen resembling a charred corn cob was once sent to my laboratory for identification; it was alleged to have been found in coal from a horizon of Pennsylvanian age. A careful examination left no doubt that it was a corn cob and one which apparently had been partially burned and, in one way or another, found its way back into the coal scuttle. I am not sure that the discoverer of the specimen took kindly to this prosaic identification but it was the only one possible.

It is perhaps evident that the only way to avoid accepting the unreal for the real is to acquire as broad a knowledge of plant form and mineral structure as possible and to examine critically any specimens that may present some reason for suspicion.

The Classification of Plants

As fragmentary as the fossil record may be it has yielded so many unique plants as to have influenced very strongly our concepts of classification. It has been evident almost since the concept of evolution became generally accepted that there are tremendous gaps in the plant kingdom as it exists today. Many species, genera, and taxa of higher orders stand apart, showing no close relationships to other plants; such discontinuities imply the extinction in past ages of large sections of the great stream of plant life. Here and there the fossil record has served to bridge a few of these gaps but it has also brought to light many plants that are unlike anything that lives today, thus adding greatly to the complexity of classification.

A classification should enable us to keep some order among the approximately one-third of a million species of living plants and, ideally, it should indicate natural relationships. We have come a long way in the past century toward achieving the latter objective, but there is a much longer road ahead. Many recent books cite what is called an "old classification" and the "new classification"; this is somewhat misleading in that our understanding of interrelationships in the plant kingdom is constantly being revised. Occasionally a botanist, perhaps more courageous than the rest, proposes somewhat more fundamental revisions; since there has never been universal agreement on any system, and very likely never will be, he must be prepared to meet a mixture of praise and ridicule.

It seems necessary at this point to present only a framework of the major divisions; further details and possible evolutionary sequences will be considered as we go along. The outline given here is patterned essentially after the classification used by Harold C. Bold in his recent *Morphology of Plants* (1957), but since the latter work deals primarily with living plants we must add a few comments.

1. Division: HEPATOPHYTA—Liverworts
2. Division: BRYOPHYTA—Mosses
3. Division: PSILOPHYTA—Psilophytes
4. (Certain early land vascular plants of unknown affinities)
5. (The preferns or plants intermediate between the earliest vascular plants on the one hand and the ferns and pteridosperms on the other; may be tentatively included in Pterophyta.)
6. Division: PTEROPHYTA—Ferns
7. Division: LYCOPODOPHYTA—Lycopods or club-mosses
8. Division: ARTHROPHYTA—Horsetails or articulates
9. Division: PTERIDOSPERMOPHYTA—Seed-ferns
10. Division: CYCADOPHYTA—Cycads and bennettites
11. Division: GINKGOPHYTA—Ginkgo and its relatives
12. Division: CONIFEROPHYTA—Cordaites and conifers
13. Division: GNETOPHYTA—Gnetum, Ephedra, Welwitschia
14. (Gymnosperms of uncertain affinities)
15. Division: ANTHOPHYTA—Flowering plants (angiosperms)

The major groups of plants, exclusive of the algae and fungi

The algae and fungi will be referred to only incidentally and are therefore not included in the outline. The angiosperms or flowering plants (the Anthophyta of Bold) are considered in some detail in Chapters 6 and 7, but for reasons given therein no attempt will be made to deal with them according to any formal taxonomic system in the way that most of the other groups are treated.

It seems to me that a new classification is justified only in so far as it reflects an advancement in our understanding of the interrelationships of plants and there is thus no point in abandoning especially useful features of the older classifications. As an example I would cite the use of the terms *pteridophyte* and *gymnosperm.* The former includes plants now assigned to the psilophyte, lycopod, horsetail, and fern Divisions. The implication of the system given here is that these four groups (there are actually probably more than four as will be explained later) have evolved independently and are not necessarily closely related. However, they have certain aspects in common and the name pteridophyte is a convenient collective one for the assemblage. Correspondingly the term gymnosperm will be used in referring to the naked-seeded plants (seed-ferns, cycadophytes, ginkgos, cordaites-conifers, and gnetophytes) even though it is evident that this assemblage includes several lines in which the seed evolved independently.

A few introductory words are especially pertinent in explanation of the three "Divisions" cited in brackets (Nos. 4, 5, and 14). Paleobotanical classification enjoys one especially exceptional problem, that of dealing with plant remains that are so different from any previously known that a precise correlation with them is impossible. Rather than immediately setting up an elaborate system of classification for such novelties it seems to me that it is less misleading, especially to the beginning student, to admit our lack of understanding. I have thus refrained from classifying some of the more problematical fossils although suggestions as to their possible affinities will be given in most instances.

It is very possible that the two major groups of the Cycadophyta, the Cycadales and Bennettitales, should be recognized as distinct Divisions. The Gnetophyta is a notoriously difficult assemblage and since their fossil record is scanty they will be dealt with only briefly.

A minor departure from Bold's system appears in the use of the Division name Lycopodophyta instead of Microphyllophyta. It is true, from the standpoint of the Greek derivation, that the word *lycopod* (wolf-foot) has no significant application but it is so well entrenched in the literature that I have chosen to retain it. It may be added that the name Microphyllophyta is not entirely satisfactory since not all of the plants in this group have small leaves (some Carboniferous forms will be considered in which the leaves are nearly a meter long) and many plants in other unrelated groups also have small leaves.

An Introduction to Wood Anatomy

It has been pointed out that compression and petrifaction fossils are especially significant in that they retain, to a greater or less degree, the cellular organization of the plant; thus in dealing with such fossils one is involved in interpreting the anatomy or internal structure of plants. It therefore seems desirable to introduce certain aspects of anatomy that will be of frequent concern. A comprehensive treatise on plant anatomy should be available for reference, two of the most useful recent accounts being those of Eames and Mac-Daniels and of Esau.

All too often it is only the more resistant woody tissues of plants that are preserved as fossils; thus this aspect of anatomy will be emphasized in the following pages. Other phases of the subject will be introduced and elaborated on as the occasion requires.

Most plants grow in length by means of a small zone of actively dividing cells located at the tips of the stem and root systems. Immediately back of this *apical meristem* new cells that are formed become structurally specialized and ultimately serve certain restricted functions; the systems thus formed are called *primary* tissues. It is a notable feature of the pteridophytic groups that their shoot system (main stem and branches) increases but little in diameter throughout the life of the plant, although the arborescent lycopods and articulates of the Carboniferous are notable exceptions. This is due to the fact that the primary tissue system becomes mature immediately back of the apical growing point and there is no mechanism which allows continued increase in girth.

In contrast to this mode of growth in the pteridophytic groups we find in most seed plants (often referred to as the "higher plants") that continued lateral or *secondary* growth is a marked feature of the plant form. It is readily observed, for example, in a pine tree that the shoot system increases progressively in diameter from the terminal twigs down through the main stem, and this increase in girth goes on throughout the life of the plant. This is accomplished by means of a layer of actively dividing cells located immediately external to the wood known as the *cambium* and the tissue systems formed are called secondary. Thus while longitudinal or primary growth is confined to the apices of stems and roots (the older parts do not increase in length) the secondary growth is general throughout these organs. There are of course exceptions and variations to this general statement; we are dealing with a very large assemblage of living things and they are often reluctant to conform to manmade rules and laws.

The following discussion will deal for the most part with stem structures since they display more variation than roots and are more often preserved as fossils. We shall also be concerned only with the wood or *xylem* (these two terms are synonymous); the wood is more resistant to decay than other stem tissues and thus more frequently preserved and it affords many diagnostic features that are useful in identification. The term *vascular* plant will refer to any in which woody tissues are present; the algae, fungi, mosses, and liverworts are notably lacking vascular tissue and it is much reduced or absent in a few of the flowering plants (certain parasites and aquatic forms).

Next, the term *stele* requires some elaboration as it will probably be used more often than any other throughout the book. This is usually defined as consisting of the wood, a central pith when present, and two living tissue systems external to the wood, the phloem and pericycle. The latter two consist of rather thin-walled cells that are highly subject to decay and thus not often preserved; consequently as applied to fossil plants the term stele will usually imply only the wood and the central pith if present.

In some of the simplest members of the pteridophyte assemblage the stele consists of a slender cylindrical core of only a few wood cells (tracheids); this is known as a *solid protostele* (Figs. 2-1B, 8-5). When, in larger axes, parenchyma cells are admixed with the tracheids, it is known as a *mixed* protostele. In other primitive plants the wood may be variously lobed forming an *actinostele* (Fig. 2-3A).

When a distinct pith of (usually) thin-walled and more or less isodiametric cells is present the vascular unit is called a *siphonostele* (Fig. 8-7), the wood itself being quite literally siphon-like or tubular. When the vascular tube is discontinuous, rather than forming a uniform unbroken cylinder, it is called a *dictyostele* (Fig. 4-12A), that is, a net-shaped stele.

Stemming out from these simple and basic types we find, particularly in the Pterophyta, a wide range of more complicated stelar patterns. In the modern tropical fern *Matonia* the vascular system in the mature stem consists of several concentric cylinders which will be termed here a *polycyclic stele*. In certain Pennsylvanian tree ferns belonging to the genus *Psaronius* (Fig. 4-4) a system of many concentric stelar units (meristeles) forms a highly complex polycyclic stele. A different degree of complexity is found in the Cretaceous fern genus *Tempskya* (Fig. 4-13A) where the "trunk" may be composed of as many as several hundred stems—not simply numerous steles within a stem but distinct stems which form an anastomosing system that is held together by thousands of slender wiry roots.

No system of characters used in identifying and classifying plants ever seems to work perfectly and vascular structures can be especially vexing to deal with. In many pteridophytic plants the mature stem possesses a stelar system that is quite constant in size and form but in some there is considerable variation within a single plant. Moreover, it has been shown that in many living pteridophytes (this is probably true for *all* living and fossil plants in this great assemblage) in the earliest stages of development the young stem starts off with a small solid protostele; as the stem grows upward the apical growing region may increase rapidly in girth and the stelar system increases correspondingly in size and complexity until the adult form is reached.

It will be useful next to examine certain developmental aspects of the stele, with particular reference to structural details of the wood cells (Fig. 1-8). In a longitudinal section through the stele of the living Whisk fern (*Psilotum*), which is protostelic in the smaller twigs (Fig. 1-7A) and siphonostelic in the larger ones (Fig. 1-7B), it will be noted that the outermost wood cells (tracheids) are small in diameter and there are conspicuous *annular* or *spiral* bands of secondary wall material laid down on the inner side of the original thin primary wall of these cells. Immediately within these wood cells there are somewhat larger ones in which the bands of secondary wall material are broader and more uniform; they are, moreover, attached to the primary wall by a slender median strand, so that this type of wall thickening is actually T-shaped when seen in section; such cells are called *scalariform* tracheids. Next in sequence are cells in which the secondary thickening is uniform over the entire primary wall with the exception of certain circular or elongate areas

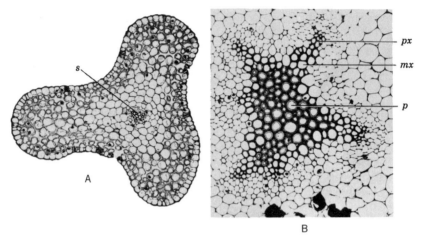

Fig. 1-7. Transverse view of the stem of the extant *Psilotum nudum,* about 15X. A. A small branch showing protostele, *s,* about 15X. The stele of a larger branch: *px,* protoxylem; *mx,* metaxylem; *p,* the central fibrous pith. (From Andrews, 1947.)

where the secondary wall forms a dome with a small orifice; these are known as tracheids with *bordered pits.*

To complete a series of useful descriptive terms, which correlate with those given previously, the smaller outermost annular or spiral tracheids are referred to as the *protoxylem* (Fig. 1-7B) while the scalariform and pitted tracheids constitute the *metaxylem.*

This account can only serve as an introduction to more detailed considerations of stelar structures that will follow, but a few variations may be noted. The entire sequence of tracheid types noted here is not found in all plants; in some of the psilophytes all of the wood is composed of annular and spiral cells; in the lycopods there are a few small annular protoxylem cells and the bulk of the wood (metaxylem) is scalariform, there being none of the circular bordered pitted type. Next, the sequence that has been given is somewhat idealized in that very small steles only a few cells thick, such as *Psilotum* itself, may not exhibit the entire series in one section. There are also transitions among the "typical" types of tracheids, so that it is not always easy to distinguish sharply between annular and scalariform cells, and there are frequently transitions between scalariform and circular bordered tracheids in which one finds circular and variously elongated bordered pits in the same cell. The drawing shown in Fig. 1-8 is taken from the primary wood of the petiole of one of the cycads; this was selected for illustration since the typical tracheid types as well as intermediates are clearly defined.

We may next turn to variations in the primary wood based on the way it matures. The annular or spiral protoxylem cells are first to mature, and the maturation process continues through the metaxylem. In a plant such as *Psilotum,* where the protoxylem is outermost, the stele is termed *exarch;* in a few rare instances, such as *Rhynia,* a fossil plant from the Devonian of Scotland, the protoxylem is centrally located and the stele is termed *centrarch* (Fig. 2-1B); when the protoxylem is innermost (but not at the center of the stem) the stele is termed *endarch;* and finally when the protoxylem is located more or less midway in the primary wood and the maturation process takes place both toward the center and toward the periphery of the stem it is termed *mesarch.*

The above discussion has referred exclusively to the primary wood; a few members of the pteridophytic groups developed sec-

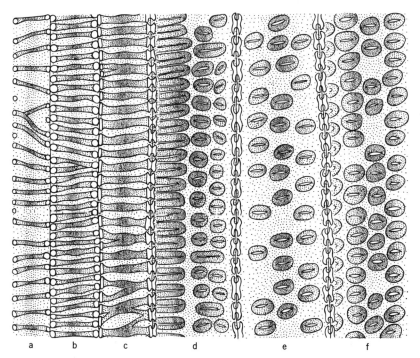

Fig. 1-8. Pitting transition in tracheids of vascular bundle in petiole of the living *Cycas revoluta: a,b,* annular; *c,* scalariform; *d,* elongate-bordered pits along left side of cell and circular to elongate pits to the right; *e,f,* circular bordered pits.

ondary wood, but it is not until we come to the gymnospermous plants that this is a constant and conspicuous element of the stem anatomy. Secondary wood is usually initiated by the division of a layer of cells, the cambium, around the periphery of the primary xylem.

By referring to Fig. 8-7, it will be noted that the primary wood can usually be readily distinguished from secondary by the radial alignment of the cells of the latter.

In the Carboniferous pteridosperms, cambial activity seems to have been something of a "new improvement," there being little secondary wood, but in some of the Permian members of the group it was formed in considerable abundance. In the Coniferophyta it seems to have been a character that was acquired early in the evolution of the group and, reaching a maximum development for the entire plant kingdom, it has enabled the sequoias to attain trunk diameters in excess of 34 feet.

There has been much skepticism cast on the potential of the fossil record but it seems to me that there is reason to be quite optimistic about the future contributions of paleobotany. There have been so few workers in the field until very recently that there has been hardly adequate opportunity for the science to display its potential. Techniques have been evolved in recent decades that enable us to obtain highly significant information from specimens that were discarded or given only the most cursory consideration fifty years ago and there is no reason to doubt that improvement in technique will continue.

It seems especially pertinent, however, to note how little of the earth's crust of sedimentary deposits have been studied intensively. There are large areas of the world in which the surface rocks have been only casually explored and paleobotanists have been limited for the most part to natural outcrops and mining operations. I have no doubt that techniques will develop in the future which will enable us to obtain much larger samples of deep-lying sediments.

REFERENCES

Abelson, Philip H. 1956. Paleobiochemistry. *Sci. Amer.,* 195: 83–92.

Andrews, Henry N., Jr. 1947. *Ancient Plants and the World They Lived in.* Comstock Pub. Co., Ithaca. 279 pp.

Barghoorn, Elso S., and W. Spackman. 1950. Geological and botanical study of the Brandon lignite and its significance in coal petrology. *Econ. Geol.,* 45: 344–357.

Berry, Edward W. 1916. The Lower Eocene floras of southeastern North America. U. S. Geol. Survey Prof. Paper, 91: 1–481.

————. 1925. A banana in the Tertiary of Colombia. *Amer. Journ. Sci.,* 10: 530–537.

Bold, Harold C. 1957. *Morphology of Plants.* Harper and Bros. 669 pp.

Bowen, R. N. C. 1958. *The Exploration of Time.* Philosophical Library, New York. Pp. 1–143.

Bower, Frederick O. 1935. *Primitive Land Plants.* MacMillan and Co., London, 658 pp.

Bradley, Wilmot H. 1946. Coprolites from the Bridger formation of Wyoming: their composition and microorganisms. *Amer. Journ. Sci.,* 244: 215–239.

Brown, Harrison. 1957. The age of the solar system. *Sci. Amer.,* 196 (4): 81–94.

Brown, Roland W. 1957. Plantlike features in thunder-eggs and geodes. *Ann. Rept. Smithsonian Institution,* 1956. pp. 329–339.

Eames, Arthur J., and L. H. MacDaniels. 1947. *An Introduction to Plant Anatomy.* McGraw-Hill Book Co., 427 pp.

Esau, Katherine. 1953. *Plant Anatomy.* John Wiley and Sons, 735 pp.

Huertas, Gustavo, and Thomas van der Hammen. 1953. Un posible banano (*Musa*) fosil del Cretaceo de Colombia. *Revista Acad. Colombinan Ciencas Exactas, Fisicas Nat.,* 9: 115–117.

Hurley, Patrick M. 1959. *How Old Is the Earth?* Doubleday and Co., New York. Pp. 1–160.

Jongmans, Willem J. 1949. Het Wissenlend Aspect van het Bos in de Oudere Geologische Formaties (from: W. Boerhave Beekman: *Hout in alle Tijden*) Deel, Deventer, Holland.

Ladd, Harry S. 1957. Chapter I. Introduction to "Treatise on Marine ecology and paleocology." *Geol. Soc. Amer. Mem.,* 67: 1–29.

Moore, John A., and Henry N. Andrews, Jr. 1936. Transitional pitting in tracheids of *Psilotum. Ann. Missouri Bot. Gard.,* 23: 151–156.

Petrunkevitch, Alexander. 1950. Baltic amber spiders in the Museum of Comparative Zoology. *Bull. Mus. Comp. Zool.,* Harvard Univ. 103: 257–337.

Rodin, Robert. 1951. Petrified forest in south-west Africa. *Journ. Paleont.,* 25: 18–20.

Sahni, Birbal. 1941. Indian silicified plants. I. *Azolla intertrappea* Sah. and H. S. Rao. *Proc. Indian Acad. Sci.,* 14: 489–501.

————, and H. S. Rao. 1943. A silicified flora from the intertrappean cherts round Sausar in the Deccan. *Proc. Nat. Acad. Sci. India,* 13: 36–75.

Traverse, Alfred. 1955. Pollen analysis of the Brandon lignite of Vermont. *U. S. Bureau of Mines, Rept. Investigations,* 5151: 1–107.

Weigelt, Johannes. 1935. Some remarks on the excavations in the Geisel Valley. *Research and Progress,* 1: 155–159.

Wetzel, Otto. 1953. Résumé of microfossils from Upper Cretaceous flints and chalks of Europe. *Journ. Paleont.,* 27: 800–804.

Zeuner, Frederick E. 1950. *Dating the Past.* Methuen and Co., London. 474 pp.

2

the PSILOPHYTA
and certain other early plants

Preface to Chapters 2–7

A word of explanation is in order concerning the arrangement of the chapters that follow. There seems to be a broad stream of evolution, with many specialized side branches, beginning in the psilophyte complex and continuing through the "preferns" in the Middle and Upper Devonian to true ferns and seed-ferns in the Carboniferous; and in turn the seed-ferns appear to be the most likely source of the flowering plants. Imperfect as our knowledge may be it seems desirable to take advantage of what is known of the evolutionary sequence in this great assemblage of plants. A return is then made to other pteridophytic groups (Lycopodophyta and Arthrophyta); these seem to represent distinct lines from their earliest recognizable members and, following their own independent paths, they expanded into flourishing groups in the Paleozoic and then rapidly declined. An attempt has been made, however, to present the subject material throughout so that the order in which the various chapters are studied is reasonably flexible.

Introduction—the Contents of This Chapter

The search for the earliest vascular plants constitutes one of the most exciting chapters in paleobotany. The problem of their origin and early trends of evolution on the land has long concerned botanists; it is assumed that they originated from the algae, for want of a better hypothesis, but we are still very much in the dark as to how this transition took place. The oldest unquestioned record of these early vascular plants dates from the upper part of the Silurian, and by Lower Devonian times a rather diverse assemblage is known to have existed. Progressing upward through the Devonian one encounters an increasing variety of plants, some of which are fairly clearly defined as early forms of the lycopod, articulate, and fern groups, whereas the affinities of others are uncertain or wholly obscure.

There is also a considerable accumulation of fossil remains which suggest that vascular plants existed long before the Silurian; the evidence is controversial and will probably continue to be so for some time, but it is such controversy that inspires men to devote their lives to the study of plant evolution. Finally, one cannot avoid wondering about the less intricate forms of life far below the level of even the simplest vascular plants. Many books and articles have been written in recent years which speculate on the ultimate origin of life on the earth; much of this is beyond the scope of paleontology, but it seems appropriate to note something of what is known of the earliest records of life at the cellular level.

It is the purpose of this chapter to consider some of these problems and, in order to present the subject matter in as clear a fashion as possible, it will be taken up under several subheadings as follows:

1. *The Psilophyta.* These are simple land vascular plants that display some evidence of being closely related and which may be the progenitors of later pteridophytic groups.

2. *Some Vascular Plants of Uncertain Affinities.* The fossils described here are classified by some authors in the Psilophyta; however, they show marked divergences from that group and are certainly not all closely related.

3. *Pre-Silurian Plant Records.* This section deals with problematical pre-Silurian reports of vascular plants as well as notations on some of the earliest known nonvascular plants.

4. *A Miscellany of Devonian Plants.* Included here are, for the most part, larger and more complex plants than those dealt with in the first two sections above; they reveal some of the great diversity of plant life in the Devonian.

THE PSILOPHYTA

There is considerable disagreement among paleobotanists as to the limits of the Psilophyta; certainly a number of fossils have been included in the group for want of any other taxonomic category in which they might be deposited. For reasons that will be explained it seems to me that this has led to serious misconceptions concerning the evolution of early vascular groups; thus the psilophytes are treated here in a narrow or restricted sense.

The Rhynie Plants

In 1917 Kidston and Lang presented a description of *Rhynia gwynne-vaughani* from a Middle Devonian chert bed near Rhynie, Aberdeenshire, Scotland. There are few plants, fossil or living, that have been as influential in botanical evolutionary thought. Unlike previously described early land vascular plants *Rhynia* and its chief associates, *Horneophyton* and *Asteroxylon,* were preserved with exquisite perfection and essentially in their entirety. *Rhynia* became established in the minds of many botanists as *the* primitive vascular plant; it is an interesting plant and a primitive one, but it is by no means certain that it should be regarded as a focal point from which all later land plants evolved.

In view of the great importance of the Rhynie deposit a few remarks may be of interest concerning the site itself. The petrified remains were first discovered in chert specimens in a stone wall and when their botanical importance was recognized the rocks were traced to their source and a considerable amount of the petrified plant material was excavated. It is worth looking to the original account for a description of the origin of the chert bed:

> The whole history of the formation of the Rhynie Chert Zone, at least of that portion from which our specimens were taken, can be clearly read. One can in imagination see a land surface, subject at intervals to inundation, covered with a dense growth of *Rhynia Gwynne-Vaughani.* By the decay of the underground parts of *Rhynia* and the falling down of withered stems (for this plant had no leaves) a bed of peat was gradually formed varying from an inch to a foot in thickness. The peat was then flooded and a layer of sand deposited on its surface. Again the *Rhynia* covered the surface, and this process of the formation of beds of peat, with the deposition of thin layers of sand, went on until a total thickness of 8 feet had accumulated. (Kidston and Lang, 1917, p. 764)

*Rhynia gwynne-vaughani** was a small plant (Fig. 2-1A) of about 8 inches in height and grew in dense aggregations with *Horneophyton* and *Asteroxylon.* It lacked leaves and the shoot system was not clearly differentiated into stems and roots although the under-

---

* Throughout this book species names will be decapitilized regardless of their etymology, other than in direct quotations.

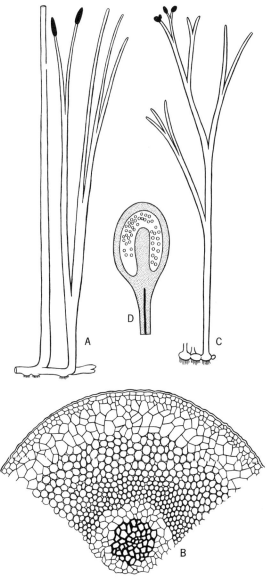

Fig. 2-1. A, B. *Rhynia gwynne-vaughani;* A. restoration; B. transverse sector of stem showing small protostele and broad cortex. C, D. *Horneophyton lignieri;* C. restoration; D. diagrammatic longitudinal section of sporangium. (A and C from Kidston and Lang, 1921.)

ground rhizome bore numerous rhizoids. The upright, gradually tapering shoots in some cases bore terminal sporangia.

Anatomically, the upright shoots, which are about 2 mm in diameter (Fig. 2-1B) consist of an epidermis with rather sparsely distributed stomata which are lacking in the rhizome, an outer and inner cortex, and a minute central stele. The outer cortex is a narrow band about four cells thick; the inner cortex is quite broad, the cells are smaller and rather large intercellular spaces are present. It extends to the surface in the vicinity of the stomates and is thought to have been the assimilating tissue of the plant. In the center is a small cylinder of annular tracheids, the number varying with the size of the shoot; the innermost tracheids are somewhat smaller than the surrounding ones, presumably representing the protoxylem. Surrounding this is a narrow zone of cells that has been interpreted as phloem by virtue of their position and elongate nature, although sieve plates have not been observed. No endodermis or pericycle is recognizable. One other feature of some interest may also be noted: small hemispherical bulges appear on the upright stems which were the seat of development of lateral branches; some of their cells developed into rhizoids and the vascular supply did not connect with the axis on which they were borne. They appear to have been readily detached and may have served the function of vegetative propagation.

The sporangia are terminally borne, attained a length of 3 mm and a diameter of 1.5 mm, although considerable size variation was found. The wall consists of a cuticularized epidermis followed by several layers of thin-walled cells and the enclosed chamber contains a large number of spores averaging about 40 μ in diameter. No specialized mode of dehiscence has been noted. It is also of interest that spores were observed in two instances in unmodified stem tips; since it seems likely that sporangia evolved here by a modification of the stem apex these may be regarded as primitive or rudimentary sporangia.

An associated species, *R. major,* has been recognized in which the aerial stems attained a diameter of 6 mm; the stele is larger and the sporangia reach a length of 12 mm, with spores about 65 μ in diameter. Hemispherical projections and adventitious lateral branches are lacking.

In the now rather voluminous literature dealing with it, *Rhynia* has been referred to as a plant that is *simple, primitive,* or *reduced.* Since these terms will be frequently encountered it may be helpful to consider their useage at this point.

That *Rhynia* is *simple* as a land vascular plant stands unchallenged. The term refers to its construction and in the lack of differentiation into stems and roots, the absence of appendages that may be called leaves, the uniform dichotomy of its branching, the vascular system composed of only a slender strand of annular tracheids, the sporangia which very possibly are only modified branch tips, it is, by comparison with any other vascular plant, living or fossil, one of the least complicated.

That *Rhynia* is *primitive* has been questioned and this necessarily involves the third term, *reduced*. *Rhynia* comes from a Middle Devonian horizon and although some geologists have questioned this, thinking that it may actually be of Lower Devonian age, this does not materially affect our discussion. Regardless of the precise age we have these rather vexing facts to explain: Land vascular plants of greater complexity are known to have existed in the Silurian. One plant (*Asteroxylon*) of significantly more complex construction lived in association with *Rhynia*.

There seem to be at least two possible explanations: First, *Rhynia* may have attained its status as a simple land vascular plant at a much earlier date and has evolved no further for many millions of years; in which case it is truly primitive. We know that many plants (the pines are a good example) have carried on with no significant changes for scores of millions of years, so this explanation is within the realm of possibility. Second, it may be that *Rhynia* has been evolved ('devolved' may be more appropriate) or reduced from more complex land plants.

This is an opportune time to insert a few notations on *Psilotum,* a living plant that presents some similarity with *Rhynia* and has been the subject of much discussion as to whether it is primitive or reduced. It is rather widely distributed in the tropics, coming as far north as Florida in the western hemisphere and Hawaii in the Pacific region. *Psilotum* is easily grown under warm humid conditions, and as a botanical oddity it is often found in greenhouses. It is a densely tufted plant not attaining a height of much over 1 foot; the rhizome system is densely covered with hairs, and both rhizome and upright shoot system branch dichotomously. The leaves are very small and certainly play no significant role in photosynthesis. The vascular strand of the smaller shoots is a minute protostele, whereas in the larger ones it is siphonostelic, often with a fibrous pith; the outermost tracheids are annular or spiral, whereas those toward the center are scalariform or have elongate to round-bordered pits. The spore-bearing organs are nearly globose, three-chambered structures borne in the axil of a two-pronged bract. This is usually referred to as a *synangium,* a term applied to a sporangiate organ which is thought to have originated from the fusion of two or more sporangia. The synangium of *Psilotum* and its tracheids with bordered pits present a degree of complexity that is not found in *Rhynia.*

Psilotum has been regarded by some botanists as a very simple land plant, possibly a very ancient type that has managed to survive for several hundreds of millions of years. Others view its simple organization as the result of degeneration from a more complex ancestor. Thus the former viewpoint regards it as *primitive* and the latter as *reduced.* Nothing is known of its fossil record and there are no related modern plants which offer much aid in settling the problem.

Returning to the Rhynie plants, we note that *Horneophyton lignieri* (Fig. 2-1C) differed from *Rhynia* in that its rhizome was a lobed parenchymatous body with numerous rhizoids but with no continuous vascular strand of its own. It gave rise to upright shoots rather similar to those of *Rhynia;* the vascular supply, which ended blindly in the upper part of the rhizome lobe, consisted of an inner core of small tracheids and a peripheral zone of somewhat larger ones.

The sporangia (Fig. 2-1D), borne on the tips of ordinary branches, were 1 to 2 mm in diameter and 2 or more mm long. Their most striking feature is the presence of a columella, a central core of slender, elongate cells continuous with the stele below but not composed of tracheidal cells. The spores are about 50 μ in diameter. The rhizome of *Horneophyton* has been compared with an early stage in the development of the sporophyte of certain species of *Lycopodium* known as the protocorm; the protocorm itself is not vascularized but bears on its upper surface leaves that show no evident orderly relation to the apex, which are called protophylls. Each protophyll has a vascularized strand which ends in its base.

The third member of the Rhynie trio is *Asteroxylon mackiei,* a somewhat larger plant of perhaps a half meter in height (Fig. 2-2). The leafless underground rhizome gave rise to a more or less monopodial upright shoot system; that is, in contrast to the equal dichotomy of *Rhynia* there is a strong central stem from which branches depart and these tend to divide by more or less equal dichotomies. The larger stems are about 1 cm in diameter and the numerous cuticularized leaves are about 5 mm long.

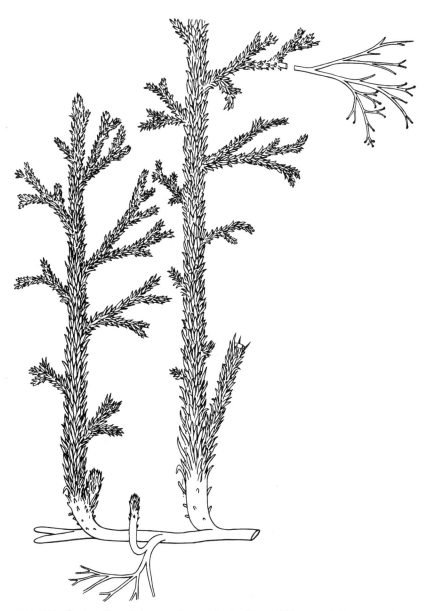

Fig. 2-2. Restoration of *Asteroxylon mackiei.* (From Kidston and Lang, 1921.)

The rhizome stele is a simple cylindrical strand of tracheids with no evident protoxylem and in the transition to the leafy shoot it becomes stellate, and strongly so in the larger stems (Fig. 2-3A); thus branches of different sizes display considerable variation in the complexity of the stele. Traces depart from the arms although they penetrate only into the base of the leaf. The middle cortex may possess rather conspicuous, radially elongate, intercellular spaces, a tissue that is characteristic of plants growing in moist places.

Associated with these leafy shoots are slender, branched, leafless axes bearing terminal pear-shaped sporangia which probably represent terminal fertile branches. The sporangia are about 1 mm long.

Psilophyton

In 1859 J. W. Dawson described an early land plant, *Psilophyton princeps,* from the Lower Devonian of the Gaspé Peninsula in eastern Canada. Since that time other species have been assigned to the genus from widely scattered localities. In view of its historical interest, apparent abundance in early Devonian times, and problems of interpretation that still enshroud it, *Psilophyton* is of more than passing interest.

As a result of later collections Dawson gave a more detailed account of the plant in 1871. He now had available three kinds of fossils: leafless dichotomizing shoots which it was thought might represent the rhizomatous part of the plant; spiny, apparently upright dichotomizing axes; and a third type bearing terminal sporangia and with markedly fewer spines. Since the fertile specimens have never been found attached to the more profusely spiny ones, they have been given the name *Dawsonites arcuatus* Halle (Fig. 2-5D). Recent critical studies of certain specimens that had been presented by Dawson to several British museums have demonstrated spores in the sporangia. The spines, which range from ⅓ mm to 2 mm in length, sometimes have a dilated tip with dark contents suggesting the possibility that they were glandular. A vascular strand has not been found in the spines nor do they contain stomates although these are present in the stem.

On a recent trip to the Gaspé I had an opportunity, through the courtesy of several Canadian paleobotanists, to collect at locations that Dawson had visited. We observed at one place a great abundance of fine specimens of the spiny type with a few fertile ones; at another place the spiny and spineless shoots were found at the

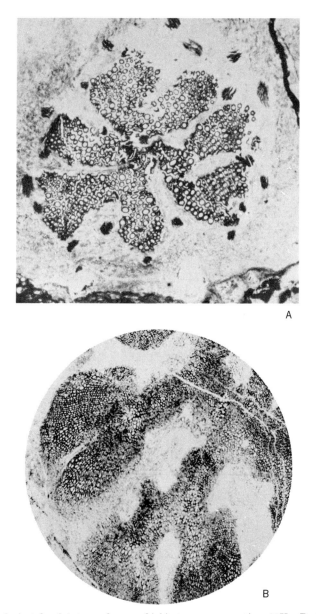

A

B

Fig. 2-3. A. A stele of *Asteroxylon mackiei* in transverse section, 30X. B. A portion of the stele of *Schizopodium davidii* showing the radially aligned cells in the peripheral region of an arm of the wood, 15X.

same horizon; and at a third locality a bed several inches thick was encountered which contained a great mass of the spineless fossils. There is still no clear-cut evidence as to the full meaning of these three shoot types although it is my impression that the smooth ones probably represent rhizomes, whereas the spiny and fertile ones are the aerial parts. Continued exploration for new collecting sites and more complete plants should ultimately result in a satisfactory restoration of this interesting plant.

It is by no means certain that all of the fossils that have been assigned to *Psilophyton* are closely related to the Gaspé species. One that does admit favorable comparison is *P. wyomingense* Dorf from the Lower Devonian of Wyoming which has been selected for illustration because of its fine preservation (Fig. 2-4). The upright

Fig. 2-4. *Psilophyton wyomingense.* A. A rhizome with three upright spiny shoots. B. A single upright shoot at a higher magnification. (From Dorf, 1933.)

spiny shoots arise from what was apparently a rhizome, the latter being more sparsely clothed with appendages.

The botanical world seemingly was not ready to accept so strange a plant in 1859 and little attention was accorded Dawson's discovery until the Rhynie fossils stimulated an interest in such early floras and left no doubt that the plants of that time were quite different from any modern ones.

Some Other Psilophytes

In view of the variety of plants that have been assigned to the Psilophyta it seems essential to refer to Kidston and Lang's original diagnosis of the group which they described as follows:

It is therefore necessary to recognize another group of Pteridophyta, of equivalent value to those mentioned, to include *Rhynia Gwynne-Vaughani* and certain of the specimens described under the name *Psilophyton princeps.* This Class is characterized by the sporangia being borne at the ends of certain branches of the stems without any relation to leaves or leaf-like organs. For this Class we propose the name Psilophytales. (1917, p. 779)

In dealing with the earliest land vascular plants we are of course working along the fringes of the unknown and one would expect that concepts of the taxonomic entities involved must change as our knowledge increases; however, for a family, order, or class to have any significance or serve a useful purpose it must have at least tentative limitations. It may be of interest to keep in mind the original concept as given above as we consider other fossils that seem to be related.

Among the oldest and smallest of vascular plants is the genus *Cooksonia,* known from two species, *C. pertoni* Lang and *C. hemispherica* Lang. They come from rocks of Downtonian age which have been variously considered as uppermost Silurian, basal Devonian, or transitional. *Cooksonia* (Fig. 2-5C) consists of slender, leafless, dichotomizing shoots up to 6.5 cm long and not exceeding 1.5 mm wide. A thin vascular strand of annular tracheids penetrates the delicate branch system; the terminal sporangia in *C. pertoni* are wider than long, whereas they are nearly spherical in *C. hemispherica.*

Hedeia corymbosa Cookson (Fig. 2-5E) was originally described from the Upper Silurian of Australia and more recently specimens have been found in the Lower Devonian. All that is known of the plant is a terminal aggregation of more or less dichotomously forking branches that formed a three dimensional, corymbose system;

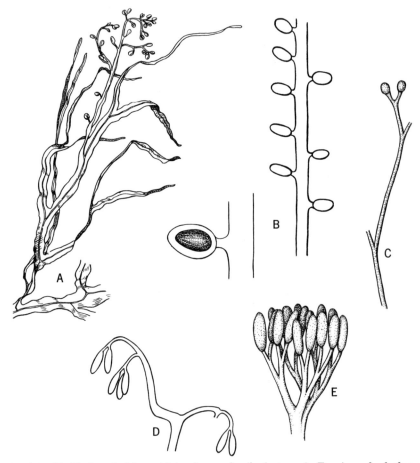

Fig. 2-5. Fertile terminal branchlets of several psilophytes. A. *Taeniocrada decheniana*. B. *Taeniocrada langi*. C. *Cooksonia* sp. D. *Dawsonites* sp. E. *Hedeia corymbosa*. (A from Kräusel and Weyland, 1930; B from specimen in Brussels Natural History Museum; C from Croft and Lang, 1942; D from specimen in British Museum of Natural History; E from photographs by Cookson.)

the terminations were mostly fertile but a few sterile tips were associated.

Taeniocrada decheniana (Goeppert) Kräusel and Weyland (Fig. 2-5A) is a Lower Devonian plant from the Rhineland with a dichotomously forking, ribbon-shaped shoot system, a central vascular strand, and terminally borne sporangia. An approximate idea of the over-all size of this plant may be gained from the fact that the branches are about 1.5 cm broad and the sporangia are 3 to 7 mm long. Specimens preserved in the Geological Institute at Cologne display tremendous quantities of the plant at a particular horizon and suggest that it probably existed as a dense growth like some modern aquatics such as the cat-tail (*Typha*). As additional evidence of its aquatic habit stomates have not been detected although the epidermis is fairly well preserved.

Other specimens have been described under the name *T. langi* Stockmans from the Lower Devonian of Belgium (Fig. 2-5B) in which the shoots bear short-stalked, laterally arranged, ovoid bodies about 3 mm long and 2 mm broad. These have been given the probationary status of sporangia, but spores have not been found in them to settle the point; if this should prove to be the case the quesion arises as to whether this plant should be included in the genus *Taeniocrada*. The possibility may also be entertained that these are specialized vegetative reproductive structures, or even food storage organs such as are found in the living horsetail, *Equisetum arvense*.

PLANTS OF UNCERTAIN AFFINITIES—DOUBTFULLY REFERABLE TO THE PSILOPHYTA

The plants described below have been referred by some authors to the Psilophyta. It does not seem to me appropriate to do so but, in any event, they reveal some of the divergence in morphology of Silurian and early Devonian plants.

Associated with the Gaspé fossils that he described as *Psilophyton princeps,* Dawson encountered several of stouter construction to which he gave the name *P. robustius.* Recent investigation of a specimen preserved in the Hunterian Museum collections in Glasgow has revealed a plant of very distinctive branching pattern which is now known as *Trimerophyton robustius* (Dawson) Hopping. The main axis is about 1 cm wide and the portion preserved is almost 9.5 cm long (Fig. 2-7E). Lateral branches depart in a spiral pattern and almost immediately trifurcate; each branch again divides into three but this time unequally. Two of these secondaries assume an ascending position and dichotomize twice, each ultimate branch being terminated by a closely compacted cluster of two or three sporangia; the third one continues out a little farther, dichotomizes, and likewise is terminated by sporangial clusters. Spores have been found in the sporangia, leaving no doubt as to the identity of these organs.

Zosterophyllum has been depicted as having a dichotomously forking, leafless shoot system with a vascular strand of annular tracheids; it is possible that the plant depended on water for support. The sporangia are aggregated into a terminal spike, which, in *Z. rhenanum* Kräusel and Weyland, from the Middle Devonian of the Rhineland, attained a length of 5 cm. In *Z. fertile* Leclercq from the Lower Devonian of Belgium it is clear that the sporangia, which were rather few in number on the spike, were borne on a short stalk which is recurved so that the sporangia tend to point back toward the axis. In *Z. australianum* Lang and Cookson (Fig. 2-6) the spike was in excess of 4.5 cm long and bore several dozens of sporangia; the reniform sporangia were attached by short stalks and were rather large, attaining a width of 4 mm and opened by means of a slit along the distal edge.

Several other early Devonian plants bear a resemblance to *Zosterophyllum* in having their sporangia aggregated together into some sort of spike-like cluster. The sporangia of *Bucheria ovata* Dorf (Fig. 2-7B) from the Lower Devonian of Wyoming were arranged in two rows; thus the spike presents a bilaterally symmetrical habit in contrast to the spiral arrangement in *Zosterophyllum;* the sporangia were sessile and apparently dehisced longitudinally.

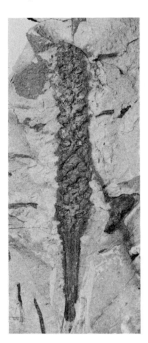

Fig. 2-6. The fertile spike of *Zosterophyllum australianum.* (From Cookson, 1949.)

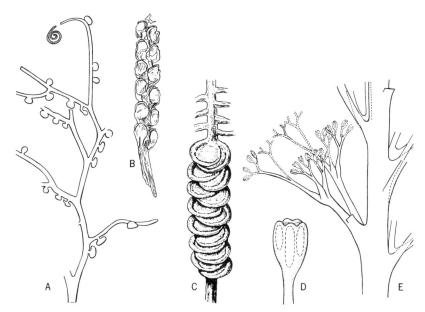

Fig. 2-7. Spore-bearing organs of certain early vascular plants. A. *Gosslingia breconensis.* B. *Bucheria ovata.* C. *Pectinophyton bipectinatum.* D. *Yarravia oblonga.* E. *Trimerophyton robustius.* (A from specimen in British Museum of Natural History; B from Dorf, 1934; C from Ananiev, 1957; D from photographs by Lang and Cookson, 1935; E from Hopping, 1956.)

Protobarinophyton obrutschevii Ananiev is known from dichotomously forking, leafless shoots containing a strand of annular tracheids; at the terminations of the shoots are spikes consisting of two rows of sporangia which are sessile, apparently elongate-oval, and dehisced along an apical slit. *Pectinophyton bipectinatum* Ananiev (Fig. 2-7C) is known from leafless shoots with a rather distinct monopodial type of branching, there being a prominent central axis with side branches which present a combination of monopodial and dichotomous forking. Some of these are terminated by spikes with two rows of short appendages. Each one tends to fold inward making a nearly complete ring, the inner margin of which bears a flat sporangium. The incurved appendage is considered as the stalk of the sporangium and may have also functioned as a dehiscence aid by straightening out at the time of maturity. Both these fossils come from the Lower Devonian of western Siberia and, although it is hoped that future discoveries will shed further light on the structure of the plants as a whole, they are of particular interest in that they add to the variety of plants with terminally clustered sporangia. It is conjectural as to just how closely related these plants may be.

Yarravia oblonga Cookson (Fig. 2-7D) is a member of the Silurian flora of Australia and is based on axis fragments up to 5 cm long which bear a terminal synangium of five or six sporangia.

One of the most puzzling of the Devonian fossils is *Gosslingia breconensis* Heard, a strange plant from the Lower Devonian of Wales which is known from small, nearly equally dichotomizing fragments, the largest of which is shown in Fig. 2-7A. The main axis is about 2 mm wide and the ultimate tip was circinately coiled. The most novel feature of *Gosslingia* is the distribution of the supposed sporangia which are scattered along the branch system. Each one was borne on a short stalk not more than 1 mm long; the sporangium itself was about 2 mm wide and 1 mm long and apparently opened into two valves. A fine strand of tracheids is present in the slender branches.

Quite a number of Devonian fossils are known from petrified stem remains in which the reproductive organs are lacking; their exact relationships are thus difficult or impossible to determine. Nevertheless some of them present anatomical features of considerable interest. *Schizopodium davidi* Harris is a stem from the Devonian (probably Middle) of Queensland with a stele that is similar in its general form to that of *Asteroxylon*. Stems up to 15 mm in diameter have been found and, at least in the portions preserved, they lack leaves. In the smallest specimens the wood takes the form of a small cross, whereas in larger ones it is more complex, being rather profusely lobed; some of the lobes may become separate entities resulting in a dissected protostele. Of particular interest is the arrangement of the tracheids. The scalariform protoxylem cells are located toward the outer part of the arms; the tracheids toward the center are metaxylem cells with bordered pits; the peripheral tracheids are, however, unique. In transverse section (Fig. 2-3B) they are seen to be arranged in regular radial rows suggesting secondary origin, but several features indicate that this is not the result of cambial growth of the kind found in most higher plants. The tracheids vary considerably in structure, some being vertically elongate, some nearly cubical, and others are radially elongate. In well-preserved stems there is, moreover, no evidence of crushing of the phloem or cortical tissues. It is therefore likely that these peripheral tracheids constitute a tissue that is intermediate between typical primary and secondary wood, having been formed by a general meristematic zone rather than a narrowly defined cambium.

The cambium is one of the greatest achievements of plant evolution and a contributing factor to the success of the plants in which it is present. *Schizopodium* seems to offer a clue to the origin of this tissue.

PRE-SILURIAN PLANT RECORDS

In 1935 Cookson and Lang described a flora from the Silurian of Australia which stands as a landmark in paleobotany, for before that time only the most fragmentary vascular plant remains had been found from horizons below the Devonian. Two members of this flora, *Hedeia* and *Yarravia,* have been cited and in Chapter 8 a third is described, the lycopod *Baragwanathia,* a plant of respectable size with stems several centimeters in diameter. Thus it is evident that vascular plants of some diversity existed at the time, which initiated considerable speculation as to how much farther back in the Paleozoic they might be traced. Intriguing clues were not long in making their appearance.

In recent years several Indian paleobotanists have described spores and wood fragments from the Middle and Upper Cambrian of Kashmir, Spiti, and other Indian localities. Several scores of spore types have been recorded which are thought to represent pteridophytic groups and even the seed-ferns, although the latter seems questionable. These discoveries have been supported by reports of large assemblages of spores from the Lower Cambrian blue clay of the Estonia-Latvia-Lithuania area by Naumova in 1949. Dr. Naumova has informed me that she has found many other Cambrian spores since that time.

This evidence from spores has been supported by a few fragmentary macrofossil remains. Probably the most significant of these is *Aldanophyton antiquissimum* Kryshtofovich, a supposedly lycopodiaceous plant from the Middle Cambrian of the Aldan Mountain range in Siberia. The plant is represented by shoots up to 13 mm wide and 8.5 cm long which are covered with microphyllous leaves attaining a length of 9 mm. Although these specimens leave much to be desired it seems possible that they do represent land vascular plants, but since reproductive organs are unknown they cannot be identified with certainty as lycopods. Association of the plant remains with trilobites, which are considered to be an index fossil for the horizon, is the principal evidence for their Cambrian age.

Although this discussion is designated as one dealing with pre-

Silurian remains it seems worth mentioning the discovery in Silurian rocks of minute plant fragments which seem to represent pieces of tracheid cells. The large bordered pits that characterize the specimens (Fig. 2-8B) are quite unlike those found in any really primitive land plant and suggest fragments from the tracheids of large forest trees such as a pine or sequoia. It is not implied that they do represent these plants but rather that they are tracheidal fragments of arborescent plants.

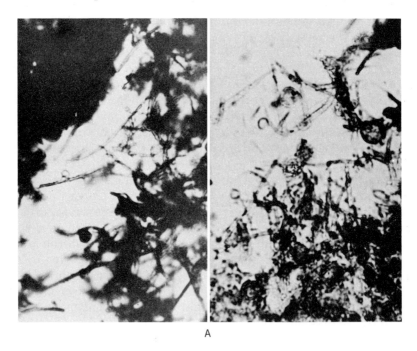

A

B

Fig. 2-8. A. The oldest known structurally preserved fossil plants from the Gunflint iron formation, Ontario. B. A tracheid fragment isolated from Silurian sediments of New York State. (A from Tyler and Barghoorn, reprinted from *Science* with permission; B courtesy of William Evitt.)

In summary it may be noted that, combined with the clearly recognized Silurian vascular plants, the suggestive evidence from earlier horizons indicates that vascular plants may have occupied the land as early as the Cambrian. Many paleobotanists are reluctant to accept this as yet and it is an exciting paleontological frontier that will be explored with increasing interest in the years to come.

It seems pertinent to consider at this point certain evidence pertaining to the earliest records of plant life of any kind on the earth. There are many reports of fossils referred to as algae from pre-Cambrian rocks, but very few show any recognizable cellular structure. Thus the discovery of well-preserved plants in rocks in southern Ontario that are dated as being 1700 million years old stands as a significant landmark in our knowledge of the earliest forms of plant life. These fossils are tentatively assigned to the blue-green algae and "simple fungi" (Fig. 2-8); it has been possible to separate them from the rock matrix so that there is no question of their organic nature. Supposed algal remains have been recorded from Rhodesia in rocks which have been dated as not less than 2600 million years old. These do not, however, display the clear-cut structural features of the Ontario fossils. Carbon deposits have also been dealt with in some detail by Rankama who has presented evidence in support of their organic origin. Some of these, reported from Finland, take the form of rings or tubes and have been given the name *Corycium enigmaticum* Sederholm. It does not seem appropriate to this writer to assign a binomial to deposits that are lacking any recognizable plant form. Rankama has also described carbonaceous slates from Canada, some of which are also believed to be biogenic, with an age cited as 2.5 billion years.

As a matter of convenience in reference some of this evidence pertaining to the earliest known land vascular plants and the earliest evidence of thallophytic plant life is tabulated below:

Earliest thallophytes	*Earliest vascular plants*
1. Structurally preserved fossils from the Gunflint iron formation, Ontario: 1700 m.y.	1. *"Baragwanathia"* flora from the Silurian of Australia: about 340 m.y.
2. Algae reported from Rhodesia: 2600 m.y.	2. *Aldanophyton* from the Cambrian of Siberia: about 520 m.y.

Based on these records, some simple arithmetic reveals a gap of about 1200 million years between the earliest record of the thallophytes and that of vascular plants. It is of course hardly necessary to point out that we are dealing in very "round" figures. Perhaps the most obvious answer to explain this great gap is that land vascular plants did exist prior to the Cambrian but, if preserved, they have not been found thus far; in turn it would have to be admitted that the Ontario fossils represent a significant period of prior evolution. Using the facts at hand it appears probable that thallophytic plant life existed for a very long time before vascular plants ventured forth on the land.

Although it is entirely speculative, one may suppose that long before the appearance of vascular plants the thallophytic forms occupied a wide range of habitats such as marine waters of varying degrees of salinity, fresh waters, as well as the land. The group that we call the algae today includes an extremely diversified and versatile assemblage of plants; they may well have been so long before the Cambrian and were able to indulge in numerous vascular plant experiments.

A MISCELLANY OF DEVONIAN PLANTS

Thus far certain fossils have been introduced which hold together as a natural, if somewhat narrowly defined, group, the psilophytes. Others have been described that might be considered referable to the psilophytes, but with less confidence. In this final section there are gathered together some of the real problems; it seems to me misleading to classify them, other than in a very tentative way, in the present state of our knowledge yet they speak vividly for plant evolution in Devonian times. There is no significance to the order in which they are described.

Nematothallus

One of the least spectacular of early Devonian plants, yet perhaps deserving more serious consideration than has been conferred on it, is *Nematothallus pseudovasculosa* Lang. In several of its characters it is a really primitive vascular plant, yet quite unlike any cited above. The fossils consist of thalloid compressions that range from minute fragments up to specimens 6.5 by 1 cm and 4.5 by 2.5 cm, although even these are evidently parts of a larger plant. Lang's investigation of the thallus has revealed a reticulum of fine tubes 2 to 4 μ in diameter which are associated with larger ones up to 25 μ

in diameter; the latter had annular thickenings and appear to have been abundant in the interior, forming rudimentary conducting tracts. Toward the outside the slender tubes form an increasingly close network and are covered by a cuticularized epidermallike layer described as having a "pseudocellular pattern." Scattered among the tubes are spores which are also believed to have been cuticularized. These curious plant fragments present some of the minimum essentials for life on the land: specialized conducting cells of a sort, a cuticle as protection against excess water loss, and resistant spores. If one can put aside for a moment the traditional concept of a primitive vascular plant in which we seek root, stem, leaf, and a spore-bearing organ we may see in *Nematothallus* a uniquely unspecialized plant in the early stages of becoming established in a land environment.

Prototaxites

The year 1859 stands as a prominent one in the records of problematical fossil plants, for in the same publication in which he first presented *Psilophyton,* Dawson also gave a brief description of the unique *Prototaxites* which he found along the south shore of the Gaspé. Since that time at least 13 other species have been described from horizons ranging from the Upper Silurian to the Upper Devonian. If measured in terms of its enigmatic nature and controversies that have arisen over it *Prototaxites* occupies a very special niche in paleobotanical literature. The Gaspé specimens are trunks up to 3 feet in diameter and 7 feet long, although the total length of the plant is not known. Since their preservation is mediocre the description given here is of *P. southworthii* Arnold from Ontario. The trunk or stem consists of a matrix of interlaced filaments (Fig. 2-9) which are aligned essentially lengthwise of the trunk; these are about 5 to 9 μ in diameter and are occasionally septate. Associated with them are larger filaments or tubes up to 50 μ in diameter; these are quite conspicuous and have much thicker walls than the smaller filaments. Another distinctive feature is the presence of fairly uniformly spaced spots about 0.5 mm in diameter which seem to represent areas of specialized filament organization, but the cellular structure is poorly preserved.

Although the name *Prototaxites* was given to these plants by Dawson because of a supposed resemblance to the wood of the yew (*Taxus*) it has been clearly demonstrated since that they are in no way related to the conifers, and most paleobotanists have been in-

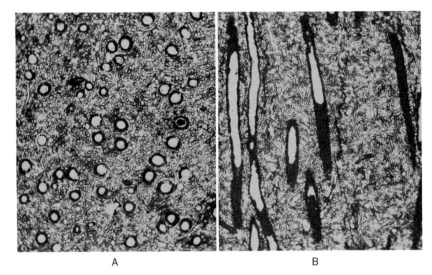

A   B

Fig. 2-9. *Prototaxites southworthii;* A. transverse section; B. longitudinal section; both about 100X.

clined to favor affinities with the algae, especially the large brown seaweeds. However, the notion of an alga with a stipe 3 feet in diameter has not been universally accepted with complete confidence! The cellular structure allows no close comparison with the woody tissue of true vascular plants and the habitat that it occupied has met with controversy, opinions ranging from an aquatic to a semiaquatic, to a land environment.

Protosalvinia and Foerstia

At a rather restricted horizon in the uppermost Devonian black shales of the east central states there is a dense concentration of fossil plants that present an unusual assortment of problems. They have been assigned to two genera, *Protosalvinia* and *Foerstia* (Fig. 2-10). The latter consists of dichotomously forking bodies which superficially resemble the terminal ends of the common seaweed, *Fucus,* whereas *Protosalvinia* is based on more or less disc-shaped structures which are, however, not infrequently lobed. *Foerstia* specimens occasionally bear tetrads of spores in deep cavities along the inner margin of the forked tip and similar tetrads are found in the *Protosalvinia* discs. The spores and at least the outer layer of cells in both are highly resistant to macerating chemicals, being impregnated with a waxy substance.

These fossils, known collectively as sporocarps, present a number of unusual and unsolved problems. Although they extend over a wide geographical area, at least from Tennessee to Michigan, they are stratigraphically confined to a few meters of shale and, in abundance, to a little more than one meter. The word abundance is an understatement, for at certain localities they make up a significant portion of the shale; there may be several hundreds exposed on a square foot of rock. The types are intermingled and it is questionable as to whether they represent two distinct genera. As to their

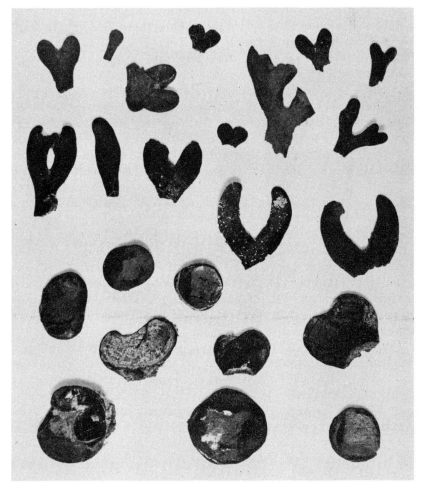

Fig. 2-10. A selection of fossils belonging to the *Foerstia* (upper half of photo) and *Protosalvinia* complex, about 2X.

affinities, they have been regarded by different investigators as referable to the red algae, as an independent class of thallophytes, or as having reached a low bryophytic level. Their great abundance seems to be best accounted for by a free floating habit and rapid vegetative reproduction; yet the tough, waxy nature of the cell walls suggests an adaptation to a dry land habitat. They show no evidence of having vascular tissue.

Some Problematical "Megaphyllous" Devonian Plants

Distributed through the Devonian from Lower to Upper and from widely scattered localities through the northern hemisphere plant fossils have been found which have been referred to the genera *Platyphyllum, Cyclopteris, Ginkgophyllum, Psygmophyllum, Germanophyton,* and *Enigmophyton.* There is no reason to assume that these were closely related, for about the only common feature they share is the presence of leaflike organs that were rather large, elongate, fan-shaped structures. Although reproductive organs are known in only a few, the plant form they present is quite in contrast to that of the psilophytes, nor are they referable to the other more clearly defined groups that are considered in later chapters.

Germanophyton psygmophylloides (Kräusel and Weyland) Høeg from the Lower Devonian of Germany is a plant with a branching axis which bore rather deeply dissected, fan-shaped leaves approximately 30 to 40 cm long. Its anatomy is described as consisting of a parenchymatous ground tissue through which tubelike cells ramified; it was first described as a species of *Prototaxites.* Chiefly because of the parenchyma, which is not found in *Prototaxites,* and the lack of evidence that the parts of the fossil that are known were borne on large trunks, it was transferred to a new genus, *Germanophyton.*

Psygmophyllum gilkineti Leclercq, from the Middle Devonian of Belgium, is a branching stem fragment some 21 cm long which bore 22 leaves. Each leaf consisted of an elongate petiole up to 33 cm long which expanded into a wedge-shaped blade 10 to 14 cm long and of about the same breadth. Frequently dichotomizing veins ramified from the petiole out into the blade.

Enigmophyton superbum Høeg is somewhat better known than the plants considered above but no less problematical. Its geographical origin adds to the interest of the plant, for it was found at Mimerdalen in Spitsbergen, is of Middle or Upper Devonian age, and is representative of the extensive floras that have thrived in the

Arctic in past geological ages. *Enigmophyton* is thought to have attained a height of 1 meter or more; the smooth stems were 5 mm in diameter and bore lateral branches about 1 to 2 mm thick. The leaf developed from a narrow base into a partially dissected fan-shaped structure (Fig. 2-11) some 16 cm long and 12 cm broad. Although woody tissue has not been observed in the stems the impressions of regularly spaced lines, which apparently indicate veins in the leaves, allow little doubt that it was a vascular plant.

Associated with the vegetative remains are fructifications which are thought to have been borne as shown in the restoration; the two are not, however, known in organic connection. It is a bifurcating spike 2 to 3 cm long bearing numerous sporophylls; there is apparently but one sporangium on each sporophyll, some of which contain megaspores up to 250 μ in diameter, whereas others contain much smaller spores 60 to 85 μ in diameter. If these fructifications and vegetative remains are correctly associated one may see evidence of lycopodiaceous affinities in the spore-bearing organs, but certainly nothing to suggest such a relationship as far as the rest of the plant is concerned.

Fig. 2-11. *Enigmophyton superbum,* a problematical fossil from the Devonian of Spitsbergen. (From Høeg, 1942.)

SUMMARY

It is my impression that the only unqualified statement one can make about the early land floras is that they included a considerable variety of plant types and that we are just beginning to understand them. The diversity is actually greater than is indicated by this chapter partly because it is not possible to consider them all and partly because the earliest representatives of the lycopods and articulates are dealt with in later chapters.

The discovery of structurally preserved plant remains, referable to algal and funguslike plants (as reported by Tyler and Barghoorn) from rocks in Ontario which have been dated at about 1700 million years by the potassium–argon technique is one of the great paleontological advances of this century. To know with reasonable certainty that plants have lived on the earth for one and three-quarters billions of years is exciting in itself and would have been considered as fantastic only a few decades ago. Certainly, an immense array of thallophytic plant forms evolved between that early date and the latter part of the Silurian period when we catch a first glimpse of reasonably well-preserved land vascular plants. Now we have tantalizing bits of evidence that vascular plants existed as far back as the Cambrian and perhaps in late pre-Cambrian times. The present intense interest in these discoveries will probably discredit or confirm them within the next few years, although even the most optimistic searchers can hope to find only vestiges of such early plant life. A reasonable explanation lies in the theory that a great many upland forms lived in the early Paleozoic which were not preserved, or the fossil record has been destroyed by erosion.

The problem of how to evaluate the Psilophyta is one that will also be subject to much debate in coming years. Should we reserve the category 'psilophyte' for a few plants which conform closely to Kidston and Lang's diagnosis or should it be evaluated in a broader sense to include plants such as those described in the second section of this chapter? One writer suggests a 'psilophytic stage' through which all of the major groups of vascular plants evolved. However, regardless of the particular classification one may see fit to follow it is clear that there were many novel forms of vascular plant life in the early Paleozoic; a few of these survived to expand into numerous pteridophytic lines in the Carboniferous while many apparently faded out in blind alleys.

The study of fossils confirms what might be suspected from a study of living plants alone, namely, that evolution is a continual fanning process; a line of racial development radiates out in many directions producing families, genera, and species that become adapted to an almost infinite variety of environments. In some of the larger groups, such as the ferns and flowering plants, the diversification becomes so great that authorities cannot agree as to whether they originated as a single line (monophyletic) or whether they are polyphyletic, that is, whether several or many independent lines served as a source.

REFERENCES

Andrews, Henry N., Jr. 1960. Evolutionary trends in early vascular plants. *Cold Spring Harbor Symposia,* 24: 217–234.

————, and Karen S. Alt. 1956. A new fossil plant from the New Albany shale with some comments on the origin of land vascular plants. *Ann. Missouri Bot. Gard.,* 43: 355–378.

Ananiev, A. R. 1957. Now Lower Devonian fossil plants from the southeast of western Siberia. *Akad Nauk USSR (Botany),* 42: 691–702.

Arnold, Chester A. 1940. Structure and relationships of some Middle Devonian plants from western New York. *Amer. Journ. Bot.,* 27: 57–63.

————. 1952. Observations on fossil plants from the Devonian of eastern North America. VI. *Xenocladia medullosina* Arnold. *Univ. Michigan, Contrib. Mus. Paleont.,* 9: 297–309.

————. 1952. Fossil sporocarps of the genus *Protosalvinia* Dawson, with special reference to *P. furcata* (Dawson) comb. nov. *Svensk. Bot. Tidskrift,* 48: 292–300.

————. 1952. A specimen of *Prototaxites* from the Kettle Point black shale of Ontario. *Palaeontographica,* 93B: 45–56.

Axelrod, Daniel I. 1959. Evolution of the Psilophyte Paleoflora. *Evolution,* 13: 264–275.

Cookson, Isabel C. 1935. On plant remains from the Silurian of Victoria, Australia, that extend and connect floras hitherto described. *Phil. Trans. Roy. Soc. London,* 225B: 127–148.

————. 1949. Yeringian (Lower Devonian) plant remains from Lilydale, Victoria, with notes on a collection from a new locality in the Siluro-Devonian sequence. *Mem. Nat. Mus. Melbourne,* No. 16: 117–130.

Croft, W. N., and W. H. Lang. 1942. The Lower Devonian flora of the Senni Beds of Monmouthshire and Breconshire. *Phil. Trans. Roy. Soc. London,* 231B: 131–163.

Dawson, J. William. 1859. On fossil plants from the Devonian rocks of Canada. *Quart. Journ. Geol. Soc. London,* 15: 477–488.

————. 1871. The fossil plants of the Devonian and Upper Silurian formations of Canada. *Geol. Survey Canada,* pp. 1–92.

Dorf, Erling. 1933. A new occurrence of the oldest known terrestrial vegetation from Beartooth Butte, Wyoming. *Bot. Gaz.,* 95: 240–256.

————. 1934. Lower Devonian flora from Beartooth Butte, Wyoming. *Bull. Geol. Soc. Amer.*, 45: 425–440.

Ghosh, A. K., and A. Bose. 1950. Microfossils from the Cambrian strata of the Salt Range, Punjab. *Trans. Bose Research Institute, Calcutta*, 18: 71–77.

————. 1955. Did vascular plants exist in Cambrian times? *Nat. Inst. Sci. India Bull.*, No. VII, Symposium on organic evolution, pp. 298–303.

Harris, Thomas M. 1929. *Schizopodium davidi* gen. et sp. nov.—a new type of stem from the Devonian of Australia. *Phil. Trans. Roy. Soc. London*, 217B: 395–410.

Høeg, Ove A. 1937. The Devonian floras and their bearings upon the origin of vascular plants. *Bot. Rev.*, 3: 563–592.

————. 1942. The Downtonian and Devonian flora of Spitsbergen. *Norges Svalbard og Ishavs-Undersokelser*, Nr., 83: 1–228.

Hopping, C. A. 1956. On a specimen of *"Psilophyton robustius"* Dawson from the Lower Devonian of Canada. *Proc. Roy. Soc. Edinburgh*, 66: 10–28.

Jacob, K., C. Jacob, and R. N. Shrivastava. 1953. Evidence for the existence of vascular land plants in the Cambrian. *Current Science*, 22: 34–36.

Kidston, Robert, and William H. Lang. 1917–1921. On Old Red Sandstone plants showing structure, from the Rhynie chert bed, Aberdeenshire. Parts I-V. *Trans. Roy. Soc. Edinburgh*, vols. 51–52.

Kräusel, Richard, and Hermann Weyland. 1930. Die Flora des deutschen Unterdevons. *Abh. Preussischen Geologischen Landesanstalt, Berlin*, 131: 1–92.

Krishtofovich, African N. 1953. Discovery of Lycopodiaceae in the Cambrian deposits of eastern Siberia. (Russian) *Doklady Acad. Sci. USSR*, 91 (6): 1377–1379.

Lang, William H. 1931. On the spines, sporangia, and spores of *Psilophyton princeps*, Dawson, shown in specimens from Gaspé. *Phil. Trans. Roy. Soc. London*, 219B: 421–442.

————. 1937. On the plant remains from the Downtonian of England and Wales. *Phil. Trans. Roy. Soc. London*, 227B: 245–291.

————, and I. C. Cookson. 1930. Some fossil plants of early Devonian type from the Walhalla series, Victoria, Australia. *Phil. Trans. Roy. Soc. London*, 219B: 133–163.

————. 1935. On a flora, including vascular land plants associated with *Monograptus*, in rocks of Silurian age, from Victoria, Australia. *Phil. Trans. Roy. Soc. London*, 224B: 421–449.

Leclercq, Suzanne. 1954. Are the Psilophytales a starting or a resulting point? *Svensk Bot. Tidskr.*, 48: 301–315.

Naumova, Sofia N. 1949. Spores of the Lower Cambrian. *Acad. Sci. USSR Bull. Geol.*, No. 4: 49–56.

Rankama, Kalervo. 1948. New evidence of the origin of Pre-Cambrian carbon. *Geol. Soc. Amer. Bull.*, 59: 389–416.

————. 1954. Early Pre-Cambrian carbon of biogenic origin from the Canadian shield. *Science*, 119: 506–507.

Read, Charles B. 1936. A Devonian flora from Kentucky. *Journ. Paleont.*, 10: 215–227.

————, and G. Campbell. 1939. Preliminary account of the New Albany shale flora. *Amer. Mid. Nat.*, 21: 435–453.

Stockmans, François. 1940. Végétaux éodévoniens de la Belgique. *Mem. Mus. Roy. Hist. Nat. Belgique*, 93: 1–90.

Tyler, Stanley A., and Elso S. Barghoorn. 1954. Occurrence of structurally preserved plants in Pre-Cambrian rocks of the Canadian shield. *Science*, 119: 606–608.

3

the PTEROPHYTA
—early ferns of the Devonian
and Carboniferous

The Classification of Early Fernlike Plants

By mid-Devonian times an assemblage of plants had begun to evolve that, by virtue of their size and complexity, may be interpreted as transitional between some of the psilophytes of an earlier age and the fern and seed-fern complexes of the Carboniferous. This chapter is devoted to these plants and they will be referred to collectively as *preferns;* the difficulties encountered in their classification are more than compensated by their fascinating morphology and the light they shed on vascular plant evolution.

The *preferns,* as well as the plants that are considered in the following two chapters, are based on a variety of fossil remains, both compressions and petrifactions; some are remarkably well known, whereas others are fragmentary and offer only a hint as to what their affinities may be. There are many beautifully preserved compression fossils which reveal much about the general morphology of the plant and the spore-bearing organs, whereas others are known from well-preserved petrifactions with excellent cellular preservation.

The stratigraphic sequence in which we find the preferns, ferns, and seed-ferns is of limited use in understanding the various evolutionary lines, for there is considerable overlapping in time of the groups, and it is by no means easy to determine which are primitive and which advanced. The seemingly primitive preferns extend into the Upper Carboniferous and the generally more complex seed-ferns are found well down in the Lower Carboniferous. As to the true ferns, one extant family, the Schizaeaceae, appears to have had its origin in the Lower Carboniferous and two others, the

Marattiaceae and Gleicheniaceae, are well established in the Upper Carboniferous. This "stratigraphic mixing" is not uncommon and usually means that certain members of a group carried on essentially unchanged for long periods of time, competing successfully for a place in the sun with the more advanced lines.

A word of caution is important here concerning the use of sterile foliage in classification. The true ferns are a diverse assemblage themselves and very possibly stemmed from several distinct lines of preferns and in this sense are polyphyletic. There can be little doubt that fernlike foliage has evolved independently several times from the prefern complex. Moreover, several leaf types, once considered to represent ferns, have been found with seeds attached and are thus members of the Pteridospermophyta.

The following classification will be followed, although it must be regarded as tentative.

PTEROPHYTA
    Protopteridales ⎱ Preferns
    Coenopteridales ⎰
        Anachoropteridaceae, Botryopteridaceae, Stauropteridaceae, Zygopteridaceae
    Cladoxylales
    Marattiales
        Marattiaceae
    Filicales—True ferns
        Schizaeaceae, Osmundaceae, Gleicheniaceae, Matoniaceae, Dipteridaceae (several modern families in which the fossil record is very sparse are not considered)

A few introductory notes on the terminology used in describing ferns and fernlike plants seem appropriate. The leaves (*fronds*), although extremely varied in size and form, are often large and highly divided. The main axis of the frond is referred to as the *rachis,* and the first and second order branches are called primary and secondary *pinnae,* and so on, to the ultimate terminations which are known as *pinnules* or leaflets. The proximal portion of the rachis, between the first primary pinna and the stem, is called the *petiole.* Leaves of this general type are called *megaphylls,* which means that they are large leaves, in contrast to the *microphylls* or small leaves of certain other groups, notably the lycopods. In this discussion of the preferns the term *frond* will be applied to structures that are regularly arranged on a stem and appear to be homologous with the fronds of the true ferns.

THE PROTOPTERIDALES

This group was established by Høeg and defined as follows:

Plants of a more or less fern-like habit, the foliar shoots having a tendency to flatten in one plane, but with little differentiation between axes and leaves. The ramifications of the foliar shoots may resemble pinnules, but no broad laminae are found. Sporangia borne terminally on pedicels, forming panicles or clusters. Sporangium, as far as is known, with apical dehiscene. Homospory, or in some species probably heterospory. (1942, p. 178)

Professor Høeg referred the following genera to the Protopteridales (to which I have added *Archaeopteris*): *Svalbardia, Protopteridium, Aneurophyton, Eospermatopteris, Rhacophyton,* and *Archaeopteris.*

Svalbardia

Svalbardia polymorpha Høeg is from the upper part of the Middle Devonian of Spitsbergen and seems appropriate to begin with since it might be considered as an advanced psilophyte or a simple prefern. It is based on a considerable number of fragments of various sizes and degrees of preservation, some of which reveal very clearly the significant details of the general branching pattern as well as the ultimate sterile subdivisions (leaves) and spore-bearing parts. The basal portion of the plant is not known; the largest axis found is 1.5 cm in diameter and the longest branch fragment 45 cm long; it is supposed that the plant attained a height of as much as 2 meters. Branching throughout, except for the ultimate foliar organs, is monopodial; that is, there is a strong central or main axis in contrast to the equal dichotomous branching in *Rhynia.* The primary branches often depart in nearly opposite pairs, but it is possible that as many as three or four branches departed at a node. It is certain that the branches were not arranged in one plane; thus the entire shoot system seems to have been three-dimensional.

The ultimate sterile branches (Fig. 3-1) consist of a fairly distinct main axis which bears the "leaves"; the latter may consist of simple undivided appendages but more usually are dichotomous-forking structures which divide two to four times. As many as six veins have been observed in the proximal portion (petiole) and the ultimate divisions seem to have but one. The leaf as a whole is about 2.5 cm long. It is significant to note that there is a certain amount of variation in the form of these leaves; some are nearly filiform throughout and others are laminated to some degree.

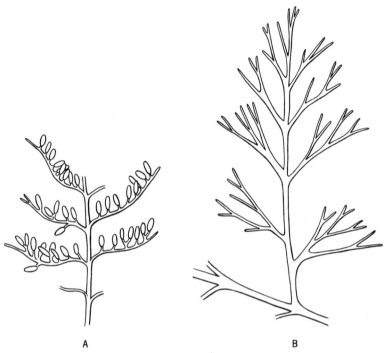

A   B

Fig. 3-1. *Svalbardia polymorpha,* probably a transitional type between the psilophytes and coenopterid ferns.   A, fertile branch; B, sterile branch; both somewhat restored. (From specimens in the Geological Museum, Oslo.)

The fertile branches bear nearly opposite, ultimate divisions on which the sporangia are arranged.   The latter are 1.5 to 2 mm long, nearly cylindrical, and borne, singly or in pairs, on a short stalk.   *Svalbardia,* although larger and more complex than psilophytes of the *Rhynia* type, is quite clearly a plant in which the differentiation of leaves as distinct organs had not progressed to a very advanced degree.   The general organization of the fertile branches is very similar to that of *Archaeopteris* and although most species of this genus possess distinct laminate leaflets, *A. fissilis,* from the Upper Devonian of Ellesmere Island and the Donetz region, has leaflets that are finely dissected and closely approach those of *Svalbardia.*

Archaeopteris*

Archaeopteris is a characteristic Upper Devonian plant (Fig. 3-2A) and is represented by 15 or more species which have been

* See footnote on page 412.

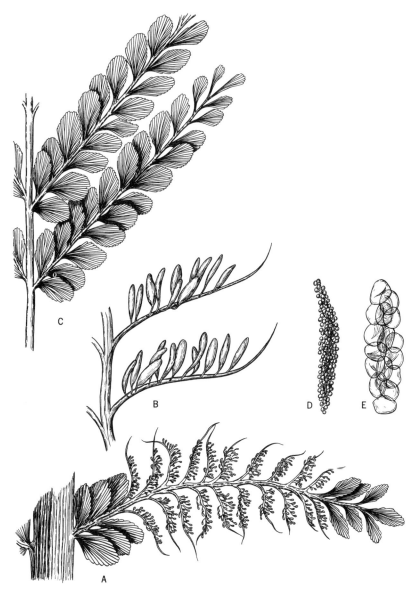

Fig. 3-2. A. *Archaeopteris hibernica,* a pinna with fertile pinnules in the central region. B–E. *Archaeopteris latifolia:* B, C, partially restored fertile and sterile fragments of fronds; D, E, micro- and megasporangial masses. (A from Schimper, 1874; B–E drawn from photographs by Arnold, 1939.)

recorded from numerous localities in New York, Pennsylvania, and eastern Canada, from Great Britain, Ireland, Belgium, Germany, the USSR (Donetz), and from high latitudes in Ellesmere Island and Bear Island.

Archaeopteris hibernica (Forbes) Dawson, the type species, was found at Kilkenny, Ireland, a little over a century ago and specimens from there are rather widely distributed in European museums. The fronds attained a length of over 80 cm and are bipinnate; that is, the strong central rachis bore primary branches on which the pinnules are arranged; the latter are more or less wedge-shaped but vary in different species, some having entire margins whereas others are lobed to fimbriate. The fertile pinnae are found toward the basal part of the frond; a pinna may be entirely fertile or partially so, with sterile pinnules toward the proximal end. The sporangia are borne in clusters which seem to be homologous to the sterile pinnules.

Archaeopteris latifolia Arnold (Fig. 3-2B–E) from the Upper Devonian of Port Allegany, Pennsylvania, is of special interest as it was in all probability a heterosporous species. The fertile pinnae have been shown to bear two kinds of sporangia, both of which are about 2 mm long but some are slender (about 0.3 mm in diameter) and others notably broader (about 0.5 mm in diameter). Although the sporangia actually attached to the pinna rachis had shed their spores, it has been demonstrated that the numerous associated ones contain spores of two kinds. There are 8 to 16 spores, which average about 300 μ in diameter, in the larger sporangia, whereas the others contain a hundred or more spores of about 35 μ diameter. It is thus virtually certain that this was a heterosporous plant although all other species, so far as is known, were homosporous.

Rhacophyton

Professor Leclercq's study of *Rhacophyton zygopteroides* has added a notable chapter to our understanding of the protopterids; it was a plant with a fairly stout stem (Fig. 3-3) and probably attained a height of a meter. A dozen or more sterile fronds were borne near the base of the stem and above this were one or more fertile ones.

The sterile fronds bore two rows of pinnae, and these in turn bore two rows of "pinnules" which consist essentially of a branch system that dichotomizes several times, there being no appreciable development of a lamina.

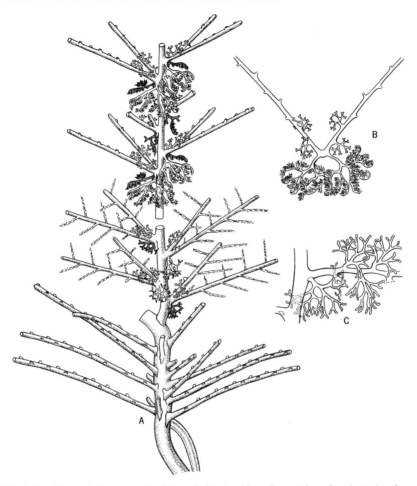

Fig. 3-3. *Rhacophyton zygopteroides.* A. Restoration of a portion of a plant showing the basal parts of several sterile fronds and a single distal fertile frond. B. Basal portion of a fertile pinna showing two sporangial masses. C. Portion of a sterile pinna with several "pinnules." (From Leclercq, 1951.)

The fertile frond is larger and presents several points of particular interest. The restoration figure is based on a specimen on which only one fertile frond was found, and since the stem was broken above its point of departure it is more than likely that there were others. The fertile frond has pinnae in two rows, but each forks immediately after its departure from the rachis; thus at first glance the frond gives the appearance of having four rows of primary pinnae. Each pinna branch divides to form secondary pinnae and pinnules of a pattern very similar to that of the sterile leaves. In the middle to upper portion of the fertile frond each pinna-pair bears on its under side, close to the rachis, two rather profusely dichotomizing branch systems which terminate in sporangia. The sporangia are about 2 mm long and exannulate, there being a rather large number in each cluster.

Eospermatopteris

By Upper Devonian times the land flora included plants of rather large size. The Lycopodophyta was represented by respectable forest trees and *Callixylon,* possibly an early cordaite relative, is known from trunks in excess of 5 feet in diameter. The Pterophyta, not to be outdone in this advance toward arborescent forms, is nobly represented by *Eospermatopteris* from eastern New York state. As long ago as 1869 large stump casts of this plant were exposed near the town of Gilboa as the result of a flood; later intensive search and quarrying operations brought to light several scores of them and occasional trunk fragments, a few of which bore the basal portions of petioles. The plant has thus been regarded as a tall unbranched tree "fern" which attained an estimated height of as much as 40 feet, with a loose crown of fronds 6 to 9 feet long (Fig. 3-4). The stumps are bulbous in shape and quickly tapered to a more slender trunk; for example, one of the largest measures 3.5 feet in diameter and tapers to a breadth of a little less than 2 feet at a height of 22 inches. Numerous slender roots radiated out from the base of the stump to a length of 9 feet. The fronds that are believed to have been borne by these trunks were fernlike in their gross morphology, but they lacked a distinct lamina; that is, the ultimate divisions were dichotomously forking terminations of the branch system. Some of these terminated in ovoid bodies up to 6 mm long and 3 mm in diameter; although originally thought to have been seeds it has been shown that they were sporangiate organs.

The stumps have been found at several horizons indicating a succession of forests which existed over a considerable period of time; they were periodically inundated and partially destroyed by invasions of the sea. Nothing is known of their internal structure other than that some stumps display a coarse reticulate pattern suggesting the cortical fiber strands found in the Carboniferous seed-ferns.

Fig. 3-4. Restoration of *Eospermatopteris* from the Upper Devonian of eastern New York. (From Goldring, 1924, courtesy of New York State Museum and Science Service.)

Aneurophyton germanicum Kräusel and Weyland from the
Upper Devonian of Germany has been considered to be closely
related to, or possibly identical with, the Gilboa plant. It is
represented by thrice pinnate fronds with ultimate ramifications
that are little more than dichotomizing terminations of the branch
system. Some fronds were apparently wholly sterile whereas
others bore clusters of sporangia in the proximal region. In its
anatomy the rachis or main axis consists of a central triangular
core of primary wood which is surrounded by secondary wood composed of tracheids with numerous series of circular (probably bordered) pits and tall, uniseriate rays. The primary pinnae were
probably borne in three rows, or at least departed so as to build up
a three-dimensional branching frond. Suggestions of possible affinities of this fossil have been drawn with the Lower Carboniferous
Stenomyelon which is considered to be the stem of a pteridosperm,
as well as the fronds of *Eospermatopteris*.

Tetraxylopteris

A partially petrified fossil plant, *Tetraxylopteris schmidtii* Beck,
from the Upper Devonian of New York offers several points of
special interest as a transitional type between the psilophytes and
the more advanced vascular plants of the Carboniferous. The
plant as a whole is believed to have attained a height of about 3
meters. The stems reach 2.5 cm in diameter, occasionally dichotomized, and bore regularly arranged appendages which are interpreted as fronds. These are three-dimensionally branching organs
with the primary divisions decussately arranged; that is, they were
given off in nearly opposite pairs and the successive pairs alternate,
the arrangement being similar to that of the distribution of leaves
on a horsechestnut twig. The ultimate divisions terminate as
short dichotomized branchlets which are thought to have been
terete in cross section; there was certainly very little flattening and
consequently little or no lamina.

The sporangia, which were borne in rather dense clusters toward
the apical portion of certain fronds, are 2.5 to 5 mm long and 0.5
to 0.8 mm in diameter; there is no evidence of an annulus and, although spores were not found, the plant is tentatively considered
to have been homosporous.

The first and second order branches of the frond have a central
cruciform primary xylem strand, whereas that of the ultimate
branchlets was terete; the rachis was not as well preserved but is

thought to have been cross-shaped also. The rachis, as well as the first order branches, and the basal part of the second order ones, developed secondary xylem in which the tracheids have several regular rows of crowded, circular-bordered pits. The wood rays are uni- to multiseriate and vary considerably in height. A narrow band of elongate, apparently fibrous cells, was present in the outer cortex.

In the branching pattern of its shoot system *Tetraxylopteris* displays an early but distinct phase in the differentiation of organs that can be designated leaves; the secondary wood and cortical fiber cells are characters suggestive of the pteridosperms. It would be most helpful if more information could be obtained concerning the reproductive organs.

COENOPTERIDALES

It is with some reluctance that the plants described here under this ordinal name are separated from the Protopteridales; the latter are known largely from compression fossils whereas the coenopterids are known from petrifications. This is obviously an unsatisfactory basis on which to segregate fossil plants. It is more than likely that there is overlapping of the plants included in the two orders but, in the light of our present knowledge, this seems to be the most expedient way of dealing with them.

The coenopterids are characterized by protostelic stems, fronds that are three-dimensional in their branching pattern and usually lack a lamina, and with terminally borne sporangia. There are exceptions to these features in the plants that are included within the group. A particular problem lies in the fact that in certain ways, such as size, complexity of their fronds, and stem anatomy (where known), they are of somewhat simpler organization than the Protopteridales, yet the coenopterids are predominantly Carboniferous with some members found in the Permian. It is possible that some of them are reduced rather than truly primitive and the smaller size of many of them may simply mean that we are dealing with fragments of larger plants.

Botryopteridaceae

Botryopteris is one of the better known genera of coenopterids and is represented by several species ranging from the Lower Carboniferous to the Permian. The stem of *Botryopteris trisecta* Mamay and Andrews (Fig. 3-5A) is about 8 mm in diameter, with a small terete protostele surrounded by a broad cortex, a portion of which is distinguished by a conspicuous band of dark sclerotic or secretory cells. The stele produces, at 120° intervals (Fig. 3-13A), an oval-shaped vascular strand which gives rise to two laterals; the central strand of these three becomes the trace of the rachis and the two laterals each trisect again but in a plane that is essentially at right angles to the previous division. The proximal member of this trisection assumes a terete form, resembling closely the structure of the stem stele.

In the figure the lateral or "primary pinna" at the left shows the division of the wood into three segments; in the lateral to the right the strand is just starting to divide; the central rachis trace is clearly M-shaped. It may also be noted that the next leaf trace has just departed from the stem and is in an insipient stage of division itself.

The primary pinna continues to divide and apparently terminated in slender cylindrical branchlets with a very small central xylem strand. However, some specimens of the main rachis have been found with subdivisions that terminate in distinct pinnules.

We may then visualize *B. trisecta* as a small plant, possibly epiphytic, with three-dimensional fronds (that were at least partially laminated) arranged on the stem in a close spiral.

Some of the preferns show a tendency to produce their sporangia in clusters; this is carried to an extreme in *Botryopteris* which renders it unique in the plant kingdom. *Botryopteris globosa* Darrah is a species based on such a sporangiate cluster; specimens have been found at the same Pennsylvanian locality in Illinois from which *B. trisecta* was derived and it is very possible that the two represent one plant; other specimens have come from Iowa and Kansas. *B. globosa* consists of a massive globose aggregation of thousands of sporangia, about 5 cm in diameter (Fig. 3-5B, C). A profusely divided branch system ramifies the fructification with a central strand containing the characteristic *Botryopteris* M-shaped rachis trace. The densely packed, pear-shaped sporangia are attached in clusters at the terminations of the ultimate ramifications of this branch system.

Aside from its tremendous size the chief peculiarity in this organ lies in the modification of the peripheral sporangia (Fig. 3-5B). In most cases the annulus covered the distal three-quarters of the sporangium with the exception of a narrow longitudinal band of

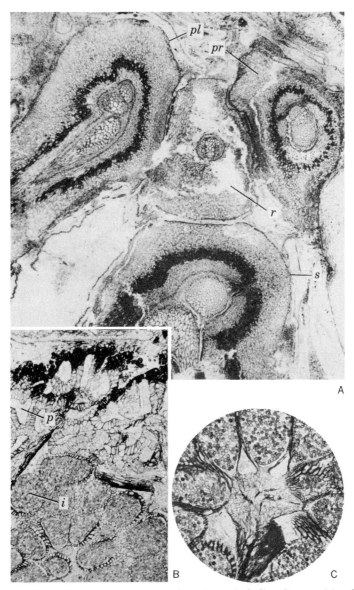

Fig. 3-5. *Botryopteris.* A. A cross section of *B. trisecta* including the stem (s) and basal portion of a frond; the central rachis (r) presents the characteristic M-shaped trace; the strand of the primary pinna at the left (pl) has trisected and the one at the right (pr) is in the initial stages of doing so. B. A small portion of the sporangial aggregate of *B. globosa* showing the peripheral sporangia (p) with their large annulus cells and the interior sporangia (i) which are filled with spores. C. A terminal branchlet of the sporangial aggregate. (A from Mamay and Andrews, 1950.)

thin-walled cells that functioned as the line of dehiscence. The outermost two or three layers of sporangia are, however, quite different in that the annulus cells are tremendously enlarged. Many years ago Bernard Renault described this transition in a French species, *B. forensis,* and came to the conclusion that the peripheral sporangia were sterile ones that served a protective function. Recent studies based on the American material described here suggest that the difference in structure is a maturation process; the enlargement of the annulus cells is gradual through the peripheral region of the entire sporangial aggregate and, although the outermost sporangia, with their tremendous annulus cells, have dehisced, some retain a few spores. If the spores in the vast bulk of the interior sporangia were to have been dispersed, it would seem as though the fructification as a whole must have expanded (for which there is no evidence) or the outermost sporangia progressively matured, shed their spores, and then fell away. A final problem lies in the mode of attachment of the massive sporangial aggregate; the central strand with its M-shaped trace attests to its correct assignment to *Botryopteris* but we do not know how it was borne.

Adding to the complexity of this genus, Surange has described in three species (*B. antiqua, B. ramosa,* and *B. elliptica*) a relationship in which an organ with a bilaterally symmetrical stele (shallowly M-shaped) produced appendages with a radially symmetrical stele (protostele) which in turn gives off spirally arranged structures with the M-shaped strands. This is interpreted as a trailing organ (dorsiventral stem) which gives rise to a radial upright stem that in turn produces petioles. Regardless of the names one applies to this sequence of morphological entities, its real significance seems to lie in the fact that we are dealing with plants in which the differentiation between stem and leaf was not as clear-cut as in most modern plants.

Anachoropteridaceae

The plants assigned to this family have protostelic stems and abaxially turned petiole traces, that is, the trace is curved with the convex side facing out.

Tubicaulis

This is a genus of about five valid species with terete protosteles and with petioles in which the C-shaped strand presents the unusual abaxial orientation; note, for example, the contrast in this character between *Tubicaulis* and the osmundas (see page 114).

In *Tubicaulis scandens* Mamay from the Upper Pennsylvanian of Illinois the trace departs from the periphery of the stem stele as a bar-shaped strand which gradually assumes the form of a shallow C and bears two rows of pinna traces. It grew as an epiphyte on a *Psaronius* plant; a *Tubicaulis* specimen from Chemnitz, Germany, has been reported which occupied a comparable habitat.

Tubicaulis stewartii Eggert (Fig. 3-6), also an Illinois species, was probably an upright plant with stems 1 cm in diameter and xylem in which a considerable amount of parenchyma was associated with the tracheids. Of particular interest is the conspicuous lacunar middle cortex of the stems and petioles; the presence of such air passages suggests rather strongly an aquatic or semiaquatic habitat for the plant.

In contrast to the species cited above, which seem to have been quite diminutive, *T. solenites,* the type species described by Bernhard Cotta in 1832, was a large upright plant, the known fragment being 40 cm high and with a basal diameter of 14 cm.

Apotropteris

Apotropteris minuta Morgan and Delevoryas from the upper Pennsylvanian of Illinois has "inverted" traces like those of *Tubicaulis* but differs in several significant ways. The stem contains a minute protostele with a centrally located protoxylem and

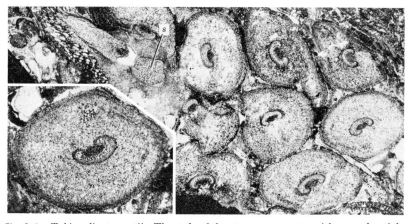

Fig. 3-6. *Tubicaulis stewartii.* The stele of the stem appears at *s* with several petioles toward the right, 3X; the characteristic aerenchymatous tissue may be seen in the enlarged petiole (inset).

the leaf traces depart at irregular intervals in contrast to the closely compacted phyllotaxy of *Tubicaulis*. At the node the stem stele is bisected, leaving a deep indenture by the departing petiole strand, but regains its terete form above the node.

Anachoropteris

The indecisive stem-leaf structure of *Botryopteris* is not confined to that genus of coenopterids. In *Anachoropteris clavata* Graham (Fig. 3-7) from the Upper Carboniferous of Illinois a "petiole" has been observed to produce on one side of its C-shaped strand a "stem" with a terete stele which in turn bears appendages with C-shaped strands, although appreciably smaller than those of the first order. It has been suggested that some petioles of the plant may have functioned as stolons producing new stems in somewhat the same way some modern plants do.

Grammatopteris

A possible connecting link between the coenopterids and the Osmundaceae is found in *Grammatopteris baldaufi* (Beck) Hirmer. The fossil was originally found in a Lower Permian deposit at Hilbersdorf, near Chemnitz; it was cut into several pieces and ultimately scattered among several private and public European collections. These were later located through the efforts of Professor Sahni from whose work the following description is taken. The plant was a small tree-fern with a stem that was enclosed in a heavy armour of cylindrical petioles 6 to 8 mm in diameter. The stem has a protostele about 4 mm in diameter composed chiefly of elongate, pitted, or scalariform tracheids but with an admixture of broad short ones in the central portion. The middle cortex is characterized by crowded nests of sclerotic cells.

Features that suggest affinities with the Osmundaceae are: the structure and closely compacted orientation of the petiole bases; "incisions" in the outer part of the stele which may be interpreted as rudimentary leaf gaps; "parenchymatous" tracheids in the center of the stele which suggest the earliest origin of a pith (a somewhat more extensive differentiation is found in *Thamnopteris*).

Stauropteridaceae

Stauropteris burntislandica P. Bertrand from the Lower Carboniferous of Pettycur, Scotland is of special importance as the only known heterosporous coenopterid. The central shoot of the plant

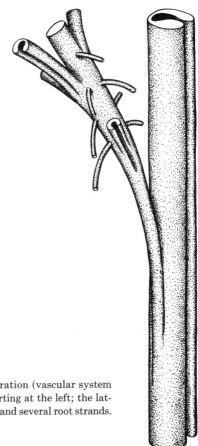

Fig. 3-7. *Anachoropteris clavata.* Restoration (vascular system only) of a primary petiole with a stem departing at the left; the latter is giving off four secondary petiole traces and several root strands. (From Delevoryas and Morgan, 1954.)

is usually regarded as a rachis and the stem that it is presumed to have been borne on has not been found; the petiole is small, measuring about 2 mm in diameter (Fig. 3-8) and its vascular strand is distinctively four-lobed. Branch traces depart at intervals of about 2 mm from opposite sides of the rachis strand; these almost immediately divide into two, resulting in a four-ranked branching system, or perhaps it is more accurately described as consisting of two double rows.

The discovery of heterospory in this plant by K. R. Surange constitutes one of the notable advances in our knowledge of the coenopterids. The megasporangium (Fig. 13-5) is a slightly asymmetric, ovoid organ with a tapered tip and measures about 1.3 mm long and 0.5 mm in maximum diameter. The basal two-thirds is

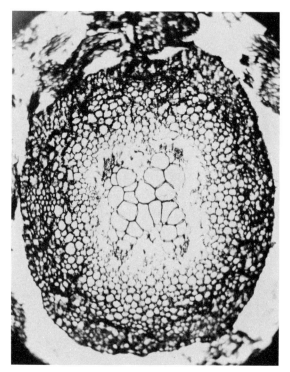

Fig. 3-8. *Stauropteris burntislandica,* transverse section of central rachis, 50X.

parenchymatous with a very slender, central vascular strand. The distal part consists of a cavity containing two relatively large megaspores.

The history of the investigation of this plant is an interesting one and the final chapter or chapters remain unwritten. The structures that have been demonstrated to be megasporangia were first investigated by R. Scott in 1908 who suggested that they might be megasporangia but discarded this concept in favor of a glandular nature; Surange's studies revealed the undoubted presence of two large megaspores and more recently W. G. Chaloner found two aborted spores associated with them, thus demonstrating the presence of a tetrad in which only two spores mature. The latter author has also discovered closely comparable spore tetrads from Carboniferous (Dinantian and Namurian) coals in Ireland, Scotland, and England and more recently still they have been reported by J. M. Schopf from an Upper Mississippian coal in western Kentucky.

A consideration of the possible significance of these curious megasporangia in the evolution of the pteridorperm seed is taken up in Chapter 13.

The microsporangia are also terminally borne and are nearly spherical structures about three-quarters of a millimeter in diameter. The numerous microspores measure 40 to 50 μ in diameter in contrast to approximately 170 μ for the megaspores. In the material of *S. burntislandica* that I have personally had available to study the microsporangia are quite scarce although the megasporangia are abundant.

The Upper Carboniferous *Stauropteris oldhamia* Binney appears to have been homosporous since sporangia, closely comparable with the microsporangia of *S. burntislandica,* are the only reproductive organs that have been found on the plant.

Zygopteridaceae

Ankyropteris

Eight species of *Ankyropteris* have been described, several of which display features of particular interest. *A. glabra* Baxter (Fig. 3-9) from the Pennsylvanian of Indiana is known from a stem specimen about 29 cm long which gave off four leaves through this distance; the internodes are thus quite long. The stem is about 12 mm in diameter and contains a lobed, mixed protostele. In the broad cortex several dozens of minute vascular traces can be observed passing out to peripheral scale-like appendages which have been designated as aphlebiae or protective scales.

The nodal anatomy of *Ankyropteris* is worth considering in some detail since it seems to give us a clue to the origin of axillary branches, one of the distinguishing morphological features of most of the higher plants. The origin of the axillary branch stele in *A. glabra* is somewhat variable; at some nodes a peripheral lobe of the stem stele departs and quickly assumes the characteristic cross-sectional (⊣) shape of the petiole (also referred to as a *phyllophore*). A few millimeters above this point on the stem a terete strand departs which presumably supplies an axillary branch. In other specimens from the same locality it has been observed that a "common trace" departed from the stem stele at the node, and this quickly divided to form the strand of the petiole (phyllophore) and axillary branch.

Axillary branching is one of the most characteristic features of the modern conifers and angiosperms, whereas in the extant ferns it has been reported in the Hymenophyllaceae and the Ophioglos-

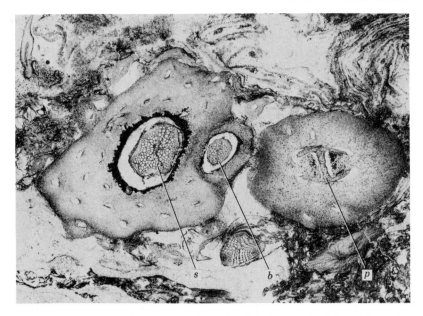

Fig. 3-9. Transverse section through the node of *Ankyropteris glabra*, 6X; *s*, stem stele; *b*, branch stele; *p*, strand of petiole. (From Baxter, 1951.)

saceae. In *Botrychium lunaria* the branch stele departs from the leaf trace shortly after the latter has left the stem stele, whereas in *Helminthostachys* the branch stele originates directly from the stem stele. In these living ferns the branch develops from vestigal buds only if the apical meristem of the stem is destroyed.

On the basis of the little evidence that is available it is my supposition that the axillary branch originated as a modification of a basal "pinna" and probably at a time when the megaphyllous leaf itself was evolving. It is evident, however, that much more research is needed to elucidate this important morphological structure.

An early Pennsylvanian species, *A. hendricksi* Read, has been found in Oklahoma which was probably a rather massive upright tree fern. The only known specimen is a fragment of a trunk including a stem about 3 cm in diameter with a heptagonal stellate stele; several petioles or "phyllophores" are associated with the stem and are held together in a dense matrix of epidermal hair and adventitious roots. The specimen measures about 12 × 8 cm in cross section, but the position of the stem on one edge indicates that this is only a portion of the original trunk.

Zygopteris

The nomenclature of some of the fossils assigned to this family of coenopterids is involved because several names have been given to different parts of the same plant by earlier investigators who had only fragmentary material available. The plants in question present several novel features and add appreciably to our knowledge of the group; it is thus necessary to introduce an unfortunate confusion of names. In 1832 Cotta described *Zygopteris primaria* as the trunk of a small tree-fern from the Lower Permian of Flöha, near Chemnitz. The specimen at his disposal was from the terminal part of the plant and displayed only leaf bases. As a result of searching through several European museums and performing a clever bit of correlation Professor Sahni demonstrated the existence of other portions of the specimen which gave a more complete picture of the entire trunk. When the scattered pieces were fitted back together it was shown to be a trunk about 20 cm in diameter, consisting of a stem only 1.5 cm in diameter surrounded by a thick armor of leaf bases. The xylem of the stem is pentagonal in cross section with a stellate core of primary wood only 1 mm thick which is surrounded by radially aligned secondary tracheids.

The name *Botrychioxylon paradoxum* next enters into the story as one given by D. H. Scott for petrified stems found in the Lower Coal Measures of Oldham, England. His original description was based on a dichotomously branching stem, the stele of which consists of a small mixed protostele surrounded by a zone of secondary wood; the petiole trace, although not well preserved, was described as zygopterid and was accompanied at first by secondary wood as is the case in *Z. primaria*. In describing the stem structure of *Z. primaria* Sahni noted a close similarity with *Botrychioxylon* and indicated that the latter genus probably would never have been created had *Z. primaria* been fully understood at the time. He also noted the identity of the petiole genus *Etapteris* with the petiole of *Z. primaria*.

Some years ago the present writer described a stem from an American coal ball under the name *Scleropteris illinoiensis* which has been shown by R. W. Baxter to be very similar to the English *Botrychioxylon,* and he has also demonstrated the attachment of *Etapteris*-type petioles and has given the fossil assemblage the name *Zygopteris illinoiensis.* It may be noted, however, that this species probably had a trailing stem with widely spaced nodes and was thus quite different from the erect treelike *Z. primaria*.

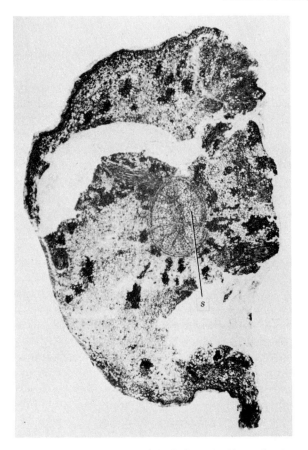

Fig. 3–10. *Zygopteris illinoiensis;* the stele includes a significant development of secondary wood (*s*), 8X.

Austroclepsis

Another arborescent coenopterid, *Austroclepsis australis* (Osborn) Sahni, from the Carboniferous of New South Wales attained a height of 9 to 12 feet with a diameter of 1 foot. The trunk was constructed of several dichotomously branching stems, each with a stele similar to that of the English *Ankyropteris grayi,* as well as numerous leaf bases and many roots. This plant should perhaps be considered as a probable coenopterid; if correctly assigned to that group it extends the known range into the southern hemisphere.

Biscalitheca

It has been pointed out that the aggregation of sporangia into dense masses characterized at least some of the protopterids and coenopterids. *Biscalitheca musata* Mamay is based on a sporangial aggregate which measured $10 \times 5 \times 2$ cm and the unique complexity of the sporangium wall is quite unlike that found in any other plant. Each sporangium is banana-shaped, being 4 mm long by 1 mm in diameter. A conspicuous multiseriate (several cells wide but one cell in thickness) annulus extends along each side. The tissue between the annuli consists, on the upper surface, of long, narrow cells, whereas those of the lower surface are more nearly isodiametric; distributed among these "interannular" cells are tiny groups of smaller and somewhat thicker walled cells. The sporangium wall is, therefore, constructed of three strikingly different cell types and is one of the most complex in the entire plant kingdom. There is a fairly close comparison between *Biscalitheca* and sporangia borne on *Etapteris lacattei* and it is primarily on this evidence that the plant is included in the Zygopteridaceae.

The sporangium contained large numbers of spores and many were observed in which gametophyte development had progressed,

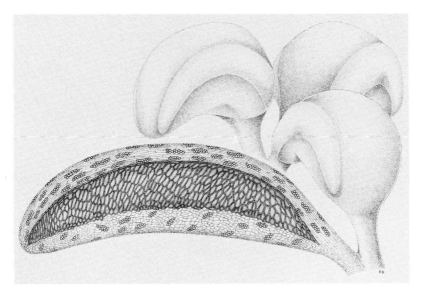

Fig. 3-11. Restoration of the sporangium of *Biscalitheca musata*. (From Mamay, 1957.)

within the spore, up to a 10-cell stage. In a few the spore coat had ruptured and papillate protrusions indicate the initiation of exosporal growth.

Cladoxylales

With the exception of some of the Carboniferous lycopods and articulates there are few pteridophytic plants which develop very much secondary vascular tissue; some, however, attain considerable size through the proliferation of the primary xylem system. The fossils described below display considerable variety of stelar form, and it is very possible that when the plants as a whole become better known they will be revealed as a rather diverse assemblage.

Cladoxylon

Cladoxylon is a genus of numerous species of stems, with complex stelar system, that range from mid-Devonian through the Carboniferous. The most comprehensive account is Paul Bertrand's study of the cladoxylons of Saalfeld and it is another interesting example of the problems involved in fitting together isolated plant parts.

Cladoxylon radiatum (Unger) Bertrand (Fig. 3-12) from the basal part of the Carboniferous is a small stem about 1 cm in diameter with a stelar system consisting of about 18 slender, sheetlike xylem segments which tend to radiate out from the center; these consist of primary wood only. As in all of the cladoxylons there are conspicuous peripheral "loops" near the distal end of each stelar segment (meristele); these were probably occupied by the protoxylem elements admixed with thin-walled parenchymatous cells. Rather large branches were initiated by the division of the distal part of several of the meristems. *C. taeniatum* (Unger) Bertrand (Fig. 3-12) is a larger stem, up to 5 cm in diameter. The stelar system consists of five small, central meristeles that are cylindrical in cross section; they are surrounded by numerous peripheral ones that are radially elongated to various degrees. In contrast to *C. radiatum,* each of the primary steles is surrounded by a band of secondary wood; its tracheids are scalariform and interspersed with parenchymatous rays. Appendages are given off in which the stelar segments tend to be arranged in a bilateral symmetry; these had been described previously from fragmentary material under the generic names *Hierogramma* and *Syncardia.*

A Middle Devonian species, *C. scoparium,* from Germany, has been described as having a stelar system that was apparently simi-

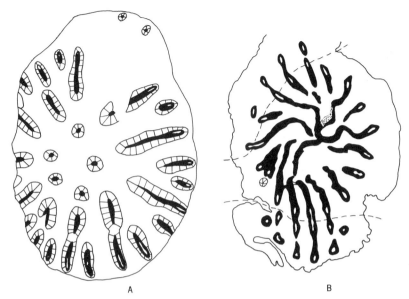

A   B

Fig. 3-12. Diagrams of the stems showing the stelar systems of: A. *Cladoxylon taeniatum* (about 2X) and B. *C. radiatum* (about 6X); solid black indicates primary wood. (From Bertrand, 1935.)

lar to that of *C. radiatum*. This description was based on compression fossils, and in their restoration Kräusel and Weyland show it as an irregularly branching plant with two kinds of small appendages, both sterile ones that were several times forked and probably photosynthetic as well as others that were dissected, fan-shaped structures which probably bore sporangia along the margin. Better preserved specimens are needed to substantiate fully the restoration of this plant and Bertrand held the opinion that it was not closely related to the Saalfeld fossils.

It seems likely that these fossils represent more than one genus and their curious and varied appendages cannot be closely identified with the usual leaf-stem categories of the higher plants.

Xenocladia

Xenocladia medullosina Arnold from the upper Middle Devonian of Erie County, New York, is based on a stem fragment measuring 1 × 5 cm, although it is apparently only a portion of the original. It is polystelic with a central core of units that are more or less circular in cross section, and peripheral to these are more or less radially elongated ones. Each stele consists of a small central mass of primary tracheids surrounded by secondary xylem in which the tracheids have two rows of bordered pits.

Steloxylon

This is another Upper Devonian genus of polystelic stems, somewhat similar to *Xenocladia,* but the numerous anastomosing steles, which are surrounded by a parenchymatous tissue, are all nearly cylindrical. In *S. irvingense* Read and Campbell the individual strands range from 0.8 to 1.7 mm in diameter, whereas in *S. sanctaecrucis* R. and C. they reach 3.8 mm in diameter.

Pietzschia

Pietzschia polyupsilon Read and Campbell from the Upper Devonian of Kentucky is known from stems about 25 mm in diameter in which the vascular system consists of a single peripheral ring of about 54 radially elongate strands of primary wood only. In the development of the vascular supply of an appendage four of the strands pinch off a segment of xylem and these fuse to form a strand with the form of an inverted U.

THE EVOLUTION OF THE MEGAPHYLLOUS LEAF— THE TELOME CONCEPT

Thus far we have dealt with plants of a very ancient lineage which in many cases cannot be described with the same terminology that is used with extant ones. Prior to the discovery of the Rhynie fossils it had long been accepted that vascular plants were composed of roots, stems, leaves, as well as reproductive organs which were generally regarded as modified leaves of one sort or another. This older philosophy of plant form was based largely on studies of the flowering plants, which did not make their appearance in any abundance until mid-Mesozoic times, and grew out of the ideas expressed by the German poet-philosopher Goethe in his treatise on the metamorphosis of plants or, as Sporne aptly expresses it, as "a slight misrepresentation of those ideas."

If the concept of organic evolution is to be applied to the higher plants we might expect that in earlier geologic periods we should find traces of the origin of the root-stem-leaf differentiation that does characterize most of the living pteridophytes and seed plants. This is precisely the contribution that the psilophyte and prefern assemblages of fossils are beginning to make. Although the information at hand is only a beginning, significant gaps have already been filled in and we are able to catch glimpses of how modern plants evolved. An especially significant aspect of this evolution is the light that is shed on the development of the megaphyllous leaf, that is, the leaf type that is found in most true ferns and seed-ferns.

The sketches given in Fig. 3-13 illustrate some of the apparent stages in the evolution of such a leaf. As a starting point a simple, dichotomously forking shoot system such as *Rhynia* (Fig. 2-1) may be taken. Here there is no distinction between stem and leaf and the ultimate units of the system are now generally referred to as sterile or fertile *telomes,* depending on whether or not a terminal sporangium is present. This morphological concept originated from the work of the able German botanist, Walter Zimmermann; a brief and readable account is found in the reference cited at the end of the chapter. It is a most useful term with the primitive vascular cryptogams where the terminology of living plants proves inadequate; its application to the higher plants seems to me unfortunate since it is by no means clear as to what telomic units are in all cases; consequently, their homology with telomes in the lower groups is obscure.

One of the first steps in the evolution of the leaf seems to have been the differentiation of the shoot system into a central axis with secondary units that are arranged in a regular fashion. These units or appendages were at first three dimensional in their branching pattern and probably nonlaminate; *Botryopteris trisecta* may be taken as an example (A).

Confining our attention to a secondary unit or appendage the next stage that may be recognized is known as *planation.* The three-dimensional branch system is now flattened into a two-dimensional one and as an example a portion of the frond of *Telangium affine* is shown in diagram B. (It will be evident that some liberty is taken with the strict rules of geometry in calling a flat leaf two-dimensional since it of course does have some thickness!) It will be noted here that there is still little evidence of the development of a blade or lamina.

Most modern fern fronds are monopodial in their branching; that is, they possess a strong, straight, central rachis from which the primary branches depart. This originated through a modification known as *overtopping;* instead of equal dichotomies one fork of each branch is conspicuously stronger than the other. This may be observed occasionally in certain living fern leaves. As an example a fragment of the Carboniferous *Lonchopteris bricei* is given in Fig. C.

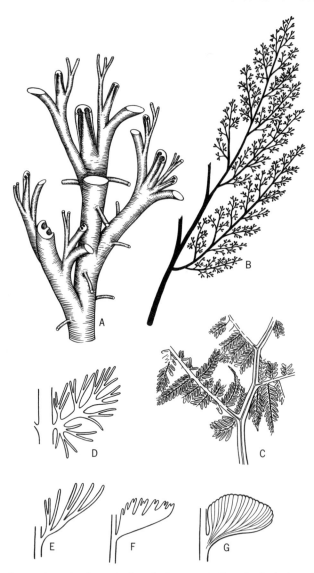

Fig. 3-13. Some phases in the evolution of the megaphyllous leaf. A. A portion of the stem of *Botryopteris trisecta* with the basal parts of three fronds. B. A representative part of a leaf of *Telangium affine;* note two-dimensional form, dichotomous branching, and lack of distinct pinnules. C. A frond fragment of *Lonchopteris bricei* with unequal dichotomies. D–G. Pinnules of *Rhacopteris* species: D, *R. geikiei;* E, *R. inaequilatera;* F, *R. flabellata;* G, *R. lindseaeformis.* (B from Miller, 1857; C from Kidston, 1911; D–F from specimens in the British Museum of Natural History; G from Kidston, 1923.)

The development of a distinct blade (pinnule) appears to have resulted from the webbing together (called *syngenesis*) of a simple telomic system. As an example the pinnules of four species of *Rhacopteris* are shown in Figs. D–G. It should be emphasized that this evolutionary "series" from a completely undifferentiated shoot system such as that of *Rhynia* to a fern or seed-fern in which the fronds are monopodial, flat, laminated organs arranged in a regular pattern on an axis (stem) is outlined only in a very general way. Many other examples might, however, be given; for example, in *Rhacophyton* and *Tetraxylopteris* there is a distinction between stem and leaf in that we find morphological units arranged in a regular fashion along a main axis. The "leaves," however, are best described as telomic branch systems since they are three-dimensional and nonlaminate. This type of leaf-stem relationship seems to have been characteristic of many of the protopterids and coenopterids.

There is now no reasonable doubt that megaphyllous leaves evolved along this general path but it was a kind of evolution that was going on in many different subgroups in slightly different ways and at different rates. The examples selected reveal only the trend as a whole.

REFERENCES

Arber, Agnes. 1946. Goethes' Botany. The Metamorphosis of Plants 1790. *Chronica Botanica,* 10: 67–115.

Arnold, Chester A. 1939. Observations on fossil plants from the Devonian of eastern North America. IV. Plant remains from the Catskill delta deposits of northern Pennsylvania and southern New York. *Univ. Michigan, Contrib. Mus. Paleont.,* 5: 271–314.

————. 1952. Observations on fossil plants from the Devonian of eastern North America. VI. *Xenocladia medullosina* Arnold. *Univ. Michigan, Contrib. Mus. Paleont.,* 9: 297–309.

Baxter, Robert W. 1951. *Ankyropteris glabra,* a new American species of the Zygopteridaceae. *Amer. Journ. Bot.,* 38: 440–452.

————. 1952. The coal-age flora of Kansas. II. On the relationships among the genera *Etapteris, Scleropteris* and *Botrychioxylon. Amer. Journ. Bot.,* 39: 263–274.

Eggert, Donald A. 1959. Studies of Paleozoic ferns. The morphology, anatomy and taxonomy of *Ankyropteris glabra. Amer. Journ. Bot.,* 46: 510–520.

————. 1959. Studies of Paleozoic ferns. *Tubicaulis stewartii* sp. nov. and evolutionary trends in the genus. *Amer. Journ. Bot.,* 46: 594–602.

Beck, Charles B. 1957. *Tetraxylopteris schmidtii* gen. et sp. nov., a probable pteridosperm precursor from the Devonian of New York. *Amer. Journ. Bot.,* 44: 350–367.

Bertrand, Paul. 1935. Contribution a l'étude des Cladoxylées de Saalfeld. *Palaeon-tographica,* 80B: 101–170.

Chaloner, William G. 1958. Isolated megaspore tetrads of *Stauropteris burntislandica. Ann. Bot..* n. s., 22: 197–204.

Delevoryas, Theodore, and Jeanne Morgan. 1952. *Tubicaulis multiscalariformis:* a new American coenopterid. *Amer. Journ. Bot.,* 39: 160–166.

————. 1954. An anatomical study of a new coenopterid and its bearing on the morphology of certain coenopterid petioles. *Amer. Journ. Bot.,* 41: 198–203.

————. 1954. A further investigation of the morphology of *Anachoropteris clavata. Amer. Journ. Bot.,* 41: 192–198.

————. 1954. Observations on petiolar branching and foliage of an American *Botryopteris. Amer. Mid. Nat.,* 52: 374–387.

Goldring, Winifred. 1924. The Upper Devonian forest of seed ferns in eastern New York. *N. Y. State Mus. Bull.,* 251: 50–72.

Holden, H. S. 1930. On the structure and affinities of *Ankyropteris corregata. Phil. Trans. Roy. Soc. London,* 218B: 79–114.

Lang, William H. 1913. Studies in the morphology and anatomy of the Ophioglos-saceae I. *Ann. Bot.,* 27: 203–242.

————. 1915. Studies in the morphology and anatomy of the Ophioglossaceae III. *Ann. Bot.,* 29: 1–54.

Leclercq, Suzanne. 1951. Etude morphologique et anatomique d'une fougére du Dévonien Supérieur. *Ann. Soc. Geol. Belgique, Mem.,* 40, 9: 1–62.

Mamay, Sergius H. 1952. An epiphytic American species of *Tubicaulis* Cotta. *Ann. Bot.,* n. s., 16: 145–163.

————. 1957. *Biscalitheca,* a new genus of Pennsylvanian coenopterids, based on its fructification. *Amer. Journ. Bot.,* 44: 229–239.

————, and Henry N. Andrews, Jr. 1950. A contribution to our knowledge of the anatomy of *Botryopteris. Bull. Torrey Bot. Club.,* 77: 462–494.

Read, Charles B. 1935. An occurrence of the genus *Cladoxylon* Unger in North America. *Journ. Washington Acad. Sci.,* 25: 493–497.

————. 1938. A new fern from the Johns Valley shale of Oklahoma. *Amer. Journ. Bot.,* 25: 335–338.

Renault, Bernard. 1875. Recherches sur les végétaux silicifies d'Autun et de Saint-Étienne, étude du genre *Botryopteris. Ann. Sci. Nat., Botanique, Paris,* ser. 6, 1: 220–240.

Sahni, Birbal. 1928. On *Clepsydropsis australis,* a zygopterid tree-fern with a *Tempskya*-like false stem, from the Carboniferous rocks of Australia. *Phil. Trans. Roy. Soc. London,* 217B: 1–37.

————. 1932. On the genera *Clepsydropsis* and *Cladoxylon* of Unger, and on a new genus *Austroclepsis. New Phyt.,* 31: 270–278.

————. 1932. On a Palaeozoic tree-fern, *Grammatopteris baldaufi* (Beck) Hirmer, a link between the Zygopterideae and the Osmundaceae. *Ann. Bot.,* 46: 863–877.

————. 1932. On the structure of *Zygopteris primaria* (Cotta) and on the relations between the genera *Zygopteris, Etapteris* and *Botrychioxylon. Phil. Trans. Roy. Soc. London,* 222B: 29–46.

Scott, Rita. 1908. On *Bensonites fusiformis,* sp. nov. a fossil associated with *Stauropteris burntislandica,* P. Bertrand and on the sporangia of the latter. *Ann. Bot.,* 22: 683–687.

Scott, Dukinfield H. 1905. The sporangia of *Stauropteris oldhamia. New Phyt.,* 4: 114–120.

Sporne, Kenneth R. 1959. On the phylogenetic classification of plants. *Amer. Journ. Bot.,* 46: 385–394.

Surange, K. R. 1952. The morphology of *Stauropteris burntislandica* P. Bertrand and its megasporangium *Bensonites fusiformis* R. Scott. *Phil. Trans. Roy. Soc. London* 237B: 73–91.

————. 1952. The morphology of *Botryopteris antiqua* with some observations on *Botryopteris ramosa. The Palaeobotanist* 1: 420–434.

————. 1954. *Botryopteris elliptica* sp. nov. from the Upper Carboniferous of England. *The Palaeobotanist,* 3: 79–86.

Zimmermann, Walter. 1952. Main results of the "Telome Theory." *The Palaeobotanist,* 1: 456–470.

Appendix—Carboniferous Fernlike Foliage

Fern—like foliage constitutes one of the most conspicuous features of Carboniferous plant-bearing rocks. Numerous genera and hundreds of species have been recognized since plants of that age began to be collected and studied seriously at the start of the nineteenth century. In the earlier days efforts were made to identify such fossils with living ferns, but it soon became apparent to the more critical investigators that they were dealing with extinct plants and in the latter decades of the last century the added suspicion began to take form that many of them were indeed not ferns at all.

The problem of identification of these fossils centers around the facts that a vast majority of specimens are sterile, that is, they bear no reproductive organs, sporangia or seeds, and some species have been founded on fragmentary or poorly preserved specimens. Many of these "fossil ferns" do have a certain esthetic attraction, with the result that museum collections have been swollen with quantities of material that is of little biological importance. One good fertile specimen of a given species will tell far more than any quantity of sterile ones. It is true, however, that a few workers have become sufficiently clever in distinguishing the various types to use them effectively in stratigraphic studies.

Regardless of one's opinions concerning the relative importance of such fossils their abundance makes it impossible to ignore them. As a result numerous *form-genera* have been established; in the case of the fernlike foliage a form-genus may be defined as one including species that are similar in their gross morphological features but in which reproductive organs are lacking and consequently the natural affinities are unknown. This definition should perhaps be qualified in that some of the genera, notably *Neuropteris* and *Alethopteris,* include species which are known to have borne seeds and it is now highly probable that all of them are leaves of pteridospermous plants; the genus *Pecopteris* includes several species that have been found to bear marattiaceous fructifications, and are thus true ferns, whereas seeds have been found on other pecopterids.

The general problem is complicated by the fact that the form-genera are not in all cases sharply defined as far as their gross morphological characters are concerned. A similar intricacy is encountered in the identification of cycadophyte and ginkgophyte foliage (see Chapters 10, 11) in Mesozoic rocks. Studies of cuticular characters of the leaves of these groups have aided greatly in their classification; very few such investigations have been undertaken with the Carboniferous fernlike leaves.

It is the objective in this brief treatment to simply introduce a few of the more common and representative form-genera, particularly ones that must be referred to in later chapters on true ferns and pteridosperms.

Adiantites. The fronds are three to four times pinnate with slender rachises; the pinnules are obovate or wedge-shaped with a truncate or rounded apex.

Alethopteris. The fronds are several times pinnate, with pinnules that are attached by a flaring base and usually make an acute angle with the pinna midrib. Most, if not all, of the *Alethopteris* species were borne on stems of the medullosan pteridosperms (see Chapter 5.

Alloiopteris. These are presumed to have been fronds of considerable size in which the ultimate pinnae are elongate, nearly linear, and divided to varying degrees in different species. The linear pinnae and characteristic mode of lobing seem to distinguish this group quite clearly.

Diplotmema. The rachis dichotomizes into two equal branches below which no pinnules are attached; the latter are deeply and finely divided and merge into pinnules of the *Sphenopteris* type, although the fronds as a whole were much smaller than in that form-genus.

Linopteris. This is a rather uncommon but distinctive type. The pinnules are similar in shape to those of *Neuropteris* but have a fine-meshed net venation in contrast to the open dichotomous venation of *Neuropteris*.

Lonchopteris. This is another uncommon type in which the pinnules are closely comparable to those of *Alethopteris* but differ in their net venation.

Fig. 3-14. Pinnule morphology of some Carboniferous fronds. A. *Alethopteris lonchitica* Schlotheim; B, *Neuropteris heterophylla* Brongniart; C. *N. scheuchzeri* Hoffmann; D. *Mariopteris bellani* P. Corsin; E. *M. leharlei* P. Corsin; F. *Linopteris* sp.; G. *Lonchopteris rugosa* Brongniart; H. *Pecopteris rarinervis* P. Corsin; I. *P. volkmanni* Sauveur; J. *P. platoni* (Grand'Eury); K. *Alloiopteris* sp.; L. *Sphenopteridium dissectum* (Goeppert); M. *Diplotmema furcatum* Brongniart; N. *Rhodea smithi* Kidston; O. *Sphenopteris flabellifolia* Kidston; P. *S. elegantiformis* Stur; Q. *S. kayi* Arber; R. *Adiantites antiquus* (Ettingshausen); (D, E from Corsin, 1932; I, J from Corsin, 1951; L, O, P, Q, R from Kidston, 1923, by permission of the Controller of Her Britannic Majesty's Stationery Office.)

Pecopteris. This group consists of many species of large fronds that were several times pinnate. Many have been found with fern sporangia and consequently have been segregated to distinct genera. The pinnules are erect or nearly so, attached along their entire base and with a sparsely forking vein system.

Neuropteris. This is another common group of many species, most, or all of which are pteridosperm leaves; several have been found with seeds attached. The pinnules have a contracted base, being attached by a very slender stalk or petiolule.

Mariopteris. This includes a large assemblage in which specific delimitations are especially difficult. The pinnules are attached along their entire base and it perhaps is not amiss to describe them as somewhat intermediate between those of *Pecopteris* and *Sphenopteris;* the lobing is similar to the latter but usually not as pronounced. In some species the fronds divide by uniform dichotomies comparable to a *Gleichenia,* and Kidston was of the opinion that most *Mariopteris* species were vines with relatively slender stems and depended on surrounding vegetation for support.

Rhacopteris. The fronds are but once pinnate, with alternate, often overlapping flabelliform pinnules which range from entire to very deeply dissected (Fig. 3-13, D–G).

Rhodea. The pinnules are dissected to the point where there is essentially no lamina present; it seems questionable, however, whether there is a real generic distinction between the species of this form-genus and certain ones in *Sphenopteris* and *Diplotmema.*

Sphenopteridium. The fronds are twice pinnate and the rachis dichotomizes in the basal part; the pinnules are more or less wedge-shaped and variously divided.

Sphenopteris. The fronds are two to four times pinnate; the pinnules have a contracted base and vary from nearly entire to dissected types of all sorts, and with blunt or pointed lobes. In his "Fossil plants of the Carboniferous rocks of Great Britain" Kidston recognized 77 species; the total number for all areas and ages would probably amount to several hundred.

REFERENCES

Corsin, Paul. 1951. Bassin Houiller de la Sarre et de la Lorraine. I. Flore Fossile. 4me fasc. Pécoptéridées. *Études des Gîtes Minér. de la France.* pp. 175–370.

———. 1932. Bassin Houiller de la Sarre et de la Lorraine. I. Flore Fossile. 3me fasc. Marioptéridées. *Études des Gîtes Minér. de la France.* Pp. 110–173.

Crookall, Robert. 1929. *Coal Measure Plants.* Edward Arnold Co., London. 80 pp.

Danzé-Corsin, Paule. 1953. Contribution a l'étude des Marioptéridées. Les Mariopteris du Nord de la France. *Service Géologique des H.B.N.P.C.*

Danzé, Jacques. 1956. Contribution a l'étude des Sphenopteris. Les fougéres Sphenoptéridiennes du Bassin Houiller du Nord de la France. *Service Géologique des H.B.N.P.C.,* Lille, 568 pp.

Janssen, Raymond E. 1939. Leaves and stems from fossil forests. *Illinois Popular Science,* ser. 1: 1–190. Springfield.

Kidston, Robert. 1923-25. Fossil plants of the Carboniferous rocks of Great Britain. *Mem. Geol. Survey Great Britain.,* vol. II, pts. 1–6.

Miller, Hugh. 1857. The Testimony of the Rocks. Gould and Lincoln, Boston, 502 pp.

Noé, A. C. 1925. Pennsylvanian flora of northern Illinois. *Illinois State Geol. Surv. Bull.* 52: 1–18.

Schimper, W. P. 1874. *Traité de Paleontologie végétale.* Atlas. J. B. Bailliere et Fils, Paris. 110 pp.

4

the PTEROPHYTA
—Marattiales and Filicales

Most of the plants that are dealt with in this chapter are referred to as the true ferns, a large and exceptionally varied assemblage. Many of them are plants of great beauty and have long attracted the interest of both amateur and professional botanists. Numerous systems of classification have been proposed and recent studies have tended to split them into an increasing number of taxa. For example, in his excellent three-volume work *The Ferns,* F. O. Bower, in 1923 recognized 14 families (excluding the "Polypodiaceae") whereas Pichi-Sermolli, in a recent treatise on the classification of the pteridophyta, recognizes 17 orders and 44 families. This number of families seems to me excessively high, but it does point to the clearly evident diversity of the, approximately, 10,000 species of extant ferns.

Only those fern families with significant fossil records are described; there is also included, at the close of the chapter, a consideration of the Noeggerathiales, a problematical and interesting group of pteridophytes. The order of subject material is, therefore, as follows:

PTEROPHYTA
  Marattiales
    Marattiaceae
  Filicales
    Schizaeaceae, Gleicheniaceae, Matoniaceae, Dipteridaceae, Osmundaceae, Tempskyaceae
    (Some ferns of uncertain affinities, possibly transitional between the coenopterid-protopterid complex and the true ferns)
    Noeggerathiales—relationships to other pteridophytes unknown

A few additional introductory comments may be helpful in outlining the nature of the fossil records and reasons for the particular selection of subject material. The Marattiaceae is represented by many beautifully preserved sporangiate organs in the Upper Carboniferous which are associated with stems of complex stelar anatomy that probably bore the fertile fronds; the group was well established at that time and undoubtedly had had a long previous history of which little is known. The Gleicheniaceae likewise come in rather suddenly at about the same time. The Schizaeaceae has been traced back to the Lower Carboniferous and there is a particularly significant series of compression fossils which sheds light on the evolution of the sporangia of this family, but virtually nothing is known of the anatomy of these plants. By contrast the earliest members of the Osmundaceae, which appear in the Permian, are known from a series of petrified stems which attained a rather modern aspect by mid-Mesozoic times. The Tempskyaceae is described in some detail because of its unparalleled trunk structure and the fact that it was an abundant floral element of late Mesozoic landscapes. Under the heading "ferns of uncertain affinities" a few fertile ferns are introduced because of their transitional nature, relative to the position of the sporangia, between the preferns and true ferns.

MARATTIALES

Marattiaceae

This family is represented today by seven genera; all are tropical with four confined to the Malayan region and only one, *Marattia,* native to both eastern and western hemispheres. The fronds of many members of the family are large; particularly distinctive are the sori, which are usually arranged in two rows, one along either side of the midrib. The sorus of *Angiopteris* consists of 20 or more separate sporangia arranged in two closely aggregated rows, whereas in *Marattia* the sorus is made up of a similar number of sporangia that are fused into a single unit. *Christensenia* differs from the other genera in having a radially symmetrical synangium of about a dozen sporangia. In all, the sporangial walls are thick, multicellular structures enclosing prodigious numbers of spores, at least for members of the Pterophyta. Some of the spore numbers per sporangium are approximately as follows: *Angiopteris* 1500, *Marattia* 2500, and *Christensenia* 7000. In the genera *Angiopteris, Marattia,* and *Archangiopteris* the stem is rather short and massive and like most, if not all, modern ferns the stelar system originates in the sporeling as a single small protostele, but soon develops into a large and complicated system of several series of meristeles or stelar segments.

The Carboniferous Marattiaceae were tall, graceful tree ferns with crowns of large pecopterid fronds; the trunks were unbranched, gradually tapering columns and heavily buttressed with adventitious roots in the basal region. The leaves, with their sporangia, and the stems are often remarkably well preserved.

Among the more frequently encountered fossils in American coal ball petrifactions are leaf fragments bearing sporangiate organs, often exquisitely preserved, which have been considered to be of marattiaceous affinities. *Cyathotrachus altissimus* Mamay (Fig. 4-1) is a rather large synangium consisting of five to nine sporangia which are united for about half their length around a central column. The sporangia are distinct morphological units consisting of a wall,

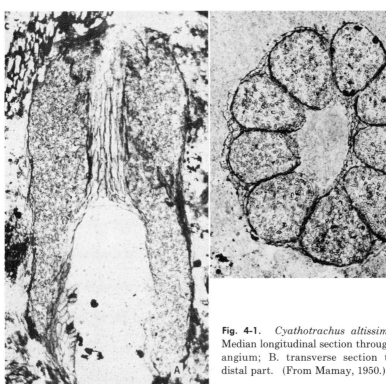

Fig. 4-1. *Cyathotrachus altissimus.* A. Median longitudinal section through a synangium; B. transverse section through distal part. (From Mamay, 1950.)

probably one cell thick at maturity, enclosing many thousands of spores; they are, however, enclosed in a synangial envelope several cells thick. As the fructification approached maturity this envelope ruptured and breaks occurred between the closely appressed walls of the sporangia so that they became separate beyond the column and dehisced by longitudinal slits along their inner wall, the spores being shed toward the central chamber. This species is abundant in Iowa coal balls and the genus is also represented by a species from Illinois and one from the Lower Coal Measures of England. The synangia of *Cyathotrachus,* which were probably borne on foliage of the *Pecopteris* type, are comparable with those of the living *Christensenia,* the chief differences being in the discharge of spores in the latter by apical pores in contrast to the longitudinal slits of *Cyathotrachus.*

Since many of the fossil plants considered in this and several following chapters come from coal ball petrifications of England and this country, the correlation chart given on page 126 is included as a reference to the position of the more important localities.

Scolecopteris is a genus of some nine species and is common in European and American Upper Carboniferous horizons. The synangium of *S. iowensis* Mamay (Fig. 4-2) from the Des Moines series of Iowa usually consisted of 6 sporangia attached to a common pedicel but free beyond this basal point; the sporangia are nearly 1 mm long and taper to bluntly acute apices. The synangia were arranged in two rows, one along either side of the midrib of the pinnule. Although the pinnule structure of some species of *Scolecopteris* appears to have been of the *Pecopteris* type this is not so in all cases. In *Scolecopteris incisifolia* Mamay (Fig. 4-3), also from the Des Moines series of Iowa, there are usually four sporangia in each sorus and the margins of the pinnule are lobed and strongly enfolded with a synangium attached at the base of each lobe.

Eoangiopteris andrewsii Mamay differs from the above in that the sporangial aggregate is distinctly elongate; there are five to eight sporangia which are shallowly immersed in a basal receptacle but are essentially separate; they were borne on *Pecopteris* type pinnules (Fig. 4-3). The term *sorus* is used to designate such sporangial units; it would perhaps be preferable to use it throughout our description of the marattiaceous fructifications, but since it is important to distinguish between those in which the sporangia are organically fused (synangia) and those in which they are not, the latter will be called sori.

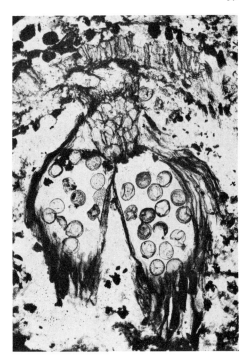

Fig. 4-2. *Scolecopteris iowensis,*
a median longitudinal section.
(From Mamay, 1950.)

As usual the problem exists here of correlating petrifaction and compression specimens. *Asterotheca* is a common genus of pecopterid fronds bearing marattiaceous sporangiate organs. In *Asterotheca parallela,* Radforth has shown that each fructification is composed of three to five sporangia joined at their bases to a common stalk; approximately 1500 to 2000 spores were produced by each sporangium. Several students of these plants are of the opinion that *Asterotheca* and *Scolecopteris* species should be referred to as a single genus. The chief difficulty lies in the fact that the compression fossils are rarely preserved well enough to allow positive comparison with the petrified ones.

Large pecopterid fronds bearing fructifications such as those described above are frequently found associated with petrified trunks referred to the genus *Psaronius,* which is characterized by a complex polycyclic stelar system. Numerous species of the genus have been recognized, chiefly from the European Carboniferous; although not rare on this side of the Atlantic they have received but little attention until recently. It has been recognized for some time that there is much variability displayed in the organization of the numerous primary stelar segments (meristeles) that make up a

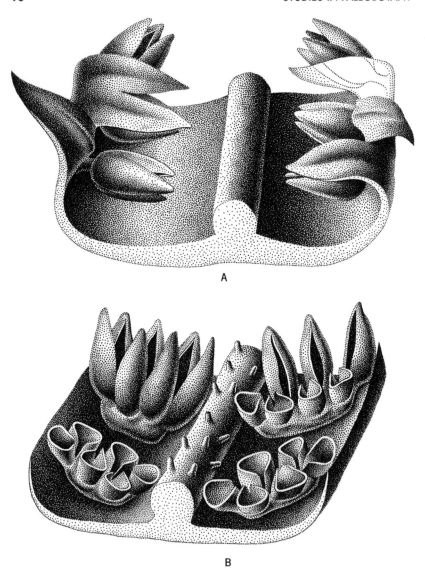

Fig. 4-3. Restorations of representative portions of the fertile pinnules of: A. *Scolecopteris incisifolia;* B. *Eoangiopteris andrewsii,* about 20X. (From Mamay, 1950.)

stem and consequently some doubt has been cast on the validity of many specific identifications. They present a problem rather akin to that of the medullosan seed-fern stems described in the next chapter, for almost every specimen that turns up differs a little from any other.

A significant advance in our knowledge of the plants has resulted from Morgan's recent study of numerous specimens from a locality in southeastern Illinois. A nearly continuous sequence of specimens was found which demonstrate clearly that the stelar system developed obconically, as, years ago, F. O. Bower demonstrated is the case with most modern ferns. That is, the stele in the basalmost part of the trunk is small and quite simple; in the case of *P. blicklei* Morgan it is about 2 cm in diameter and is described as dicyclic, the vascular system consisting of a central strand surrounded by a ring of several others. (It would be most interesting to have the sporeling stage at hand, but very probably it was a protostele.) This simple stelar system gradually proliferated as the trunk increased in height and girth to a polycyclic stele consisting of ten or more concentric cycles.

The leaf traces follow a similar pattern of increasing complexity; there are only three vertical rows of traces, representing three rows of petioles, in the basal portion of the trunk of *E. blicklei,* whereas

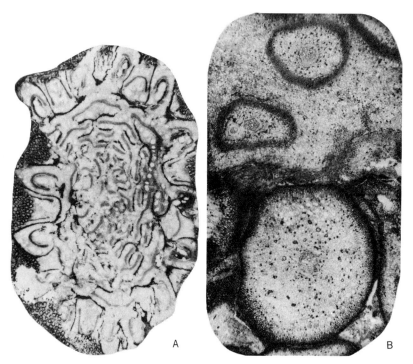

A   B

Fig. 4-4. *Psaronius blicklei.* A. Transverse view of the trunk showing numerous concentric stelar cycles; a portion of the inner root zone is intact. B. Junction of inner and outer root zones, 5X. (A from Morgan, 1959.)

specimens from the distal part of a trunk display up to 14 orthostichies.

Another characteristic feature of many *Psaronius* specimens is the mantle of roots surrounding the stem. The specimen of *P. blicklei* with the small stem cited above was found with a mantle of roots over 30 cm thick; thus the over-all diameter of the specimen (quite clearly a stump) was close to three-quarters of a meter. The root mantle is composed of two distinct zones: an inner one of adventitious roots that are embedded in a parenchymatous matrix; and an outer zone of appreciably larger adventitious, free roots. These roots, either isolated specimens or tangled clusters of them, are ubiquitous in certain American coal-ball localities and often occur to the exclusion of almost any other kind of plant remains.

The individual root has a central star-shaped stele and a broad cortex which is essentially a meshwork of chains of parenchyma cells with large intervening chambers. Such an aerenchymatous organization is usually found in the roots, stems, and leaves of aquatic or semiaquatic plants and suggests a swampy habitat for the *Psaronius* trees.

Partially petrified specimens, as well as compression or impression fossils of tree-fern trunks, have been described under the generic names *Megaphyton, Caulopteris,* and others; they are characterized chiefly by large petiole scars and distinctively shaped trace scars. *Megaphyton* was established for a specimen with only two rows of such scars, one on either side of the stem; *Caulopteris* has been used for specimens with several rows arranged in what appears to be a spiral pattern. In view of the variation that is now known to exist in a trunk of *P. blicklei* it may be understood with what hesitancy one can assign species names to impressions or poorly preserved petrifactions.

In summary, the marattiaceous tree ferns were one of the dominating elements of the Upper Carboniferous and Permian forests. Their unbranched trunks attained a height of probably 50 feet or more, and although the stem itself tapered obconically the dense mantle of adventitious roots resulted in a heavily buttressed base which tapered upward to the crown of leaves which were few in number but of great size. Some are estimated to have reached a length of 3 meters, and these in turn bore sporangiate organs quite similar to those of the living members of the family.

The evidence for such a restoration comes chiefly from the actual attachment of the sporangia to leaf fragments and the asso-

Fig. 4-5. Reconstruction of a psaroniaceous tree fern about 25 feet tall. Leaf scars are visible on the upper part of the trunk below the crown of fronds; most of the trunk is enclosed in adventitious roots which increase in thickness toward the base. (From Morgan, 1959.)

ciation of these fossils with the trunks. However, in 1888 Renault and Zeiller figured a compression specimen of a trunk with the basal portion of a frond still attached; the few fragments of the ultimate frond divisions indicated a *Pecopteris* leaf. There is thus much that remains to be learned. The abundance of petrified frond fragments in American coal balls suggests the possibility of reconstructing the entire frond with some accuracy, and basal portions might be correlated with the petiole traces that are frequently found around the periphery of the trunk.

Records of Mesozoic and Tertiary fossils of the family are not rare, but the preservation is generally unsatisfactory. *Asterotheca meriani* (Brongniart) Stur from the Triassic of Lunz, Austria, was probably comparable in habit to the Carboniferous plants. Each pinnule of the frond bore two rows of synangia which consisted of four sporangia fused in the basal portion. *Marattiopsis hoerensis* (Schimper) Thomas from the Rhaetic of Greenland is based on fertile leaf fragments with synangia 7 mm long and 1 mm broad. They consisted of 17 or more sporangia, each of which contained a spore mass 900 μ long and 350 μ wide, the spores being 28 μ in diameter. This fossil is closely comparable if not identical with modern species; in fact, Harris notes that *"Marattiopsis* is only known to differ from *Marattia* in age."* Specimens of *Marattiopsis macrocarpa* (Morris) Seward and Sahni, from the Jurassic of the Rajmahal Hills in India, are sufficiently well preserved to disclose the presence of the family at that time.

Nathorstia alata Halle, a Lower Cretaceous fossil from Lago San Martin, Patagonia, is probably marattiaceous; it is an interesting fern with slender, tapering pinnae 10 to 12 cm long with an anastomosing vein system. There is a row of sori along each side of the midrib; they are densely crowded, circular in outline, and about 1.5 mm in diameter. Each sorus is divided into 12 to 15 chambers.

FILICALES

Schizaeaceae

This family is represented today by four genera of diminutive plants, but what they lack in size they make up for in their unusual morphology and fossil history. Although the family favors warm latitudes *Lygodium,* the climbing fern, has been reported as far north as southern New Hampshire. *Schizaea* also invades temperate regions and some species are especially striking with their small grasslike leaves only a few inches high. *Anemia* is the largest genus, with over 60 species, and lives as far north as the Florida Keys and southern Texas. *Mohria* is restricted to South Africa and in its sporangial structure presents a possible link with the Osmundaceae.

For the most part the sporangia are arranged separately in two rows, rather than in groups (sori), on the pinnules and the annulus consists of a single row of cells at the distal end. The sporangia are erect in *Schizaea* and *Anemia* (Fig. 4-6) but curved in *Lygodium,* lending the effect of being attached at one side.

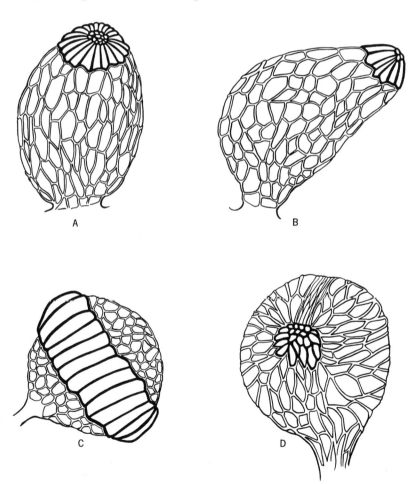

Fig. 4-6. Sporangia of some extant ferns: A. *Anemia mexicana;* B. *Lygodium palmatum;* C. *Gleichenia longissima;* D. *Osmunda regalis.*

Since petrified remains of the vegetative organs are virtually unknown, the anatomy of the living members of the family will not
be considered. The frond structure is, however, highly distinctive
and enables us to trace *Lygodium* and *Anemia* far back into the
past.

Anemia is but little changed since mid-Cretaceous times.
Anemia fremonti Knowlton (Fig. 4-7) from the Cretaceous Frontier
formation of southwestern Wyoming is very close to the living *A.
adiantifolia* found around limestone sink holes of the Florida Keys.
The fine grained shale of the Frontier formation has preserved the
fertile pinnae so well that portions of the compression can be
removed from the rock using the cellulose transfer method (see
Chapter 17). The size and general morphology of the frond with
its two conspicuous fertile pinnae, the arrangement of the sporangia
in two rows on the under side of the pinnules and the sculpturing
of the spores, are almost identical with the corresponding characters of *A. adiantifolia*. Unless the physiology of schizaeaceous
plants has changed appreciably during the past 100 million years
one may suppose that the Cretaceous landscape of what is now
southwestern Wyoming was far more lush, with a warmer and
more equitable climate, than prevails there now.

Anemia poolensis Chandler is based on fragments of fertile pinnae obtained by Miss Chandler by sifting early Tertiary sediments
of southern England. Although the fragments are small they
reveal the characteristic arrangement of the sporangia in two rows
and the conspicuous apical annulus; the spores, however, differ
from most other species, living and fossil, in being smooth. This is
far north of the present range of this tropical to subtropical genus
and fits in with the abundant evidence from angiosperm seeds and
fruits that a warm climate prevailed in southern England during
the early Tertiary. Fertile pinnules of *Lygodium* were also found
in the same deposits.

Lygodium pumilum Brown from the Upper Cretaceous of Wyoming is one of the oldest species of this genus; although known
only from sterile leaflets it is quite clearly a *Lygodium* and is considered to be allied with the living *L. palmatum,* differing from the
latter in having very small pinnules.

Lygodium skottsbergii Halle, from Eocene deposits in southern
Chile, attests to a wider distribution for the genus than is known
at present; Coronel, the locality from which it was obtained, is
some 30° south of the present limits of *Lygodium*. Macerations of
sporangia revealed spore counts of between 120 and 169; some of

Fig. 4-7. *Anemia fremonti,* an Upper Cretaceous fern from southwestern Wyoming. A. Restoration of a complete fertile frond; B. portion of a frond; C. part of a fertile pinna after removal from the rock, showing three clusters of small pinnules; D. one group of pinnules in side view; E. a single pinnule more highly enlarged with seven sporangia. (From Andrew and Pearsall, 1941.)

these were fragmentary and the number was thus very possibly 256, which is typical for modern species. Although distinct from any lygodiums that live today it was probably closest to *L. palmatum.*

There are numerous reports in the literature of fertile schizae-aceous leaf fragments assigned to the genus *Klukia,* although rather few reveal clearly defined details. Harris has described specimens from the Jurassic of Yorkshire under the name *Klukia exilis* (Phillips) Raciborski which are at least twice pinnate fronds with pinnules of the *Pecopteris* type. Each pinnule bore 6 to 14 sporangia which display the characteristic single-rowed annulus. The same species has been reported by Endo from the Upper Jurassic of northern Hondo in Japan.

Using a transfer technique which enabled him to study the sporangia in detail Radforth has demonstrated a sequence of fossil forms (Fig. 4-8) that gives a clue to the origin of the unique sporangium in the Schizaeaceae. The annulus of the Upper Carboniferous *Senftenbergia plumosa* (Artis) Radforth is apical, as in the Mesozoic and modern members of the family, but is two to three cells deep instead of only one. In *S. ophiodermatica* (Goeppert) Stur, also Upper Carboniferous, the annulus is two to three cells deep, but the arrangement is less regular than in *S. plumosa.* In *S. sturi* (Sterzel) Radforth, from a Lower Carboniferous horizon, the annulus cells are less clearly distinguished from the others of the sporangium and tend to form an irregular patch around one side of the sporangium.

The *Senftenbergia* fronds are of the *Pecopteris* type and indicate plants of larger size than most modern members of the Schizae-aceae. If all of the species mentioned above are correctly assigned to this family, there is a rather strong suggestion that the distinc-tive annulus originated from an irregular cluster of cells which be-came localized at the distal end of the sporangium; they later became oriented into two rows and finally reduced to one.

In connection with this apparent sequence of Carboniferous forms it is of interest to note, briefly, a specimen described by Rad-forth and Woods from the Lower Cretaceous of western Canada. It is probably referable to the genus *Klukia,* the ovoid, sessile sporangia being arranged in two rows on the pinnule. Although there is typically but a single row of annulus cells, occasional sporangia show a rather indefinite second row; the spore number per sporangium is approximately 390 which is higher than in most living species of the Schizaeaceae but appreciably lower than in two of the Carboniferous fossils noted above.

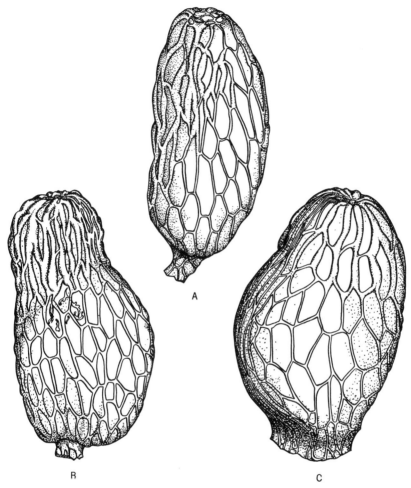

Fig. 4-8. Sporangia of certain Carboniferous species of *Senftenbergia* showing probable stages in the evolution of schizaeaceous annulus. A. *S. sturi;* B. *S. ophiodermatica;* C. *S. plumosa.* (From Radforth, 1938, 1939.)

Gleicheniaceae

The genus *Gleichenia* is a large one, including, according to the interpretations of different authorities, 80 to 100 living species, although certain subgroups of the genus may actually be of generic rank. In most gleichenias the branching pattern of the leaf is highly distinctive; it takes the form of a uniform pseudo-dichotomy because of the frequent failure of the leaf apex to develop. In some species the fronds attain a length of over 20 feet.

The sori are arranged in two rows along either side of the midrib of the leaflets. The sorus usually consists of a ring of sporangia arranged around a central receptacle; the annulus (Fig. 4-6) is a conspicuous and nearly complete ring of cells oriented in a slightly oblique equatorial position.

Gleichenia is widely scattered in the tropics and versatile of habitat; its distribution has been rather colorfully described by Seward as follows:

> *Gleichenia* occurs in the clearings of tropical forests, on the edge of jungles, on tropical alpine peaks, in the heath vegetation on the higher slopes of Ruwenzori, . . . China, India, Australia . . . in the rain forests of Mexico, Costa Rica . . . in the West Indies, along the Andes to the Falkland Islands, . . . from Natal to Table Mountain, in Madagascar, in the Island of Réunion, on Amsterdam Island, in New Guinea where Dr. Wollaston tells me that he collected several species, some growing at an altitude of over 12,000 feet. . . . (1922, pp. 224–225.)

The earliest genus attributed to the family is the Upper Carboniferous *Oligocarpia* (Fig. 4-9). It has been restudied recently by Abbott who recognizes nine species which are rather widely distributed through Europe and North America. The foliage is of the *Sphenopteris* type with fronds that are several times pinnate. The sori are rosettelike groups of (usually) four or five sporangia, although the number may be as high as 17. The sporangia are attached to a low dome-shaped receptacle and are arranged in a single ring with occasionally one or two in the center.

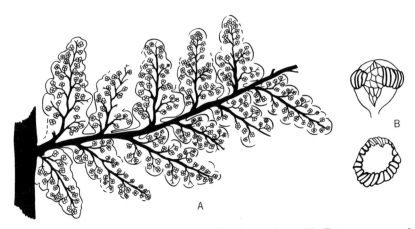

Fig. 4-9. *Oligocarpia capitata.* A. A fertile pinna, about 6X; B. two aspects of a sporangium showing the equatorial annulus. (From Abbott, 1954.)

There is sufficient variation in sporangial structure in the ferns so that a sharp delimitation of families is not always an easy matter. However, the manner of grouping of the sporangia combined with the annulus structure offers strong support of the assignment of *Oligocarpia* to this family. Some of the species appear to have been quite widely distributed; for example, *O. brongniarti* Stur is recognized from Upper Carboniferous horizons in Silesia, France, England, southeastern Canada, and Ohio.

Unlike the Carboniferous *Oligocarpia* most of the Mesozoic ferns assigned to the Gleicheniaceae display the apparent dichotomous pattern of the modern fronds and most authors have referred these fossils to *Gleichenites*. *G. rostafinskii* Raciborski from the Jurassic of Poland is one of the oldest Mesozoic species. In the Cretaceous the gleichenias were abundant and widespread; several species flourished in western Greenland and others are known from England, Germany, France, and the Balkans, and as far south as Patagonia. *Gleichenites coloradensis* (Knowlton) Andrews was associated with *Anemia fremonti* in the Frontier formation of Wyoming and compares closely with certain living species. Fossil records are rather scarce from the Eocene on, there being little doubt that the cooling climate of the early Tertiary, through what is today the north temperate zone, spelled the extinction of this family in the north.

Matoniaceae

Contrasts in present and past distribution patterns often reveal some striking stories of plant migration and few of these exceed in interest that of the Matoniaceae. Because of the unique branching pattern of the frond, it is possible to recognize the fossil ancestors with an unusual degree of certainty far back in the Mesozoic.

The family is represented today by only three or four species in the genera *Matonia* and *Phanerosorus* which are of limited distribution in the Malay region. The frond may attain a height of 8 feet and has been described as "a very perfect example of a catadromic helicoid dichopodium." In somewhat simpler language this means that the main rachis of the leaf divides into two parts, each of which bends down (catadromic) and continues to divide in a helicoid or spiral fashion. The sori consist of six to nine sporangia which are covered by a hemispherical flap, the indusium, which falls away when the sporangia are mature. The structure of the sporangium is similar to that of *Gleichenia* although the position of the annulus is not quite as regular.

During the past the matonias have wandered far beyond their present confinement. They were rather widely distributed in central Europe during the Mesozoic and have been found as far north as Greenland; they have been recorded from Australia, India and Japan, from North Africa, and from several localities in this country.

Two matoniaceous ferns referred to the genus *Phlebopteris* have been found in the Triassic of the southwest; *P. smithii* (Daugherty) Arnold (Fig. 4-10) comes from the Petrified Forest National Monument in Arizona and *P. utensis* Arnold, with smaller fronds, shorter pinnae, and a somewhat different vein pattern was found in Utah. A fine specimen of a fertile frond of *Matonidium americanum* Berry, about 30 cm in diameter, has been described from the Upper Cretaceous of Montrose County, Colorado. *Matonidium indicum* Sahni from Baroda State, India, is probably of Lower Cretaceous age and is distinguished by an upward flaring of the petiole into a funnel-

Fig. 4-10. *Phlebopteris smithii,* from the Triassic of Arizona, natural size.

shaped lamina, presenting a possible transition to the leaf type of the Dipteridaceae.

Dipteridaceae

This is a family of one living genus, *Dipteris,* with five species found in the Indo-Malayan region. They are plants with creeping rhizomes and long-stalked fronds similar in general appearances to those of *Matonia.* There is, in fact, some question as to whether the dipterids should be recognized as a separate family. The leaf plan differs from *Matonia* in being anadromic; that is, the branching of the two equal subdivisions of the leaf turn up rather than down.

The blade of the living *Dipteris lobbiana* may be almost filiform; the apex of the petiole dichotomizes into two equal parts and these in turn divide, resulting in a leaf that is deeply dissected into 24 or more narrow segments; there is a row of sori on either side of the midrib of the segments. The blade of *D. conjugata* (Fig. 4-11B), although deeply divided into two equal halves is generally much less deeply dissected.

The number of sporangia per sorus is quite variable and their orientation lacks the regular radial pattern of *Matonia* and *Gleichenia.* The annulus is an obliquely arranged, nearly complete ring of cells; the degree of obliquity is, however, rather slight and *Dipteris* was originally placed in the Polypodiaceae because of a general similarity in annulus structure.

In a comprehensive survey of the fossils referred to the Dipteridaceae, Ôishi and Yamasita have recognized six genera (*Hausmannia, Clathropteris, Dictyophyllum, Thaumatopteris, Goeppertella* and *Camptosorus*) with some 60 species; some of the genera are, however, separated on rather slender grounds.

Clathropteris meniscoides var. *elegans* Ôishi from the Mesozoic (Nariwa Series) of Japan must have been a fern of great beauty in life; the blade was divided into about 30 equal, slightly toothed segments that are separate almost to the petiole.

The family was represented by numerous species in the Rhaetic floras of Greenland and some of them were ferns with rather large fronds; for example, *Clathropteris meniscoides* Brongniart had a blade about 35 cm in diameter, the 10 or 12 divisions being about 2.5 cm broad and characterized by a net venation of rather large rectangular meshes. Several species of *Dictyophyllum, Thaumatopteris,* and *Hausmannia* also thrived in these northern climes during the Mesozoic.

Fig. 4-11. A. *Dipteris, Gleichenia* and other ferns growing along a river bank in central Viti Levu, Fiji; B. a single frond of *Dipteris conjugata*. (Photographs by O. H. Selling.)

The leaf structure in *Hausmannia* is rather different from that of the other genera. In *H. dentata* Ôishi the blade is only dissected a third of the way to the rachis and in *H. nariwaensis* Ôishi it is nearly entire, resembling superficially the leaf of the common garden nasturtium.

The fossil history of the family is briefly summarized by Ôishi and Yamasita as follows:

The conclusion arrived at from the investigation of fossil records of Dipteridaceae is that it forms a conspicuous group of ferns among the Mesozoic flora of the world. It has been known from Europe, Eastern Asia, North and

South America, Australia, Greenland and Graham Land but there are no satis-factory examples from India and Africa. Thus the occurrence of fossil Dip-teridaceae in these continents is of special interest as indicating that tropical or subtropical conditions prevailed in the regions where the fossil forms are now found. The fossil Dipteridaceae reached its maximum development, numerically and possibly biologically also, in the Rhaetic and the Liassic epochs, and it began to wane numerically towards the end of the Mesozoic, though a certain genus, namely *Hausmannia,* began to flourish in the younger Mesozoic time. (1936, p. 175)

In summary it may be noted that it seems likely that the three families Gleicheniaceae, Matoniaceae, and Dipteridaceae stem from a common ancestor in late Paleozoic times. There appears to be some blending in the Mesozoic and several authorities have stressed the difficulties of classification that are involved in working with fossil representatives. The works of Harris, Ôishi and Yamasita, Hirmer and Hoerhammer, Seward and Dale are particularly worth consulting.

Osmundaceae

Among modern plants that deserve the name of "living fossil" the osmundas occupy a place in the foremost ranks. Judging from stem anatomy the extant species and their immediate ancestors have been conspicuous and widespread elements of the earth's vegetation for well over 100 million years.

Foliage attributed to the Osmundaceae is quite abundant, par-ticularly in Mesozoic rocks, and there is an exceptional series of petrified stems that traces the family back to the Permian. The stem in the three living species of *Osmunda* that are so abundant through the eastern portion of the United States (*O. regalis,* the Royal fern; *O. cinnamomea,* the Cinnamon fern; *O. claytoniana,* the Interrupted fern) gives the appearance of being a rather mas-sive structure owing to the investing leaf bases which remain for many years. The stem proper is, however, usually not over a cen-timeter or two in diameter and contains a very hard sclerotic cortex and a cylindrical stele that is divided into numerous segments by the departing leaf traces. The stems of our American species attain a height of only a few inches; as to other extant osmundas, *Todea barbara* in South Africa may reach a height of several feet and *Leptopteris wilkensiana,* found in New Caledonia, is reported to attain a height of 10 feet.

Our knowledge of the petrified remains has been gathered to-gether in a series of contributions by Kidston and Gwynne-Vaughan and much of the following is taken from their work. One of the

youngest of the fossil stems is *Osmundites schemnitzensis* from the
Upper Miocene (or Lower Pliocene?) of Hungary which was found
associated with other plant remains including beech and sequioa.
The stem is small with a parenchymatous pith and a dissected xylem
cylinder of 17 or 18 strands and only 3 to 4 mm in diameter; in the
structure of the stele and the form of the departing leaf traces it
corresponds closely with present day osmundas.   In fact, as far back
as the Eocene the stem anatomy is essentially the same as in modern species; *Osmundites chandleri* Arnold (Fig. 4-12B) from the
Eocene of Oregon had a stem 15 mm in diameter with a small xylem
cylinder composed of about 34 strands.   *O. dowkeri* Carruthers
from the Eocene of the Isle of Wight (England) was of very similar
size and anatomical dimensions.

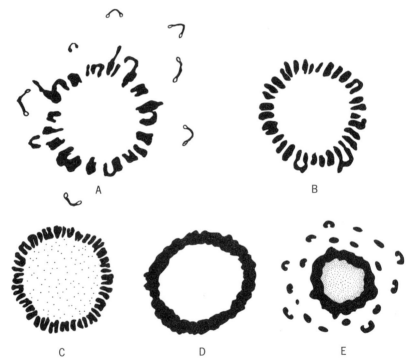

Fig. 4-12. Diagrams of the stelar systems of certain fossil osmundaceous stems.
A. *Osmundites braziliensis.* B. *O. chandleri.* C. *O. kolbei;* stippling represents scattered central tracheids. D. *O. dunlopi;* note that the stele is essentially continuous
(siphonostelic). E. *Thamnopteris schlechtendalii;* stippling represents inner tracheid
zone and solid black outer zone. (A from Andrews, 1950; B from Arnold, 1952;
C–E, from Kidston and Gwynne-Vaughan, 1907–14.)

The range of these plants extended as far north as Spitsbergen in Tertiary times. *O. spetsbergensis* Nathorst (probably late Tertiary) is known from a rhizome in which the stele had been lost, but the general organization of the leaf bases is indicative of its osmundaceous affinity. Associated with it were portions of leaves considered to be almost identical with the foliage of the living *Osmunda regalis;* preserved sporangia fragments indicate that the annulus was somewhat more strongly developed than in modern osmundas.

Some late Mesozoic fossils seem to represent the giants of the osmunda line. *Osmundites kolbei* Seward from the Wealden of South Africa is estimated to have attained a height of 90 cm and the entire "trunk" (stem and leaf bases) measured about 14 cm in diameter. The stele, although flattened prior to fossilization, measured about 19 mm in diameter and consisted of a ring of some 56 strands. Of special interest here is the presence of numerous tracheids scattered through the pith (Fig. 4-12C); these differ from the tracheids of the main woody cylinder in being irregular in size and shape but they are clearly enough tracheidal elements. *O. braziliensis* Andrews (Fig. 4-12A) from Brazil is known from a stem 8 cm in diameter which is enclosed in a closely compacted root sheath 4.5 cm thick, so that the entire trunk was close to 15 cm in diameter. It seems safe to assume that this was a tree fern of at least several feet in height. The exact age of this fossil is not known but it is probably not older than late Mesozoic.

It seems probable that the larger arborescent osmundas grew in a manner similar to the Carboniferous *Psaronius,* with the petiole bases being gradually sloughed off in the older part of the stem which was in turn enclosed in a mantle of roots, resulting in a heavily buttressed base.

A Jurassic species found in New Zealand, *O. dunlopi* Kidston and Gwynne-Vaughan, differs from the younger ones in having a stele that forms a continuous or nearly continuous cylinder.

The oldest petrified stems that are referred to the Osmundaceae come from the Permian of the southern Urals. Of six genera that have been described *Thamnopteris schlechtendalii* (Eichwald) Brongniart (Fig. 4-12E) is perhaps the most informative. The stem with its ensheathing leaf bases measures about 12 cm across and has a xylem cylinder 11 mm in diameter. The stele differs from the previously considered members of the family in that it consists of two distinct zones: an outer continuous ring of tracheids of the usual scalariform type and an inner core of irregularly shaped, somewhat thinner-walled ones. This inner portion probably formed a solid core but the central part is missing. The leaf trace buds off from the periphery of the outer xylem as a small oval-shaped mass and becomes C-shaped as it passes through the cortex. A sclerotic outer cortex and abundant, large petiole bases with C-shaped strands bear out the osmundaceous affinities of the plant.

The evolution of the osmundaceous stem seems, therefore, to have followed essentially this course: from semiprotostelic plants such as *Thamnopteris,* the Mesozoic siphonostelic forms evolved (*O. dunlopi*) and the development of large leaf gaps resulted in dictyostelic types. There is a retention of some tracheids scattered through the pith region in *O. kolbei* but they disappear in the Tertiary representatives.

As noted in the previous chapter, the Lower Permian *Grammatopteris baldaufi* may represent a link between the coenopterids (*Anachoropteridaceae*) and the Osmundaceae. Its thick armor of petioles and crowded nests of sclerotic cells in the middle cortex point toward the latter family, whereas the presence of a few short, broad tracheids in the central portion of the stele leads into the partially medullated stele of *Thamnopteris.* The general shape of the sporangia as well as their dense clustered organization in certain living species of *Osmunda* could readily have evolved from the less specialized coenopterids. In brief, it seems very likely that future research will confirm this suggested derivation of the Osmundaceae; it may be worth noting that Engler and Diels recognize 17 living species in their *Syllabus der Pflanzenfamilien* and since they have not all been studied critically significant information may be forthcoming from studies of the living plants as well as the fossils.

Tempskyaceae

The Mesozoic tree-fern *Tempskya* is one of the most bizarre creations in the plant kingdom and because it was sufficiently abundant as to have played a significant ecological role in the Cretaceous landscapes it is considered at some length. It was a plant with an unbranched columnar trunk which attained a diameter of 16 inches in the largest specimen known; this tapered slowly upward to a blunt apex, the largest ones probably attaining a height of 15 to 20 feet. The upper part of the trunk was covered with a dense mass of hundreds or perhaps thousands of small leaves (Fig. 4-14).

Fragments of the petrified trunks have been found in rather widely scattered localities: in the Karaganda river basin of the USSR, Bohemia, England, and in this country in Maryland and several of the northwestern states. Stratigraphically, the genus ranges from the upper part of the Lower Cretaceous to the mid Upper Cretaceous although there is some evidence to indicate that it extended back to the late Jurassic. About a dozen species have been recognized, but many of them are questionable. The basic organization in all seems to be very similar and the description given here is based chiefly on a large collection of specimens found in the Wayan formation of southeastern Idaho which are assigned to *T. wesselii* Arnold.

The trunk fragments are often beautifully preserved, and in a cross section taken from the middle or upper part they are found to be constructed of many stems held together in a dense matrix of small adventitious roots. It should be emphasized that these are complete *stems* and not simply steles. The trunk section shown in Fig. 4-13A is a small one 5 inches in diameter and is composed of about 190 stems; each stem (Fig. 4-13B) is siphonostelic and rather similar in its anatomy to the rhizome of certain living ferns, such as *Adiantum* (Maidenhair fern). The stems dichotomized rather profusely and gave off small petioles (about 1 mm in diameter) in two rows. It is thus evident that the trunk bore large numbers of small leaves which were distributed along the upper half, or perhaps the upper two-thirds, of its length. This is quite in contrast to most modern tree-ferns where a few large fronds are borne as a terminal crown. Although the fossil trunks have not been found with leaves attached, they occur in Idaho at a horizon in which *Anemia fremonti* fronds are also present; this association presents only a possible relationship, but the approximate size of the fronds and their distribution on the trunk, as shown in the restoration, are well established.

The most curious feature of *Tempskya* is the fact that specimens taken from the lower part of the trunk are composed exclusively, or almost exclusively, of roots. The development of the trunk was apparently somewhat as follows: in the earliest sporeling stage there was probably a single stem; as this grew and divided the resultant *trunk* increased in height and girth; at the same time large numbers of small, wiry roots were produced which held the entire mass together as a distinct unit, with a texture somewhat like that of a modern *Osmunda;* when the trunks reached a height of several feet, the stems at the base gradually began to decay and the functional stems above maintained a connection with the soil by means of the thousands of small roots. These conclusions concerning the ontogeny of the trunks were reached by the writer from a study of

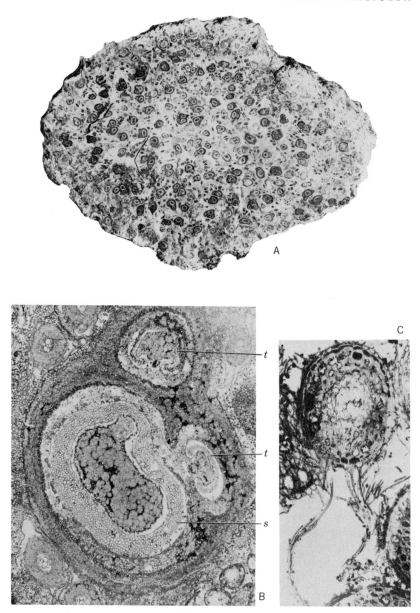

Fig. 4-13. *Tempskya wesselii* from the Upper Cretaceous of Wayan, Idaho. A. Cross section of a trunk about 5 inches in diameter composed of 190 stems. B. A single stem, 20X; *s,* siphonostele of stem; *t,* traces leaving stem stele to supply two petioles. C. A single root enlarged about 50X showing well-preserved root hairs. (From Andrews and Kern, 1947.)

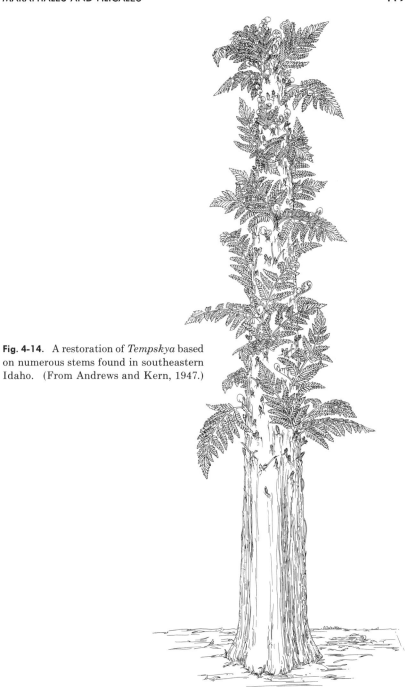

Fig. 4-14. A restoration of *Tempskya* based on numerous stems found in southeastern Idaho. (From Andrews and Kern, 1947.)

many hundreds of trunk specimens; the evidence seems reasonably conclusive, but *why* they grew in this manner is another question! As to the affinities of *Tempskya,* the anatomy of the stems is not sufficiently distinctive to offer a clue and other bits of evidence are conflicting. Seward described a specimen of *T. knowltoni* from Montana in which he found isolated sporangia among the stems and roots; they have a characteristic schizaeaceous annulus and this fits in nicely with the association of trunk specimens of *T. wesselii* and the fronds of *Anemia fremonti.* However, isolated sporangial fragments have been found in the *T. wesselii* trunks from Idaho that are quite clearly not schizaeaceous!

There are a few other ferns, living and fossil, that have a tendency to develop such compound trunks. The Carboniferous *Austroclepsis australis* was mentioned in the previous chapter, although only a very few stem branchings are involved. The extant *Hemitelia crenulata* which grows in the forests of Java is a tree-fern of considerable size, attaining a circumference of 20 cm at 30 cm above the ground. The basal 2 or 3 feet of the trunk is an aggregation of numerous branches enclosed in a dense matrix of roots, whereas above this the branches are free. A few other examples might be cited; it is clearly a mode of growth that has originated independently in several unrelated fern groups, but none compares closely with *Tempskya* in either detailed anatomy or the huge number of stems involved. *Tempskya* may well have been an important source of food for the larger herbivorous animals of the Cretaceous. Possibly future studies will be able to shed some light on its racial origin and why it disappeared from the scene so suddenly.

SOME FERNS OF UNCERTAIN AFFINITIES

It has been noted earlier that there are conspicuous gaps in our knowledge between plants of the coenopterid and protopterid types considered in the last chapter and the ferns of more modern aspect. Among the more important differences between the two assemblages are the following:

1. The fronds of the ferns are distinctly laminate or "leafy," whereas the preferns bore fronds that amounted to little more than a much divided branch system.

2. The sporangia, whether arranged singly, in clusters, or in organically fused groups, are found on the under side of the lamina in the ferns in what has been called a "superficial" position, rather than being borne terminally as is the case in the preferns.

As to point number one, evidence has been presented to show how this transition was accomplished and, at least in a general way, this is rather well understood. The fossils that are described next offer at least a sketchy answer to point number two, although it is not possible to make precise taxonomic correlations.

Acrangiophyllum pendulata (Lesley) Mamay (Fig. 4-15A) from the Pennsylvanian (Pottsville) of Alabama is known from rather small fragments of fernlike foliage with sporangia 1.5 mm long

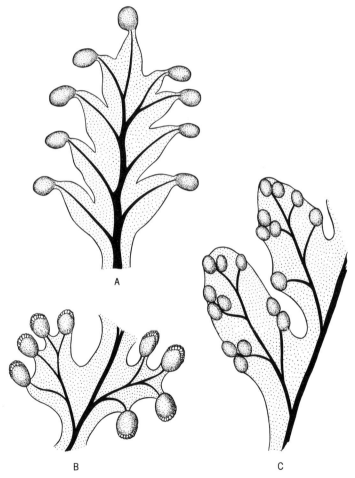

Fig. 4-15. Some Carboniferous ferns showing certain stages in the evolution of the superficial position of the sporangium. A. *Acrangiophyllum pendulata,* from the Pennsylvanian of Alabama; B. *Boweria schatzlarensis,* from the Westphalian of Yorkshire; C. *Renaultia gracilis,* from Lancashire. (A from Mamay, 1955; B, C from Kidston, 1923, by permission of the Controller of Her Majesty's Stationery Office.)

borne marginally at the tip of each lobe. Although the specimens are of the compression type, it has been shown that a single spore mass, consisting of many small spores 20 μ in diameter, is contained within the sporangium wall, which was several cells thick. It is not difficult to envisage this as a plant that evolved from one in which the sporangia were borne terminally (as evidenced by the position of the sporangia at the tips of veins) on a branch system that has become partially webbed.

Other ferns of comparable morphology might be mentioned. One other example will be noted to indicate that this evolution of the superficial position probably took place in several different groups. The pinnules of *Boweria schlatzlarensis* Kidston (from the Westphalian division of the Carboniferous) bore several marginal sporangia, each at the tip of a vein. The sporangia are smaller than those of *Acrangiophyllum,* being about 0.5 mm in diameter and having a distally located annulus of two rows of cells that are conspicuously larger than the others of the sporangium wall. The larger pinnules bore five to seven sporangia, while only one to three are found on the smaller distal ones.

Renaultia gracilis (Brongniart) Zeiller (Fig. 4-15) from the Westphalian and Lanarkian series of Britain has fertile foliage in which the sporangia are attached on the under side of the pinnule but close to the margin. They are broadly oval, being about 0.4 mm long and 0.3 mm in diameter and are arranged singly or in small groups of two or three at the end of a vein. No annulus has been observed and numerous small spores are contained in each sporangium.

Stages in the evolution of an organ in which sporangia shift from a terminal to a marginal position and finally reach the superficial one are summarized well in a series of diagrams by Zimmermann (Fig. 4-16). In Chapter 3 we dealt in some detail with certain aspects of this, namely, the problem of the evolution of a two-dimensional, laminate frond. As to the sporangia, there are now so many plants, both living and fossil, which attest to the general sequence shown in Figs. I–P as to leave no doubt that this is what took place.

I have included comparable series that Zimmermann has given for the lycopods (A–D) and the articulates (E–H). The sequence for the articulates seems probable, but there is less supporting evidence than there is for the ferns (Pterophyta); the sequence for the lycopods (A–D) is, so far as I am aware, purely speculative. The student may wish to refer back to these figures after studying the chapters on the lycopods and articulates; they constitute brilliant

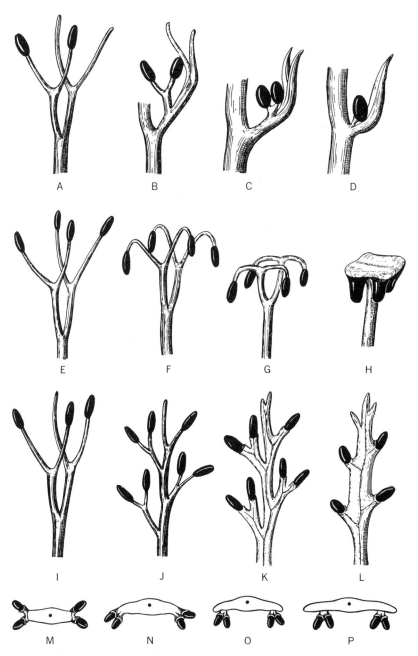

Fig. 4-16. Supposed stages in the development of sporophylls in the lycopods (A–D), in the articulates (E–H), and in the ferns (I–P). (From Zimmermann, 1952.)

hypotheses which may well be confirmed as our knowledge of the two groups expands.

NOEGGERATHIALES

In view of the distinctive nature of the spore-bearing organs the fossils that are included in this group might well merit assignment to a distinct Division. However, I am not convinced that this step should be taken until more is known about them; thus, their inclusion here is admittedly a matter of convenience. The Noeggerathiales is known at present from a small assemblage of foliage and heterosporous cone specimens found chiefly in continental Europe.

The leaves are best described as cycadlike, but the cones are quite unlike the reproductive organs of any other group of plants. The assumption that cones and leaves belong to the same plant, or closely allied plants, is based on association and the occasional presence of leaflets at the base of the cone which compare with those of the leaves proper.

Although the plants occur as compressions, some of the cones are well preserved and a recent restoration by Halle, of *Noeggerathiostrobus bohemicus* O. Feistmantel, seems to conform rather well with the known facts. The cones (Fig. 4-17) appear to have borne two rows of semicircular scales or half-discs which functioned as sporophylls; they might be compared with saucers cut in half and attached to a central axis at the middle of the cut edge. The margin of each scale is characteristically fringed and numerous sporangia, each containing 16 megaspores, are borne on the upper (adaxial surface).

In *Discinites major,* another cone genus referred to the group, Nemejc has found megasporangia with but one large megaspore (1 mm in diameter) whereas the microsporangia contain numerous microspores, each about 100 μ in diameter. The cones were at least 10 cm long and 4 cm in diameter. Another species, *D. delectus* Arnold, has been found in the Michigan Coal Basin; the mega- and microsporangia are similar in form, with the former containing 16 spores, each about 700 μ in diameter, and each microsporangium contains many microspores each about 80 μ in diameter.

These cones, if they may be called such, of the noeggerathias cannot be compared closely with the spore-bearing organs of any pteridophytic group. Although they have been referred by vari-

ous workers to the palms, ferns, and cycads it is evident that there
are no close relationships in these directions. The stand taken by
Halle seems to be the only reasonable one when dealing with such
problematical fossils, and we shall encounter others to which it
applies with equal cogency.

Fig. 4-17. A tentative restoration of *Noeggerathiostrobus*
bohemicus, about natural size. (From Halle, 1954.)

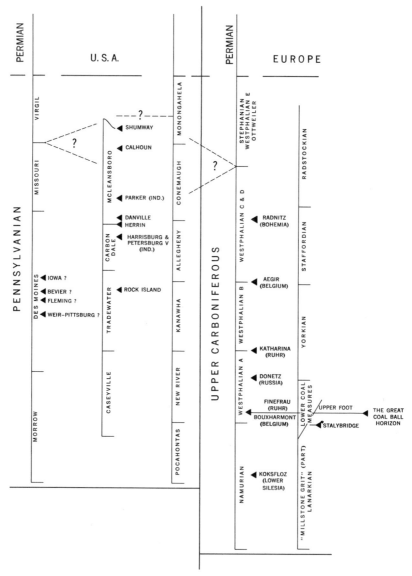

Fig. 4-18. A chart showing the stratigraphical position of American and European coal ball horizons. (From Schopf, 1941, modified.)

Halle notes:

The natural tendency to squeeze fossil forms into existing taxonomic groups has too often led to unfortunate consequences. In the present case it would seem a safer course—whatever headlines be adopted for practical purposes—to regard the Noeggerathiineae as an isolated group of *pterophyta incertae sedis.* (1954, p. 378)

REFERENCES

Abbott, Maxine L. 1954. Revision of the Paleozoic fern genus *Oligocarpia*. *Palaeontographica*, 96B: 39–65.

Andrews, Henry N., Jr. 1950. A fossil osmundaceous tree-fern from Brazil. *Bull. Torrey Bot. Club*, 77: 29–34.

————, and Kern, Ellen M. 1947. The Idaho Tempskyas and associated fossil plants. *Ann. Missouri Bot. Gard.*, 34: 119–186.

————, and Cortland S. Pearsall. 1941. On the flora of the Frontier formation of southwestern Wyoming. *Ann. Missouri Bot. Gard.*, 28: 165–180.

Arnold, Chester A. 1952. Fossil Osmundaceae from the Eocene of Oregon. *Palaeontographica*, 92B: 63–78.

————. 1956. Fossil ferns of the Matoniaceae from North America. *Palaeont. Soc. India, Lucknow*, 1: 118–121.

Bhardwaj, D. C., and Singh, Hari Pall. 1956. *Asterotheca meriani* (Brongn.) Stur and its spores from the Upper Triassic of Lunz (Austria). *The Palaeobotanist*, 5: 51–55.

Bower, Frederick O. *The Ferns (Filicales)*. Cambridge Univ. Press. Vol. 1. 1923; Vol. II. 1926; Vol. III. 1928.

Brown, Roland W. 1949. Cretaceous plants from southwestern Colorado. *U. S. Geol. Survey Prof. Paper*, 221D: 45–66.

Chandler, Marjorie E. J. 1955. The Schizaeaceae of the south of England in early Tertiary times. *Bull. Brit. Mus. Nat. Hist., Geology*, 7: 291–314.

Daugherty, Lyman H., and Howard R. Stagner. 1941. The Upper Triassic flora of Arizona. *Carnegie Inst. Washington Pub.*, 526: 1–108.

Endô, Seido. 1952. *Klukia* remains newly found in Japan. *The Palaeobotanist*, 1: 165–167.

Gillette, Norman J. 1937. Morphology of some American species of *Psaronius*. *Bot. Gaz.*, 99: 80–102.

Graham, Roy. 1934. Pennsylvanian flora of Illinois as revealed in coal balls. I. *Bot. Gaz.*, 95: 453–476.

Halle, Thore G. 1940. A fossil fertile *Lygodium* from the Tertiary of South Chile. *Svensk Bot. Tid.*, 34: 257–264.

————. 1954. Notes on the Noeggerathiineae. *Svensk Bot. Tidskrift*, 48: 360–380.

Harris, Thomas M. 1931. The fossil flora of Scoresby Sound East Greenland. Part I. *Meddelelser om Grönland*, 85 (2): 1–102.

————. 1945. Notes on the Jurassic flora of Yorkshire. 19. *Klukia exilis* (Phillips) Raciborski. *Ann. Mag. Nat. Hist.*, ser. 11, 12: 257–265.

————. 1947. The problems of Jurassic Paleobotany. *Bol. Soc. Geol. Portugal*, 6: 5–32.

Hirmer, Max, and Hoerhammer, L. 1936. Morphologie, Systematik and geographische Verbreitung der fossilen und rezenten Matoniaceen. *Palaeontographica*, 81B: 1–67.

Kidston, Robert, and Gwynne-Vaughn, D. T. 1907–1914. On the fossil Osmundaceae. *Trans. Roy. Soc. Edinburgh*, Pt. I (1907), 45: 759–780; Pt. II (1908), 46: 213–232; Pt. III (1909), 46: 651–667; pt. IV (1910), 47: 455–477; pt. V (1914), 50: 469–480.

Mamay, Sergius H. 1950. Some American Carboniferous fern fructifications. *Ann. Missouri Bot. Gard.*, 37: 409–459.

————. 1955. *Acrangiophyllum*, a new genus of Pennsylvanian pteropsida based on fertile foliage. *Amer. Journ. Bot.*, 42: 177–183.

Morgan, Jeanne. 1959. The morphology and anatomy of American species of the genus *Psaronius*. *Illinois Biological Mono.* No. 27: 1–108.

Němejc, Frantisek. 1928. Revise Karbonske a Permske Kveteny Stredoceskych Panvi Uhelnych. *Paleont. Bohemiae,* No. 12: 1–82.

Ôishi, Saburo, and Yamasita, Kazuo. 1936. On the fossil Dipteridaceae. *Journ. Facutly Sci. Hokkaido Imper. Univ.,* ser. IV, vol. III, No. 2: 135–184.

Pichi-Sermolli, Rodolfo, E. G. 1958. The higher taxa of the Pteridophyta and their classification. *Uppsala Universitets Årsskrift,* 6: 70–90.

Radforth, Norman W. 1938. An analysis and comparison of the structural features of *Dactylotheca plumosa* Artis sp. and *Senftenbergia ophiodermatica* Goeppert sp. *Trans. Roy. Soc. Edinburgh,* 59: 385–396.

————. 1939. Further contributions to our knowledge of the fossil Schizaeaceae; genus *Senftenbergia. Trans. Roy. Soc. Edinburgh,* 59: 745–761.

————. 1942. On the fructification and new taxonomic position of *Dactylotheca parallela* Kidston. *Canadian Journ. Res., C,* 20: 186–195.

————, and Woods, A. B. 1950. An analysis of *Cladophlebis (Klukia) dunkeri* Schimper, a Mesozoic fern from western Canada. *Canadian Journ. Res., C.* 28: 780–787.

Read, Charles B., and Brown, Roland W. 1937. American Cretaceous ferns of the genus *Tempskya.* U. S. Geol. Prof. Paper, 186-F: 105–129.

Remy, Winfried. 1953. Untersuchungen über einige Fruktifikationen von Farnen und Pterospermen. *Abhandl. Deutschen Akad. Wissenschaften Berlin, (Kl. Math., Naturwiss.) Jr.,* 1952, Nr. 2: 1–38.

Sahni, Birbal, and Rao, A. R. 1933. On some Jurassic plants from the Rajmahal Hills. *Journ. Proc. Asiatic Soc. Bengal,* n. s., 27: 183–208.

————, and Sitholey, R. V. 1945. Some Mesozoic ferns from the Salt Range, Punjab. *Proc. Nat. Acad. Sci. India,* 15: 61–73.

Seward, Albert C. 1922. The past and present distribution of certain ferns. *Linn. Soc. Journal, Botany (London),* 46: 219–240.

————. 1924. On a new species of *Tempskya* from Montana: *Tempskya knowltoni,* sp. nov. *Ann. Bot.,* 38: 485–507.

————, and Dale, Elizabeth. 1901. On the structure and affinities of *Dipteris,* with notes on the geological history of the Dipteridinae. *Phil. Trans. Roy. Soc. London,* 194B: 487–513.

Zimmermann, Walter. 1952. Main results of the "Telome theory." *The Palaeobotanist,* 1: 456–470.

5
PTERIDOSPERMOPHYTA

One of the most significant contributions that paleobotany has made to our knowledge of the plant kingdom is the discovery and gradual elucidation of the assemblage of fossils that is known as the pteridosperms or seed-ferns. A few of them are now sufficiently well known as to reveal a great race of plants that became extinct many millions of years ago. It is possible to set up a rudimentary framework of the subgroups that compose the Pteridospermophyta, and we are beginning to gain some insight into the evolutionary trends. There are, however, many fragmentary but interesting remains of the various plant organs which can with some confidence be called pteridospermous but beyond that their taxonomic position is uncertain.

Of the features that characterize the pteridosperms the following seem to be most notable: They possessed fernlike foliage and bore seeds, which are the features that contribute to the name *pteridosperm*. The seeds, which were borne in a variety of ways on the fronds, were in some cases partially enclosed in a distinctive outer integument or *cupule*. The stems developed secondary wood and a prominent outer cortex of longitudinally aligned fiber strands. The microsporangiate organs were terminally clustered sporangia which in many cases formed large complex synangial fructifications.

The pteridosperms appear first in the Lower Carboniferous and are sufficiently advanced as to suggest a Devonian origin, but as of this writing no undoubted representatives have been found at such a low horizon. They were abundant through the Upper Carboniferous and Permian, and in the Triassic and Jurassic we find distinct subgroups which most paleobotanists have regarded as later offshoots of the Paleozoic lines.

The large and complex assemblage of fossil plants that is now attributed to the Pteridospermophyta will be considered under the following headings:

Lyginopteridaceae
Medullosaceae
(Some seed-bearing pteridospermous foliage)
(Some early fossils attributed to the pteridosperms)
Corystospermaceae
Peltaspermaceae

LYGINOPTERIDACEAE

Although fossils that are attributed to the Pteridospermophyta range from the Lower Carboniferous to the Jurassic, it seems most expedient to introduce the group as a whole with the well-known, even classic example, *Lyginopteris oldhamia* (Binney) H. Potonié. This species has been extensively investigated by Williamson, Scott, and others, chiefly from a single important coal ball horizon (see chart p. 126) of Yorkshire and Lancashire. Since vegetative and reproductive characteristics of this species are correlated in exceptional detail at the level of the Ganister and Halifax Hard Bed coal, it is appropriate to use it as an example for the family.

The *Lyginopteris oldhamia* stems (Fig. 5-1) attained a diameter of about 3 cm; in the center there was usually a rather large pith of parenchymatous cells and scattered through this were clusters of somewhat thicker-walled cells with dark contents. The pith is, however, variable in both size and the relationship of the two cell types that compose it. Around the periphery are several rather large, mesarch primary strands (Fig. 5-1C); however, the bulk of the primary wood is centripetal (develops toward the center of the stem) so that the bundles are nearly exarch. Secondary wood was formed in some abundance and consists of large tracheids with numerous angular bordered pits in the radial walls; the pits are irregularly arranged in contrast, for example, to the regularly aligned vertical rows in cordaitean woods. This is a conspicuous feature of most of the Carboniferous seed-ferns (Fig. 5-8D) although it is perhaps questionable as to how much importance should be attributed to it. The rays of the secondary wood vary from one to a dozen cells wide (tangential dimension) and in height they range from a small fraction of a millimeter (4 or 5 cells) to over 2 cm. It would seem that this may be properly regarded as a very unspecialized type of secondary wood. It is of interest to mention that varying amounts of internal secondary wood are occasionally found in *Lyginopteris* stems (Fig. 5-1C). It apparently originated from a

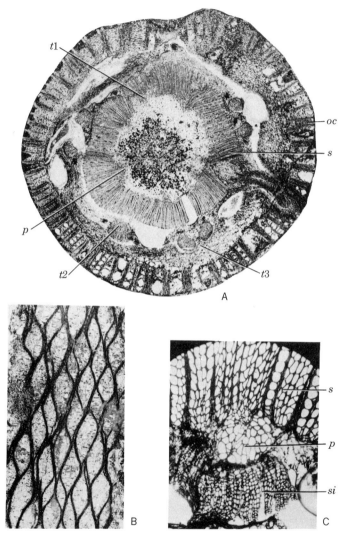

Fig. 5-1. *Lyginopteris oldhamia.* A. Stem in transverse section; *p,* pith with numerous sclerotic-secretory cells; *t1,* a leaf trace in an early stage of its departure; *t2,* leaf trace that has passed through the secondary wood but is still undivided; *t3,* leaf trace that has divided into two branches (two other traces appear in this divided condition in the secton); *s,* secondary wood; *oc,* outer cortex. B. Tangential longitudinal section through the outer cortex. C. An enlarged sector of a stele showing: *s,* normal secondary wood; *p,* primary vascular bundle; *si,* internal secondary wood (an occasional anomalous feature).

cambium just within the primary strands and developed as a mirror image of the normal secondary wood. In view of the wide variety of form taken in the formation of secondary tissues in the Upper Carboniferous and Permian medullosan pteridosperms, this may have some significance in *Lyginopteris* other than as a mere "anomaly."

A leaf trace originates by the tangential division of a primary wood strand. It passed out through a gap in the secondary wood, being accompanied by a small arc of the latter on its outer side. Shortly after leaving the secondary wood it divides into a pair of strands which traverse the cortex and ultimately unite again in the base of the petiole to form a V-shaped leaf trace. Petioles of this type were first observed and classified under a separate name, *Rachiopteris aspera.* Their identity in the northern European coal balls now seems fully established.

In the best preserved stems the secondary phloem may be distinguished immediately outside the secondary wood. This is followed by a narrow band of radially aligned cells which is referred to as a periderm; it is a distinctive feature of certain pteridosperms, particularly the medullosas. The outer cortex is especially noteworthy (Fig. 5-1B); it consists of radially elongate fiber bands which form a regular anastomosing network in their longitudinal course through the stem. The "windows" of this network are filled with parenchymatous cells. This fibrous outer cortex has been relied on as a diagnostic feature of pteridosperm stems and petioles. It is certainly characteristic of many, although the form and organization of the strands are not the same in all.

The frond is a primitive sphenopterid type, apparently identical to that of *Sphenopteris hoeninghausi,* which attained a length of about one-half meter; the rachis dichotomized 10 cm from its point of junction with the stem, and each of the two divisions was three times pinnate. Some years ago I had occasion to collect some rather fine stem and leaf impressions, tentatively referable to *L. oldhamia* (Fig. 5-3), from a horizon in southern Scotland. The fossils composed a very nearly "pure stand" in the rock, there being few other associated plants and I have wondered if, as is supposed, *Lyginopteris* actually grew as a vine depending on others for partial support, or whether it may have formed dense thickets of its own along the borders of the forests or within them. It should be noted that these specimens were identified by leaf characters and the presence of the "dictyoxylon" outer cortex which is so striking a feature of the *Lyginopteris oldhamia* stems. However, the fact that this type of cortical structure is not strictly confined to *L. oldhamia* makes is impossible, so far as I am aware, to positively identify this species in compression form. Two examples may be cited to illustrate this rather important point:

In his *Fossil Plants of the Carboniferous Rocks of Great Britain,* Part 1, Kidston figures a stem compression about 1.4 cm in diameter from the Carboniferous Limestone Series (Lower Carboniferous) of the "dictyoxylon" type that is attributed to *Sphenopteris fragilis* (Schlotheim).

Figures 5-3B, C show portions of a stem and leaf from the Carboniferous of Belgium that are attributed to *Sphenopteris bäumleri* Andrä.

In both of these examples the pinnule structure is distinct from that of the frond of *L. oldhamia* (although all are probably closely related), but I do not believe it would be possible to distinguish the three on the basis of stem impression characters.

The seeds of *L. oldhamia,* sometimes classified separately as *Lagenostoma lomaxi* Oliver and Scott, are barrel-shaped, radially symmetrical and 5.5 mm long by 4.4 mm in diameter (Fig. 5-2); they have a rather stout integument, containing nine vascular strands, which encloses the nucellus; the latter is elongate with a chamber at the distal end for the reception of pollen grains. Long has reported remarkable preservation in seeds of a closely related species, *Lagenostoma ovoides* Williamson, containing the female gametophyte; several archegonia are present near the micropylar end. In all probability the pollen grains germinated in the nucellar chamber and liberated sperms which fertilized the eggs of the archegonia in much the same fashion as in modern cycads.

The seed was partially enclosed in a lobed husk or *cupule* (Fig. 6-2A) which bears distinctive glandular emergencies identical with those present on the petioles and stems. Since most of the seeds found in English petrifactions are isolated, it is presumed that they were released from the cupule by a basal abscission mechanism.

Because the seeds present a critical phase in the restoration of the plant, a few additional notes are appropriate at this point. The name *Lagenostoma lomaxi* was first formulated by Williamson in his catalogue notes but was formally published by Oliver and Scott in 1904. Although the seeds were not found actually attached to the *Lyginopteris* leaves, the following evidence led them to believe that such was the case: the two are often associated in the same petrifaction; rachis and cupule bear distinctive, globular-

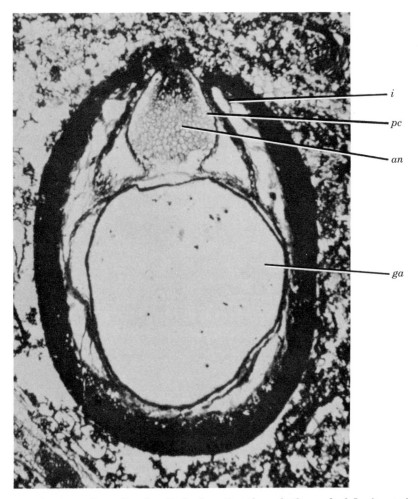

Fig. 5-2. A nearly median longitudinal section through the seed of *Lyginopteris oldhamia;* the cupule is not present; *i,* free or apical portion of the integument; *an,* apical portion of nucellus; *pc,* pollen chamber; *ga,* gametophytic area—seeds have been found in which this central region is occupied by a female gametophyte.

headed glands which are identical, and these are not known in any other plant; the vascular bundle of the cupule pedicel compares closely, even to pitting of the tracheids, with the trace of a small rachis. This evidence has been accepted by most paleobotanists as conclusive.

The male or pollen-producing organs of the plant have not been positively identified, but based on correlation with allied species of

Sphenopteris it is possible that they were of the *Crossotheca* type. The specimen shown in Fig. 5-4 is from Illinois and, as is usually the case, the fertile pinnate branch system apparently terminated part of a *Sphenopteris* frond. The ultimate branchlets (fertile pinnules?) consist of a slightly flattened lamina with a peripheral row of elongate, pendant sporangia.

The nomenclatorial problems centered around this and related plants are somewhat more complicated than has been intimated here, but it has not seemed advisable to discuss them fully since practically an entire chapter would be required. It may be added

Fig. 5-3. A. A portion of a frond from southern Scotland that was probably borne on the *L. oldhamia* stems. B, C. Frond and stem fragments of *Sphenopteris bäumleri* from Belgium. (A from Andrews, 1947.)

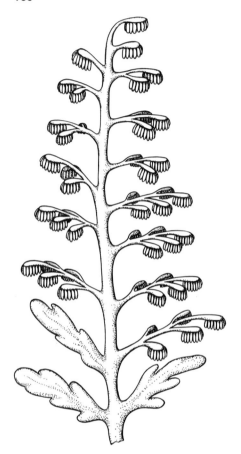

Fig. 5-4. *Crossotheca* sp.; restoration based on a specimen from the Upper Carboniferous of Illinois, about 2X.

that other species of *Lyginopteris,* based on stem anatomy, have been described by Kubart from somewhat older deposits in Lower Silesia. Furthermore, the multiplicity of species that *probably* possessed similar organization is suggested by fossils such as the one shown in Fig. 5-3B, C. However, it is only in exceptional instances that diagnostic features of a species of *Lyginopteris* and a species of *Sphenopteris* can be demonstrated in connection. It is most difficult therefore to establish a precise correlation outside of deposits like the British coal balls which are actually exceptional in their occurrence, in the preservation of included plant materials, and in the extent to which they have been studied. As a consequence, although the foliar characteristics of *Lyginopteris oldhamia* appear to be identical with those of *Sphenopteris hoeninghausi* at the Ganister coal bed horizon, the history of this correlation of diagnostic features is uncertain and debatable in younger and older deposits.

In summary, *Lyginopteris oldhamia* was a fairly common plant in the European Upper Carboniferous which bore sphenopterid foliage and, in all probability, the seeds called *Lagenostoma lomaxi*. It is important as one of the better known pteridosperms and the pioneering studies devoted to it by Williamson, Oliver, and Scott, and others form an important chapter in our understanding of the seed-ferns. However gaps still remain in our knowledge of the plant.

Some Other Fossils Attributed to the Lyginopteridaceae

Stems

Callistophyton poroxyloides Delevoryas, from upper Pennsylvanian coal balls of Illinois, is a stem that is marked by certain distinctive features and especially fine preservation. In the center is a rather large parenchymatous pith with occasional secretory cavities; the cells are, in fact, quite thin-walled and in some specimens this region has collapsed. Several mesarch primary strands are present around the periphery of the pith and these are followed by a strong development of secondary wood. In the largest stems, which attain a diameter of a little over 2 cm, this tissue makes up a considerable portion of the stem; it is quite like that of *Lyginopteris* although the rays are usually narrower.

A leaf trace originates from a peripheral primary bundle which divides to form two, each of which develops a distinct pair of protoxylem groups. Thus at first glance the trace, which is quite conspicuous, gives the impression of being composed of four strands. It passes out through the wood and cortex, being accompanied on the outer side by a segment of secondary wood. It may also be noted that axillary branching seems to have been a well-developed feature of the plant.

The secondary phloem is composed of rays that are continuous with those of the wood, parenchyma cells, and sieve cells; the latter bear simple sieve plates on their radial walls—an exceptional feature of preservation. Fiber bands are found in the outer cortex, although they are less conspicuous than those of *Lyginopteris* and they follow a parallel course. Occasional glands are present on the cortex which are similar to those of the latter genus but smaller. This fibrous outer cortex tends to be sloughed off in older stems following the development of periderm in the tissue immediately peripheral to the phloem.

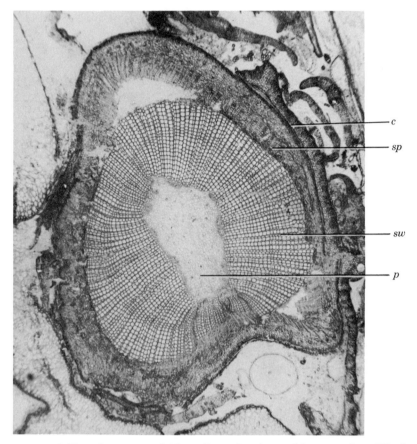

Fig. 5-5. *Callistophyton poroxyloides,* a Pennsylvanian seed-fern stem from Illinois, 6X. *c,* outer primary cortex; *p,* pith; *sp,* secondary pholem; *sw,* secondary wood. (From a preparation by T. Delevoryas.)

Schopfiastrum decussatum Andrews is a distinctive pteridosperm stem (Fig. 5-6) from the Des Moines series of Iowa. It is a little more than 2 cm in diameter and has a mixed protostele surrounded by secondary wood with narrow rays and tracheids with closely compacted bordered pits. The outer cortex is composed of parenchyma and regularly spaced, radially elongate fiber strands; these run a nearly parallel longitudinal course and thus present a distinct contrast to the netlike arrangement found in *Lyginopteris.*

Of special note are the leaf traces which are large, tangentially elongated strands given off on opposite sides of the primary wood, with one departing slightly above the other. Although this plant

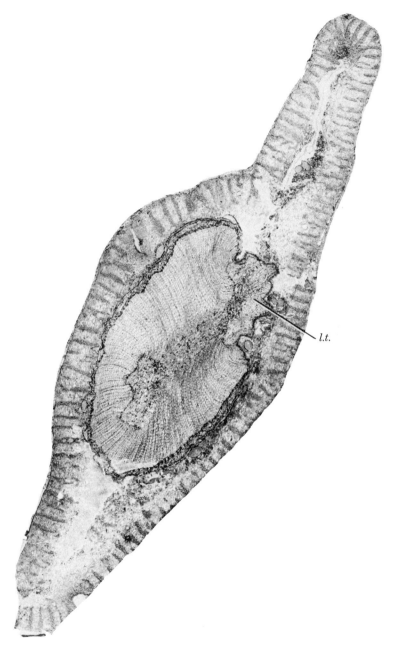

l.t.

Fig. 5-6. *Schopfiastrum decussatum,* a pteridosperm stem (somewhat crushed) from the Upper Carboniferous of Iowa; *l.t.,* a large, lobed leaf trace departing from the stele.

is known from only two small fragments, the available material suggests that successive pairs of traces alternate, resulting in four rows of leaves. It was either a rare plant or is represented by specimens that were carried in from a distant region.

Heterangium is a genus of several species from both Lower and Upper Carboniferous horizons. One of the best known as well as the oldest is *H. grievii* Williamson from Pettycur in Scotland. The stem reached a diameter of 4 cm and apparently branched only rarely. The center is occupied by a terete, mixed protostele in which the tracheids occur in groups of six to a dozen scattered quite uniformly through the parenchyma. A small amount of secondary wood is present surrounding the protostele. Radially elongate fiber bands are present in the outer cortex, whereas the inner cortex is distinguished by horizontal plates of thick-walled cells. Leaves known as *Diplotmema adiantoides* are attributed to the *H. grievii* stems. In an Upper Carboniferous species from the McLeansboro formation of Illinois (*H. americana* Andrews) appreciably more secondary wood is formed, there being as much as 5 mm in one specimen, although the extent of its development varies in different specimens. The secondary wood is similar in tracheidal pitting and ray structure to *Lyginopteris*. The vascular supply to a leaf originates as two separate strands at the periphery of the primary wood; these divide in their passage through the cortex to form an arc of four bundles in the petiole and the two central ones soon trifurcate.

Seeds

There are several features of special interest in the recently discovered seed *Tyliosperma orbiculatum* Mamay (Fig. 13-3), found in the Des Moines series of Kansas; it is, in fact, one of the few pteridosperms in which the cupule is well preserved. The seed is small, reaching a length of but 3.7 mm and a diameter of 3.5 mm; it is nearly circular in cross section. The nucellus and integument present a closely knit unit; a set of vascular strands passes up apparently in the peripheral region of the nucellus and the tissue that is identified as integument is not vascularized. Of particular interest is that the epidermis of the integument is continuous over the distal end of the nucellus, a relationship that is very difficult to explain if the integument originated as a distinct tissue and grew up around the nucellus. The distal portion of the integument is divided into seven lobes. One cannot avoid the suspicion that there is no real distinction here between integument and nucellus, but rather that they compose a unit tissue system. The cupule is quite deeply lobed, more so than that of *Lyginopteris,* and it did not enclose the seed as completely; rather stout, blunt protuberances occur on its outer surface, but they apparently were not glandular.

Physostoma elegans Williamson is a beautiful little seed about 6 mm long and 2.3 mm in diameter from the Upper Carboniferous of England. Like other "lagenostomalian" seeds the integument is fused with the nucellus except at the apex. The pollen chamber, formed by nucellar tissue, has a distinctive shape (Fig. 5-7) and may contain large numbers of pollen. Scott remarked on this in the first volume af his *Studies* and I once enjoyed the thrill of cutting into two specimens, found in an English coal ball, one of which contained many pollen grains. The distal part of the integument is divided into about ten elongate lobes or tentacles and the entire outer surface of the integument is covered with conspicuous, fat, unicellular hairs.

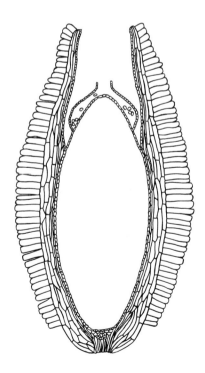

Fig. 5-7. Longitudinal aspect of *Physostoma elegans;* several pollen grains are shown in the pollen chamber.

THE MEDULLOSACEAE

The basis of this group of pteridosperms is a large and unique series of stems that has been found in Upper Carboniferous and Permian rocks; they are identified chiefly by their polystelic vascular systems which, in some species, are of unequalled complexity. Most of them have been assigned to *Medullosa* and, although it is evident that it is a "genus" of extraordinarily broad limits, it seems to be an expedient taxonomic procedure in the present light of our knowledge. The fronds borne on these stems were of the *Alethopteris* and *Neuropteris* types and in their anatomy the petioles present a striking superficial resemblance to monocot stems. The seeds that are associated with the stems and fronds are generally larger than those of the Lyginopteridaceae and the nucellus and integuments are separate; the microsporangiate organs are for the most part large, complex synangia.

Medullosa thompsoni Andrews from Iowa may be considered first as a representative Upper Carboniferous species. The stem measures about 4 cm in diameter and contains three steles, each consisting of a small central primary core of tracheids and parenchyma surrounded by secondary xylem of large, pitted tracheids and tall, slender rays. The steles range in over-all size from 7×3.5 mm in diameter for the largest to 2 mm in diameter for the smallest. The stelar system is enclosed by a continuous, narrow band of thin-walled radially aligned cells, the inner periderm. External to this is a broad parenchymatous cortex which is penetrated by numerous small leaf traces which depart from the steles. The outer cortex consists of circular to radially elongate fibrous strands which run longitudinally and occasionally anastomose; a few secretory canals are associated with them.

In Fig. 5-8A a leaf base appears near its point of departure and Fig. 5-8B shows an isolated petiole at a somewhat higher magnification. The structure of the outer cortex is essentially the same as the corresponding tissue in the stem although the secretory canals may be somewhat more abundant. The vascular system, unlike that of the stem, consists of numerous small, scattered bundles.

Over 40 species and varieties have been recognized although it is now evident that some of these actually represent variants within a species. The genus was established by Cotta as far back as 1832 and quite a number of Permian species were described during the

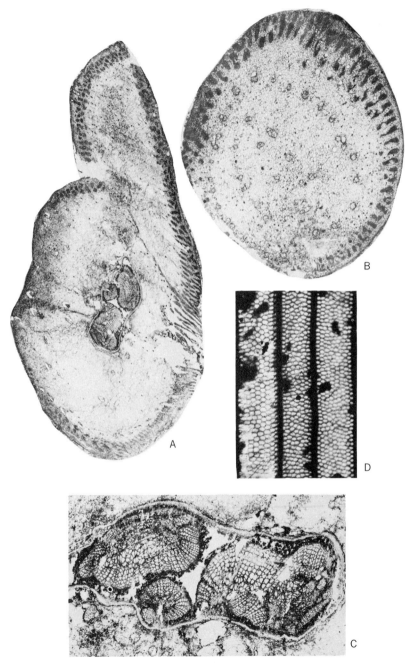

Fig. 5-8. *Medullosa thompsoni.* A. Transverse section of stem with three central steles and petiole departing above; B. a petiole; C. stelar system enlarged; D. pitting in secondary tracheids. (From Andrews, 1945.)

latter part of the nineteenth century, a comprehensive account of them being given by Weber and Sterzel in 1896. Scott described a well-preserved stem (*M. anglica*) from the English coal fields in 1899 and this was followed by the discovery of several others. With the initiation of coal-ball studies in this country a remarkable series of specimens have come to light in the Illinois, Kansas, and Iowa deposits. This work has been summarized and revised in recent studies by Stewart and Delevoryas.

The species vary tremendously in over-all size and in the nature of the stelar system. *M. endocentrica* Baxter from Illinois has a vascular system of 3 steles, 2 of which are only 5 mm in diameter and the third one is even smaller; in certain specimens of *M. grandis* Andrews and Mamay from Kansas the vascular system is composed of 18 to 20 steles with an over-all diameter of about 12 × 28 cm. In all the medullosan stems the steles tend to anastomose and divide to a greater or less degree so that the exact number is not a dependable taxonomic character. Some species vary in the uniformity of development of the secondary wood; the cambial activity was fairly uniform around the primary wood of certain species, whereas in others it was strongly endocentric, that is, secondary wood was developed almost exclusively toward the center of the stem. A particularly interesting discovery was made by Stewart and Delevoryas in dealing with a specimen of *M. primaeva* (originally described under the name of *M. heterostelica*) in which there are two steles in the internodal part of the stem and these suddenly proliferate to a total of 23 steles at the node; leaf traces depart from some of them and they quickly reassemble above the node to the two-stelar condition. A more recent survey of numerous specimens which are closely related or actually referable to this species suggest that this nodal-internodal relationship does not always hold true.

For their bizarre stem anatomy the Permian medullosas of Germany are deserving of more space than is available here; a few will be mentioned to indicate some of their diversity. Figure 5-9F is a cross-sectional diagram of a specimen of *M. stellata* Cotta that is preserved in the Swedish Natural History Museum. The stem as a whole measures approximately 24 × 34 cm and includes about 70 steles in the central portion; each one has a very small central group of primary tracheids, with perhaps a few associated parenchyma cells. The cells are surrounded by secondary xylem which makes up the greater portion of the stele. In over-all size they vary from about 3 mm to 25 mm in diameter. This entire central

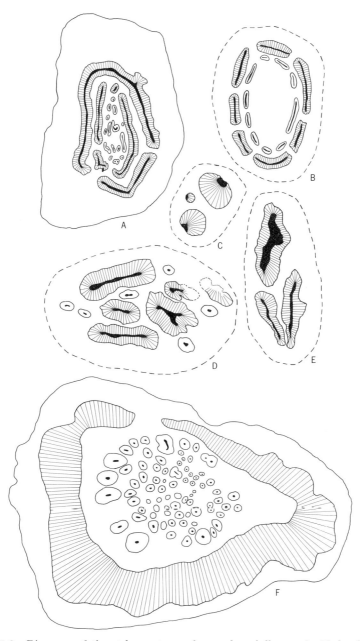

Fig. 5-9. Diagrams of the stelar systems of several medullosas. A. *M. leuckarti,* Permian of Chemnitz; B. *M. solmsi,* Permian of Chemnitz; C. *M. endocentrica,* Pennsylvanian of Illinois; D. *M. grandis,* Pennsylvanian of Kansas; E. *M. noei,* Pennsylvanian of Illinois; F. *M. stellata,* Permian of Chemnitz. (A,F drawn from specimens in the Swedish Natural History Museum.)

145

complex is, in turn, enclosed within a rather broad and nearly complete band of secondary xylem.

Several variations of *M. stellata* have been described, some of which display only 4 or 5 central steles and with a much thicker peripheral band of secondary wood than in the specimen described above. In *M. stellata* var. *typica* there are two enclosing bands of wood which are inversely oriented; that is, there is an inner band of wood with the phloem located on the inner side; this is enclosed by a parenchymatous zone with some scattered tracheids, which is followed by another zone of wood with the phloem located on its outer side. In the variety *gigantea* the stems attain a diameter of 50 cm and, in addition to the stelar system described for the variety *typica*, there is a series of several additional concentric bands of secondary wood. These undoubtedly developed from successive cambia that originated in the cortex as happens in certain living species of *Cycas*.

In *Medullosa leuckarti* Goeppert and Stenzel (Fig. 5-9A) there are several of the more or less cylindrical central steles, which are surrounded by a wavy ring of several large, tangentially elongate steles that have been called a "snake ring." In *M. solmsi* Schenk (Fig. 5-9B) there is a rather large "pith" or central region where a few inconspicuous small cylindrical steles are found, and this is surrounded by two series of tangentially flattened and elongate steles. In the variety *lignosa* of this species the secondary xylem of the outermost ring is excentrically developed; in addition there are several concentric bands of xylem that develop in the "normal" fashion.

In his Autun and Epinac flora (of France) Renault cites wood fragments of *M. gigas* Renault that are 50 cm thick. Judging from his description and illustrations this is a radial dimension and implies a stem that was in excess of 1 meter in diameter.

It is quite clear that we have here a mode of stem growth that is not comparable with that of present-day forest trees. Delevoryas has shown that in young stems the steles may be more closely aggregated than in mature ones and that the increase in trunk diameter is a result of continued growth of the parenchymatous ground tissue surrounding the steles as well as cambial activity of the individual steles; thus in older stems the steles may be farther apart than in younger ones.

The origin and evolution of the medullosan stems have been the subject of considerable speculation and will undoubtedly continue to be for some time. Several students of the group have envisaged

Fig. 5-10. Restoration of a *Medullosa* about 12 to 15 feet tall; based on specimens of *M. noei.* (From Stewart and Delevoryas, 1956.)

lines of increasing complexity, possibly originating in simpler pteridosperms of the *Heterangium* type; the single stele of the latter with its mixed primary xylem cylinders surrounded by a moderate development of secondary wood is not unlike a single medullosan stele. Delevoryas, on the other hand, regards the many-steled forms as primitive, with a phyletic fusion resulting in the few-steled Carboniferous species. A second trend, involving tangential fusion may have led to the *M. leuckarti* type with continued fusion of the "snake rings" into the *M. stellata* and *M. gigas* types in which there is a peripheral, continuous ring (or rings) of wood. I am personally inclined toward this latter theory, in which case it would seem as though one may look to late Devonian and early Carboniferous fossils such as *Xenocladia-Steloxylon-Cladoxylon* as possible ancestral types. Quite clearly, however, we do not know enough about these earlier fossils to render this more than a rough-working hypothesis. At the other end of the scale, some of the Permian medullosas begin to approach a type of stem anatomy that is not distantly removed from that of the cycads.

As a concluding comment on this great array of stems it may be stated with some certainty that *Medullosa* is a "genus" of convenience. It is difficult to draw sharp lines between the different species and we are, in fact, just beginning to learn something of the degree of stelar variation that may be present in the ontogeny of a single plant, but it seems virtually certain that *Medullosa* will eventually be divided into several distinct genera.

Associated Leaves and Reproductive Organs

In American petrifactions in which medullosan stems are abundant they are constantly associated with *Alethopteris* and *Neuropteris* type leaflets and the various branch orders of the fronds; there is little doubt that they constitute the foliage of many *Medullosa* stems. Other leaf genera which have been suspected of being medullosan, at least in part, are *Odontopteris, Linopteris, Lonchopteris, Callipteris,* and *Taeniopteris.*

Among the most interesting fossils that have been referred to the pteridosperms are the large and complex microsporangiate organs of the genus *Dolerotheca* (Fig. 5-11B); these are frequently found associated in American coal balls with medullosan stems and fronds. They are bell-shaped fructifications consisting of several hundreds of tubular sporangia embedded in a cellular groundwork.

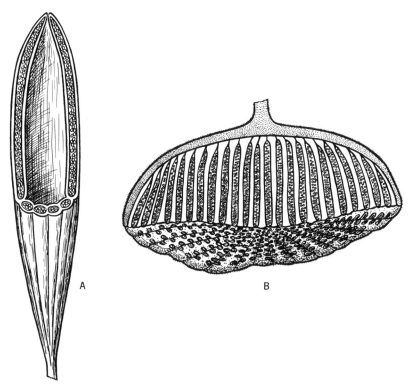

Fig. 5-11. Pteridosperm microsporangiate organs. A, *Aulacotheca elongata,* upper portion cut away to show internal structure, about 6X; B, *Dolerotheca formosa,* half of a fructification. (A from Halle, 1933; B from Schopf, 1948, modified.)

Dolerotheca formosa Schopf is nearly 4 cm in diameter and the individual sporangia are about 14 mm long and 0.8 mm in diameter; the huge ovoid microspores, of which there are several hundreds in each sporangium, are about 0.4 mm long. Since they were quite abundant in certain parts of the Carboniferous forests the big *Dolerotheca* "bells" must have presented a most picturesque sight at the time of spore dispersal.

In 1933 Halle presented a study of several presumed pteridospermous microsporangiate organs which is one of the modern classics in paleobotany. Although he dealt with compression specimens, an embedding technique was devised which made it possible to section the fructifications and prepare restorations which for the most part seem to be beyond question. The following descriptions are based largely on his study.

Whittleseya, in its compressed form, is an elongate, shovel-shaped structure that was originally described from the American coal fields and was thought to be a ginkgophyte leaf. *W. elegans* Newberry vies in size with *Dolerotheca* although its morphology is rather different. According to Halle's restoration it is a hollow, campanulate structure with a wall composed of a single ring of long, tubular concrescent sporangia. The spores are about 0.25 mm long, which is large but not as large as those of *Dolerotheca.*

In one of his last contributions the late W. J. Jongmans made a significant addition to our knowledge of the pteridosperms in discovering *Whittleseya* fructifications attached to *Neuropteris schlehani* Stur foliage (Fig. 5-12). These were found in the Limburg coal field in the southeastern corner of the Netherlands and the same collection included other specimens of *N. schlehani* with seedlike structures attached which are considered similar to seeds described in this country as *Pachytheca vera* Hoskins and Cross.

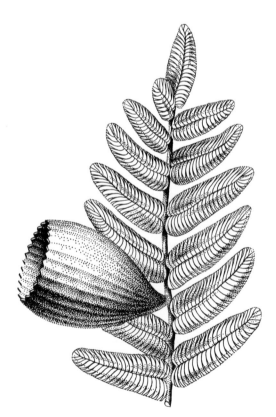

Fig. 5-12. *Neuropteris* foliage bearing a *Whittleseya media* synangium, about natural size; a restoration based on specimens from the Limburg, Holland, coal field. (From Stewart and Delevoryas, 1956.)

Aulacotheca elongata (Kidston) Halle is another synangial type
that is known from numerous specimens found in Upper Carboniferous horizons in Scotland and northern England. Enough have
been found with the spores still intact to leave no doubt as to the
validity of Halle's restoration (Fig. 5-11A). The synangium is about
2 cm long and was composed of about nine sporangia.

The synangia or capsules of *Goldenbergia glomerata* Halle were
less elongate, measuring about 8 mm long and 4 mm broad and consisting of 12 to 16 sporangia containing spores about 0.4 mm long.
The outer surface is covered with short, pointed spines.

The genus *Potoniea* is known from several species and numerous
specimens, but its structure is not as precisely understood as in the
genera cited above. As interpreted by Halle it is an open cup-shaped
organ with several dozen sporangia attached to the inside of the
cup. Specimens have been found with the spore masses intact, but
there is still some question as to whether the sporangia were actually free or whether there was an enclosing tissue.

Our knowledge of petrified seeds of medullosan affinities is based
largely on association; *Pachytesta illinoense* (Arnold and Steidt-
mann) Stewart is one (Fig. 5-13) that has been found in some numbers with the stem remains of *Medullosa noei*. The seeds vary in
size, some reaching a length of 4.5 cm. They are nearly circular in
a median cross section and become triangular toward the micropylar end. The integument is quite complex, consisting of a thin
inner layer of elongate cells with dark contents. Next there is a
layer of thicker-walled cells which develop numerous conspicuous
longitudinal ribs. The region between the ribs is occupied with
thin-walled cells of rather spongy organization and, as might be expected, not often well preserved. The integument is free from the
nucellus except at the base of the seed. Two features of the nucellus are noteworthy: a pendant skirt of tissue at the base and a
doughnut-shaped micropyle at the apex, both of which are apparent
in the restoration. About 24 vascular bundles extend up through
the nucellus and about 40 in the fleshy sarcotesta of the integument
alternating with the sclerotestal ribs.

Stephanospermum is another rather well-known genus of seeds
that may have been borne on the medullosans. *S. elongatum* Hall
from the upper Pennsylvanian of Illinois is a little under 2 cm long
and about three-quarters of a centimeter in diameter. The integument, which is free from the nucellus except at the base, is composed of the following tissues: a peripheral fleshy layer, traversed
by several vascular strands, and very irregularly lobed on the out-

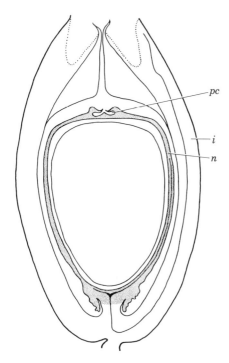

Fig. 5-13. Restoration of the seed *Pachytesta illinoense; i,* integument; *n,* nucellus; *pc,* pollen chamber. (From Stewart, 1954.)

side; next a fibrous zone, and innermost a single layer of thin-walled cells. The integument is elongated at the distal end to a length almost equal to that of the main body of the seed; one might suppose that this elongate, lobed apex could have served as a pollen-catching mechanism, but the lobing is continuous over the entire integument. The nucellus, which has an inverted bell-shaped pollen chamber at its distal end, is distinctive in the manner of its vascularization which is described as a "trachael mantle" and invests the nucellus on its inner side. In this species the mantle is two or three cells wide, the cells being elongate, slender, reticulate to scalariform tracheids. In another species, *S. stewarti* Hall, from Kansas the mantle is composed of three or four rows of tracheids which are only two or three times as long as wide.

Some Seed-bearing Pteridosperm Foliage

As the name "seed-fern" implies, this is a group of plants of fernlike aspect, yet, unlike the true ferns, they bore seeds. Enough is now known about them so that distinctive characters, as noted above, are evident in the stems, seeds, and microsporangiate organs.

However, the most critical feature of the group is the actual attachment of seeds to the organ on which they were borne. It is therefore pertinent to consider several examples of the attachment of seeds, or structures that are presumed to be seeds, to the fernlike fronds.

Sphenopteris tenuis Schenk, from the Permian of Shansi, China, (Fig. 5-14A) is based on fronds that attained a breadth of some 24 cm. About four seeds are borne on each ultimate pinna, being arranged along the midvein on the under side; they are about 4 mm long by 2 mm broad and have a protracted tip. Seeds have also been reported on specimens of *Sphenopteris alata* found in the Somerset coal field (Radstockian age) of Wales. In this species they were borne in a terminal position at the tips of apparently specialized ultimate pinnae in which little lamina was developed; the foliage otherwise consists of rather finely dissected pinnules verging on the *Diplotmema* type. The seeds are 3 to 4 mm long, 2 to 3 mm broad, and were enveloped in a cupule.

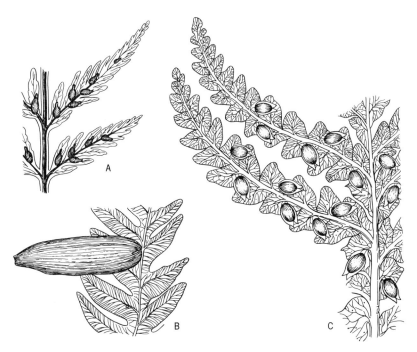

Fig. 5-14. Fertile (seed-bearing) pteridosperm foliage from the Permian of China. A. *Sphenopteris tenuis.* B. *Alethopteris norinii,* C. *Emplectopteris triangularis.* (A,B from Halle, 1929; C from a drawing supplied by Professor Halle, after Andrews, 1947.)

Several examples are now known of *Alethopteris* and *Neuropteris* type foliage that bore seeds. *Alethopteris norinii* Halle, from the Lower Permian of Shansi (Fig. 5-14B), is known from a fragment of a leaf with alethopterid pinnules and a seed 40 mm long and 12 mm broad attached to the rachis or midrib of the pinna, probably on the upper side. The portion of the frond on which the seeds are found is a normal one, with pinnules both above and below the point of attachment of the seed. Arnold has described abundant seeds associated with the foliage known as *Alethopteris grandifolia* Newberry from the Upper Carboniferous of Summit County, Ohio, and reports one specimen with the basal portion of a seed attached, apparently in much the same manner as the Shansi *Alethopteris*. As long ago as 1904, in the same year that Oliver and Scott demonstrated the association of *Lagenostoma lomaxi* with *Lyginopteris oldhamia,* Kidston described a specimen of *Neuropteris heterophylla* bearing a seed about 3 cm long. It is not quite clear as to just how the seed was attached, but it was probably joined directly to the rachis of the ultimate pinna, and enough of a cast is present to indicate that it was of the radiospermic type, that is, a seed that was radially symmetrical in cross section.

In 1911 Kidston and Jongmans described seeds attached to a fragment of a frond of *Neuropteris obliqua;* these present two points of special interest. First, the seeds are quite large, being in excess of 4.5 cm long (the tips were missing so that the entire length could not be determined) and over 2 cm in diameter and, second, they were borne on special peduncles about 1 cm long.

Specimens of *Neuropteris tenuifolia* from the Upper Carboniferous of Yorkshire are known in which the presumed seed occupies a terminal position on the pinna. It is slightly parted at the apex, suggesting the possibility that it may represent a cupule with or without its enclosed seed.

Emplectopteris triangularis Halle, from the Permian of China (Fig. 5-14C), presents several variations on the patterns described above. The proximal portion of the pinnules is net-veined and the winged seeds are attached close to the midrib of the pinna, probably on the upper surface.

One other plant from Professor Halle's collections of Chinese Permian fossils will be mentioned briefly, although its proper inclusion in the pteridosperms is by no means certain. *Nystroemia pectiniformis* Halle (Fig. 5-15) is based on portions of a branch system that has the general organization of a fern frond. The ultimate subdivisions (pinnules?) divide several times, each branchlet termi-

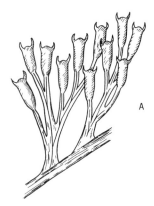

Fig. 5-15. *Nystroemia pectiniformis;* A. seed-bearing and B. microsporangiate portions of frond. (From Halle, 1929.)

nating in a small, two-horned body about 3 mm long that is presumed to be a seed. It is possible that the fossil is a specialized reproductive portion of a frond. Associated with them were slightly curved branches bearing clusters of elongate structures that may represent the microsporangiate organs.

Finally, in *Pecopteris pluckenetii,* the seeds are attached along the margins of the pinnules which are essentially normal (pecopterid) or but slightly reduced; the position of the seed also corresponds to the termination of a lateral vein.

The problem of the position of the seed

Aside from finding seeds actually attached to the fernlike foliage of the pteridosperms, the problem of how they attained the variety of positions in which they have been reported is hardly of less importance. In a summary discussion dealing with their position Halle was shown that the earlier members of the group, such as *Lyginopteris oldhamia* (lower Upper Carboniferous), bore the seeds on specialized segments of the frond; in *Pecopteris pluckenetii* (middle Upper Carboniferous) they are marginal on essentially normal pinnules; and in certain Permian representatives such as *Alethopteris norinii, Emplectopteris triangularis,* and *Sphenopteris tenuis* they are found on the surface of normal foliage leaves—in the first two, on the upper side and on the lower side in the third. There is thus a suggestion that this surface position on a laminated leaf may have come about by the "migration" of the seed from a terminal, to marginal, to surface position comparable to transitions that seem to have taken place with respect to the sori in the ferns. There is a notable difference in that the seed tends to occupy a place on the upper surface of the frond rather than the lower. Halle points out that the location of sporangia on the under surface presents an advantage for spore dispersal, whereas the position of seeds on the upper surface would be more effective for the reception of windborne pollen.

There are now enough examples known of seeds attached to fernlike foliage to leave no doubt that they were borne in a variety of positions, but a great deal more information is needed before we will have a satisfactory understanding as to how they evolved.

Some Early Pteridosperm Fossils

The pteridosperms were well established in the Lower Carboniferous although much less is known about them from that period. The fossils described here seem to be of considerable interest in that they display both primitive and advanced characters and it is possible that they all belong to a closely related group.

In 1938, W. T. Gordon described petrified stems from Lower Carboniferous rocks of East Lothian, in southern Scotland, under the name *Tetrastichia bupatides*. They are about 1 cm in diameter and contain a cruciform protostele of reticulately thickened tracheids; in some specimens there is a weak development of secondary wood, the secondary tracheids being pitted on both radial and tangential walls. The vascular traces that supply the petioles depart in opposite pairs, one on each side of the stem, with one slightly above the other, and successive pairs alternate, resulting in a four-ranked arrangement as in *Schopfiastrum*. Fiber strands are present in the outer cortex of the stem and petioles.

Leaves known as *Telangium affine* (Fig. 3-13B) also occur in the Lower Carboniferous of southern Scotland and, in their size and cortical features of the petiole, there are reasons to believe that they may have been borne on the *Tetrastichia* stems. They are common fossils in the oil shales and may be so well preserved that pieces of the leaf can be lifted from the rock without any special treatment. A complete frond would measure about 45 cm long; the "main rachis" dichotomized two or three times and, as noted previously, the frond is essentially a flattened telomic branch system, there being only a feeble development of leaflets as such. The fertile pinnae of *T. affine* bear terminal groups of six partially united sporangia, each about 3 mm long, which contain numerous spores; these were probably attached at the base of the frond as is known to be the case in an allied species, *T. bifidum*.

Some years ago the writer found a large tulip-shaped fossil (*Megatheca thomasi*) associated with the *Telangium* foliage in the oil shales of Midlothian and at about the same time Professor Walton discovered a partially petrified seed-bearing organ, *Calathospermum scoticum* (Fig. 5-16A) in a Lower Carboniferous horizon of the Kilpatrick Hills of Scotland. It is probable that the two are identical, but since *Calathospermum* is more completely preserved the following description is based on it. It is a tulip-shaped cupule 45 mm long and 20 mm broad and is partially divided into six lobes, each of which is supplied by several vascular strands. As many as 60 seeds are found within the cupule, being borne terminally on slender stalks that emanate from the basal region. The seeds are distinguished by a pollen chamber surmounted by an elongate nucellar tube which is enclosed by prolongations (usually nine) of the integument which are actually longer than the main body of the seed. It is suggested that the long stalks bearing the seeds served to extrude them for more effective pollination and that dispersal was thus facilitated.

Summary—Paleozoic Pteridosperms

The occurrence of fossils like *Calathospermum* in the Lower Carboniferous would seem to imply that the pteridosperms had already undergone a considerable period of evolution. The seeds contained in the large cupules appear to be highly specialized with respect to the elongate distal lobes of the integument. It is of interest here to comment briefly on *Gnetopsis elliptica* Renault (Fig. 5-16B), a cupulate organ from the Upper Carboniferous of St. Étienne, France; it was considerably smaller, being 6.4 mm long, not as deeply lobed and contained only two seeds. The latter are distinguished by several elongate, slender, hair-covered processes that extend out from the distal end of the integument and may have functioned as pollen-trapping mechanisms.

One cannot avoid the suspicion, on finding fossils like *Calathospermum* in the Lower Carboniferous, that pteridosperms will ultimately be recovered from Devonian horizons. Structures that might be interpreted as cupules have, in fact, been found in the

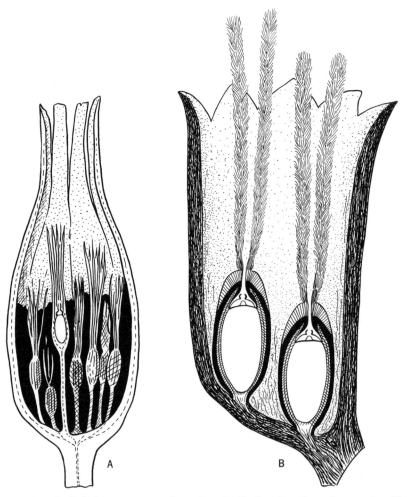

Fig. 5-16. A. *Calathospermum scoticum;* longitudinal section through a cupule with some of the enclosed seeds. B. *Gnetopsis elliptica;* a cupule from the Upper Carboniferous of France, with two seeds. (A from Walton, 1940; B from Renault and Zeiller, 1888 and Oliver and Salisbury, 1911.)

Devonian, but the evidence they afford is not conclusive; they will be considered further in Chapter 13.   The stem and leaf remains (*Tetrastichia, Telangium*) seem less advanced than *Calathospermum,* although our knowledge of the pteridosperms of this age is admittedly too scanty to allow one to be at all dogmatic about evolutionary trends.   *Tetrastichia* is similar in its anatomy to the Upper Devonian *Tetraxylopteris* which is tentatively placed in the protopterids, and *Telangium* is little more than a two-dimensional protopterid "leaf."

A seed-bearing organ as complex as *Calathospermum* does, however, afford conclusive evidence that the pteridosperms are not closely allied to the true ferns. With the exception of sporangia that are doubtfully referable to the Schizaeaceae the true ferns have not been found below the Upper Carboniferous. Thus, the evidence available at present indicates that the pteridosperms and ferns originated as independent lines (probably several independent lines in each group) from the coenopterid-protopterid complex. The term *seed-fern* is thus an unfortunate one since these plants are certainly not "ferns" that evolved into seed plants.

The pteridosperms came into their own in the Upper Carboniferous and Permian as a dominant and highly diversified assemblage, and I have no doubt that ultimately several orders and families will be recognized. The medullosas seem to form a clearly defined line and it is one that has commanded considerable attention partly because of the intrinsic peculiarity of the vascular anatomy and partly because some of the Permian species approach the kind of stem anatomy found in the cycads. There is, however, a very wide gap between the morphology of the seed and microsporangiate organs of the medullosas and the comparable structures of any Mesozoic or living cycadophytes.

It is perhaps evident that the pteridosperms, like many other fossil groups, present more riddles than solutions. Since they offer the most likely source that we have as a fountain head of angiosperm ancestry the most pressing problem centers around the seed and its enclosing cupule. In Chapter 13 a few speculations are offered concerning the nature of the cupule, the integument, and the nucellus. We do have some intriguing evidence which at least offers an evasive target for future investigations. Apparently not all of the pteridosperms had their seeds invested in a cupule—but is this really true? The positions that the seeds occupied are quite diverse; but to date there are somewhat less than a score of specimens known with the seeds actually attached to their fronds, and in most of these there is little cellular structure preserved.

Corystospermaceae

Several groups of fossils have been found at Triassic and Jurassic horizons that have been classified as "Mesozoic pteridosperms." There is sound evidence for so regarding them, but they are sufficiently distinct in many details as to allow no close correlation with the Carboniferous pteridosperms.

The Corystospermaceae (Fig. 5-17) is a unique assemblage of fossil plants, including seed-bearing and microsporangiate organs and foliage from Triassic beds of Natal, South Africa. They are

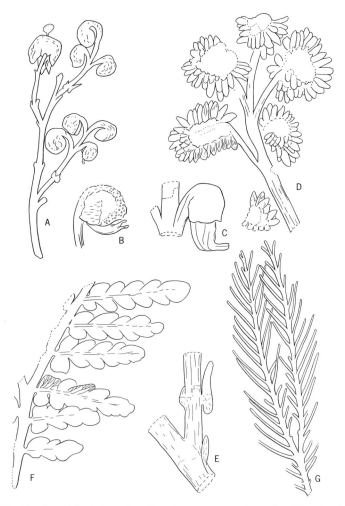

Fig. 5-17. Fossils attributed to the Corystospermaceae, from the Triassic of South Africa. A. *Umkomasia macleani,* inflorescence. B. *Pilophorosperma granulatum,* cupule showing micropyle of seed. C. *Pilophorosperma* sp., cupule with seed. D. *Pteruchus africanus,* microsporangiate inflorescence. E. *Umkomasia macleani,* portion of a main axis with base of lateral branch and subtending bract and bracteoles. F. *Dicroidium* sp. G. *Stenopteris densifolia.* (From Thomas, 1933.)

regarded as referable to a single family by virtue of close association, similarity in cuticle structure, and in one case pollen grains characteristic of a male organ have been found in one of the seeds. Two genera and several species of seed-bearing inflorescences are known. *Umkomasia macleani* Thomas is an inflorescence 3.4 cm long which produces lateral branches, in one plane, in the axils of bracts. Each of these branches in turn bears a pair of small scalelike structures (bracteoles) and several terminal cupules, each enclosing one seed. The cupule is partially divided to form a bivalved structure which is distinctly helmet-shaped; this feature is especially pronounced in *Pilophorosperma granulatum* Thomas in which the curved micropyle with its bifid tip is shown projecting from the helmet or cupule. The generic name *Pilophorosperma* implies that the helmet is lined with hairs, this being the chief distinguishing character which separates it from *Umkomasia*.

Pteruchus africanus Thomas may be considered as typical of the male or pollen-bearing organs. It is a small branch system which divides by equal or unequal dichotomies, the branches terminating in a circular or elliptical lamina. Numerous sporangia were attached over the entire under surface of the lamina and were probably pendant in life. The sporangia vary from 1 to 4 mm long and contain winged pollen.

Leaves known as *Dicroidium* and *Stenopteris* are thought to represent the foliage of these plants.

The beds from which the corystosperms were obtained are described by H. H. Thomas as having a rich fossil flora and should ultimately shed considerable light on our knowledge of seed plants and possibly those approaching the angiosperm stage. It seems to me that the "Corystospermaceae" should be regarded at present as an assemblage of pollen and seed-bearing organs that probably bore fernlike foliage. It was a fairly large and varied group; Thomas has described two species of *Umkomasia,* eight species of *Pilophorosperma,* a third seed genus *Spermatocodon* and eight species of *Pteruchus* in addition to the foliage genera *Dicroidium, Stenopteris,* and perhaps *Pachypteris.*

The helmet-shaped structure enclosing the seeds appears to be homologous with the cupule of the Carboniferous pteridosperms and the male organs are similar in form to earlier ones such as *Crossotheca.*

Another distinctive group of fossils [*Zuberia zuberi* (Szajnocha) Frenguelli] apparently of pteridospermous affinity, have been described from the Triassic of Argentina (Fig. 5-18). The dichoto-

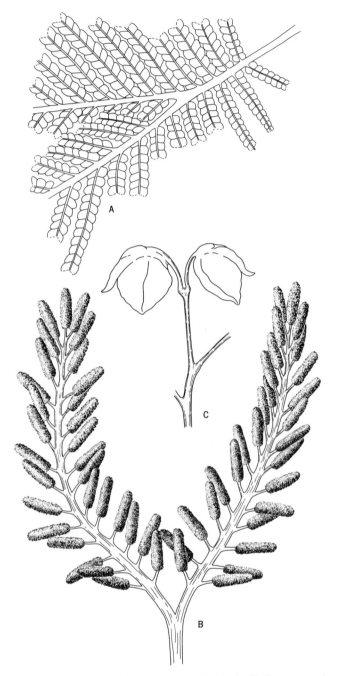

Fig. 5-18. *Zuberia zuberi.* A. A portion of a sterile frond. B. Reconstruction of the male fructification. C. Cupulate organs. (From Frenguelli, 1944.)

mously forking sterile frond bears pinnules of the *Odontopteris* type. The microsporangiate organ is a dichotomously forking branch bearing numerous appendages that consist of a pedicel beyond which many sporangia are attached at right angles. Associated with them are cupulate organs which invite some comparison with *Umkomasia.* The preservation of these fossils is by no means as perfect as one would wish for, but enough evidence is available to indicate a distinct group of plants that may tentatively be assigned to this family.

The Peltaspermaceae

This family (Fig. 5-19) is designated for seed-bearing and microsporangiate organs, as well as foliage, which have been found in Rhaetic rocks of Greenland and the late Triassic of Natal. The leaves are typically fernlike, attained a length of 30 cm, and are widely distributed and abundant in the Greenland beds. Stem

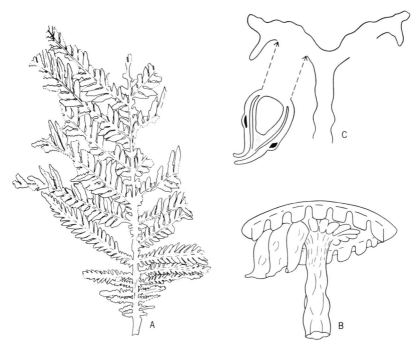

Fig. 5-19. The Peltaspermaceae. A. *Lepidopteris natalensis,* a frond about ¼ natural size. B,C. *Lepidopteris ottonis,* from East Greenland; B. restoration of cupulate disc showing three seeds still intact; C. diagrammatic longitudinal section through disc and one seed. (A from Thomas, 1933; B,C from Harris, 1932.)

fragments have been found with them which measure 5.5 cm in diameter and 10 cm long. The microsporangiate organ is a pinnately branching structure about half the size of the leaves and was probably three dimensional in its branching pattern. The sporangia or pollen sacs were borne in pairs at the tips of the ultimate branchlets; each pollen sac is about 2 mm long and 1 mm wide and dehisced longitudinally.

The seed organ is described as a cupulate disc; it was umbrella-shaped, consisting of a stalk terminated by a disc 1.5 cm in diameter. About 20 cavities are found forming a circle around the under side of the disc and are believed to represent scars where seeds were attached. In the Natal collections the cupulate discs were found attached, in a spiral arrangement, to a central axis which attained a length of 27 cm. They are only about one-third the size of the Greenland specimens and apparently only one seed per disc reached maturity.

In the Greenland specimens the seed is ovate, being about 7 mm long and 4 mm broad with a curved micropylar beak. They were sufficiently well preserved to allow observation of the inner and outer cuticle of the integument, the nucellar cuticle, and the megaspore membrane.

The stem fragment, leaf rachis, stalk of the disc, and microsporangiate organs bear characteristic swellings that are generally similar and there is a close comparison in the stomatal structure of all these organs as well as the seeds. This evidence, combined with their association in localities as widely separated as Greenland and Natal, leaves little doubt that they may be regarded as representing a single genus or a group of closely related ones. Their taxonomic position is less certain. The foliage and microsporangiate organs are pteridospermous, but the seed-bearing appendages present a striking departure.

REFERENCES

Andrews, Henry N., Jr. 1940. A new cupule from the Lower Carboniferous of Scotland. *Bull. Torrey Bot. Club.*, 67: 595–601.

――――――. 1945. Contributions to our knowledge of American Carboniferous floras. VII. Some pteridosperm stems from Iowa. *Ann. Missouri Bot. Gard.*, 32: 323–360.

――――――. 1947. *Ancient Plants and the World They Lived In.* Comstock Pub. Co., Ithaca. 279 pp.

Corsin, Paul. 1931 Fructifications de *Sphenopteris* (*Diplotemema*) *alata* (Brgt) Kidston. *Ann. Soc. Geol. Nord,* Lille, 56: 25–33.

Delevoryas, Theodore. 1955. The Medullosae—structure and relationships. *Palaeontographica*, 97B: 113–167.

————, and Jeanne Morgan. 1954. A new pteridosperm from Upper Pennsylvanian deposits of North America. *Palaeontographica*, 96B: 11–23.

Frenguelli, Joaquin. 1944. Las especies des genero *Zuberia* en la Argentina. *Anales Mus. de la Plata, Paleontologia:* n.s. sec. B, *Paleobotanica*, No. 1: 1–30.

Gordon, William T. 1938. On *Tetrastichia bupatides:* a Carboniferous pteridosperm from East Lothian. *Trans. Roy. Soc. Edinburgh*, 59: 351–370.

————. 1941. On *Salpingostoma dasu:* a new Carboniferous seed from East Lothian. *Trans. Roy. Soc. Edinburgh*, 60: 427–464.

Hall, John. 1954. The genus *Stephanospermum* in American coal balls. *Bot. Gaz.*, 115: 346–360.

Halle, Thore G. 1929. Some seed-bearing pteridosperms from the Permian of China. *Kungl. Svenska Vetenskapsakad. Handl.*, 6: (No. 8) 1–24.

————. 1933. The structure of certain fossil spore-bearing organs believed to belong to pteridosperms. *Kungl. Svenska Vetenskapsakad. Handl.*, 12: (No. 6) 1–103.

————. 1942. Some specimens of *Potoniea* from the Carboniferous (Westphalien) of Belgium. *Mus. Roy. Histoire Nat. Belgique*, Bruxelles, 18 (No. 42): 1–6.

Harris, Thomas M. 1932. The fossil flora of Scoresby Sound, East Greenland. Part 3. Caytoniales and Bennettitales. *Meddelelser om Grönland*, 85 (5): 1–133.

Hemingway, W. 1941. *Neuropteris tenuifolia* with carpons attached. *Ann. Bot.*, 5: 193–196.

Jongmans, Willem J. 1930. On the fructification of *Sphenopteris hoeninghausi* and its relations with *Lyginodendron oldhamiam* and *Crossotheca schatzlarensis. Geol. Bureau Nederland Mijngebied*, 1930.

————. 1954. Contributions to the knowledge of the flora of the seam Girondelle (Lower Part of the Westphalien A). *Mededel. Geol. Stichting*, ser., CIII No. 4: 1–16.

Kidston, Robert. 1904. On the fructification of *Neuropteris heterophylla*, Brongniart. *Phil. Trans. Roy. Soc. London*, 197B: 1–5.

———— and Jongmans, W. J. 1911. Sur la fructification de *Neuropteris obliqua* Bgt. *Archives Neerland. Sci. Exactes Nat.*, ser. IIIB, 1: 25–26.

Kubart, Bruno. 1914. Über die Cycadofilicineen *Heterangium* und *Lyginodendron* aus dem Ostrauer Kohlenbecken. *Österreichische Bot. Zeit.*, 64: 8–17.

Long, Albert G. 1944. On the prothallus of *Lagenostoma ovoides* Will. *Ann. Bot.*, 8: 105–117.

Oliver, F. W., and Scott, D. H. 1904. On the structure of the Palaeozoic seed *Lagenostoma lomaxi*, with a statement of the evidence upon which it is referred to *Lyginodendron. Phil. Trans. Roy. Soc. London*, 197B: 193–247.

————, and E. J. Salisbury. 1911. On the structure and affinities of the Palaeozoic seeds of the *Conostoma* group. *Ann. Bot.*, 25: 1–50.

Renault, Bernard, and René Zeiller. 1888. *Études sur le terrain houiller de Commentry, Atlas*. St. Etienne, 75 pp.

Schopf, James M. 1948. Pteridosperm male fructifications: American species of *Dolerotheca*, with notes regarding certain allied forms. *Journ. Paleont.* 22: 681–724.

Stewart, Wilson N. 1954. The structure and affinities of *Pachytesta illinoense* comb. nov. *Amer. Journ. Bot.*, 41: 500–508.

———— and Delevoryas, T. 1952. Bases for determining relationships among the Medulloseae. *Amer. Journ. Bot.*, 39: 505–516.

————. 1956. The Medullosan pteridosperms. *Bot. Rev.*, 22: 45–80.

Thomas, H. Hamshaw. 1933. On some pteridospermous plants from the Mesozoic rocks of South Africa. *Phil. Trans. Roy. Soc. London,* 222B: 193–265.

Walton, John. 1940. *An Introduction to the Study of Fossil Plants.* Adam and Charles Black, London. 188 pp.

—————. 1949. *Calathospermum scoticum*—an ovuliferous fructification of Lower Carboniferous age from Dunbartonshire. *Trans. Roy. Soc. Edinburgh,* 61: 719–728.

Weber, O. and Sterzel, J. T. 1896. Beitrage zur Kenntnis der Medulloseae. *Natur-wiss. Gesell. Chemitz,* 13: 44–143.

6

the ANTHOPHYTA
(Angiosperms)

Some paleobotanical problems; earliest
Angiosperms; and a Gymnospermous
Group, the Caytoniales

Introduction

The dominance of the angiosperms or flowering plants in the living flora, their use in an almost endless number of ways in our economy, and the great botanical interest in these plants have all contributed to the intense desire of many naturalists to know more about their origins and past distributions. As to origins we still know very little, but some tantalizing discoveries of pre-Cretaceous angiosperms have been made in recent years and significant studies dealing with the problem have appeared; as to former distributions there have been numerous excellent contributions and some of these will be presented in detail.

It seems desirable first to enumerate the problems that are associated with angiosperm studies so that the present approach will be clear and, perhaps of more importance, so that the student may be aware of avenues of research that may be explored to enhance our all-too-scanty knowledge.

The angiosperms are preserved in the form of leaf impression (and occasionally compressions), seeds and fruits that may be petrified or in a lignitic state, petrified or lignitic wood, pollen grains, and very rarely flowers are found as impressions or entombed in amber; in some instances various combinations of these are found together in the same deposit. As to the value of these remains, the following may be considered as an introduction which will be elaborated upon later:

1. There has been much criticism of the general validity of studies based on leaf impressions, some botanists holding that the available characters are not adequate for significant identifications. It is true that many of the earlier studies include questionable identifications and the more recent ones are admittedly not perfect. However, some of the modern works, which are based on careful comparisons with herbarium materials and extensive studies of possibly related living floras, are dependable; quite obviously in all "leaf floras" some species have more distinctive characters than others.

2. Recent investigations of fossil seeds and fruits have supplied what seems to be especially good evidence. Such studies require the availability of extensive collections of modern seeds and fruits for comparison and a knowledge of both the external form and internal structure of these organs; there is much to be done in this area.

3. Fossil angiosperm woods are abundant from Upper Cretaceous times on and have received relatively little study. Like seeds, fruits, and leaves, studies of wood must be based on a knowledge of living plants. There are still thousands of living trees and shrubs whose wood anatomy is poorly known, if at all; there are very few collections of woods that are at all comprehensive in the botanical institutions of the world, and finally, the preparation of this material for study, living and fossil, is quite time-consuming. Thus it seems fair to say that studies of fossil angiosperm woods are hardly beyond the pioneering stage.

4. Finally, fossil pollens are now receiving a tremendous amount of attention, but these studies are little more than two decades old. Ultimately they may be expected to contribute a great deal toward our knowledge of past distribution patterns.

A few introductory comments may be offered concerning the time element. In general, angiosperm remains from late Tertiary horizons can be identified with modern genera and species with a considerable degree of confidence; by "late Tertiary" is implied, rather roughly, from the upper Miocene to the present. In floras found within this time range we are dealing largely with plants whose modern equivalents may be found in the immediate vicinity or at most a few hundred miles distant. Going down through mid-to early Tertiary horizons, we find that the comparison with adjacent floras, or indeed with living species and genera, becomes progressively less distinct. As examples, the upper Oligocene Bridge Creek flora of Oregon bears close comparison with certain elements in the coastal Redwood forests; the mid-Tertiary flora of Brandon, Vermont, displays affinities with the modern flora of the extreme southeastern United States; the Eocene Goshen flora of Oregon appears to have its strongest bonds with certain modern Central American areas; the Eocene London Clay flora carries far afield, with its modern equivalents centered in the Indo-Malayan region.

As might be expected, angiosperms of the Upper Cretaceous are often difficult or impossible to correlate with modern genera or even families. Very few studies of Cretaceous angiosperm floras have been made in recent decades and most of the older ones must be revised.

In attempting to sort out the vast literature dealing with angiosperm origins and past distribution patterns, the topics outlined below seem to be of particular interest and importance.

Chapter 6

Earliest fossil evidence of the angiosperms
What is a primitive angiosperm?
   Some presumably primitive living angiosperms
   The Caytoniales
Early Cretaceous flowering plants
Some Upper Cretaceous angiosperm floras

Chapter 7

Tertiary floras of England and western continental Europe
Late Cretaceous and Tertiary floras of western North America

Earliest Fossil Evidence of Angiosperms

Fossil leaf impressions that represent undoubted angiosperms appear in mid Lower Cretaceous horizons or possibly a little earlier in the stratigraphic column, and by early Upper Cretaceous times the group appears in abundance and great diversity. In explanation of this apparently sudden appearance of the angiosperms most botanists have supposed that they actually originated much earlier than the record indicates, but for one reason or another the forerunners of the group were not preserved. Scattered scraps of evidence have accumulated which suggest rather strongly that the angiosperms did exist in the Jurassic or even the Triassic; the more important of these pre-Cretaceous records will now be considered.

As long ago as 1904 a leaf impression was found in the mid-Jurassic of Yorkshire and given the rather noncommital name *Phyllites* sp. It is a leaf 3.7 cm long and 2.5 cm broad, with three main veins. It has been compared with *Cercidophyllum,* and Seward, in his original account, apparently considered it sufficiently convincing to write the following:

> Had the specimen been found in rocks known to contain the remains of Angiosperms, there would be no hesitation in identifying it as the leaf of a Dicotyledon. (1904, p. 153)

In 1955 Kuhn described what appears to be a typical dicot leaf from the lower Jurassic of Sassendorf, near Bamberg under the name *Sassendorfites benkerti.* Of still older age, *Furcula granulifer* Harris is from the upper Triassic of Greenland; the leaves are Y-shaped or sometimes unbranched, up to 15 cm long and with a distinct network of veins similar to that in the dicots.

It must of course be kept in mind that leaves or leaf fragments with parallel or netted venation do not positively imply monocot and dicot affinities. Several ferns as well as the gymnosperm *Gnetum* have net-veined leaves; others, such as *Caytonia* which is gymnospermous, and *Glossopteris* (the affinities of which are uncertain but it is probably a gymnosperm) have leaves that fall within the range of the dicots as far as venation is concerned.

There are several reports of supposed monocot foliage from pre-Cretaceous horizons recorded in the older paleobotanical works. Most of these are based on small fragments of foliage showing parallel venation and are either unidentifiable or quite clearly misidentified. Two accounts of presumed palm leaves, however, seem to be authentic. Early in the century Lignier described impression specimens from the lower Jurassic (Lias) of France which apparently represent the basal portion of a leaf blade. Although this is a critical and distinctive portion of the leaf and several competent paleobotanists have seen fit to accept his identification of the fossils as a palm (*Propalmophyllum liasinum*), little serious attention has been given them. It was, therefore, an event of the utmost interest when palmlike leaves (Fig. 6-1), much more completely preserved, were reported by Brown from the mid-to upper Triassic of Colorado. The simple, elliptic, pleated leaves of this plant (*Sanmiguelia lewisi*) attained a length of 40 cm and a width of 25 cm and were borne in an alternate arrangement at the tip of a rapidly tapering stem. They were collected from a reddish calcarious sandstone in the Dolores formation.

Pleistocene	
Pliocene	
Miocene	Brandon, Vt. flora (?)
Oligocene	
Eocene	Goshen flora Wilcox flora at Puryear, Tenn. London clay flora
Upper Cretaceous	Lance fm., Medicine Bow fm. Fox Hills fm. Raritan fm.
	Albian stage
	Patapsco fm.
Lower Cretaceous	Aptian stage
	Patuxent fm. Greenland flora (?)
Jurassic	Caytoniales *Sassendorfites, Propalmophyllum*
	Furcula
Triassic	*Sanmiguelia*

Mesozoic-Tertiary time chart showing approximate position of certain
plants, floras, and geological horizons cited in Chapters 6 and 7.

Fig. 6-1. *Sanmiguelia lewisi,* one leaf of a presumed Triassic palm from southwestern Colorado. (From Brown, 1956.)

Palm leaves are not uncommon in the Upper Cretaceous and had *Sanmiguelia* been found at such a horizon it would have been accepted without question as a palm. There is certainly no other known group of plants with which it can be compared at all closely and I see no reason to question the author's comment that it should be "regarded tentatively but credibly as a primitive palm." Many botanists have been inclined to regard the monocots as derived from the dicots and thus of more recent origin. If this is true the presence of a palm in the Triassic would push the origin of the angiosperms back in time much farther than has previously been supposed, consequently additional evidence bearing on this exciting discovery is anxiously awaited.

Pollen grains presumed to have been derived from plants belonging to the water lily family have been reported from a Jurassic coal in Sutherlandshire, Scotland, and from the early Jurassic at Pålsjo in Sweden. Recently, a restudy of the Scottish pollens by Hughes and Couper indicates that these are gymnospermous rather than angiospermous.

What is a Primitive Angiosperm?

In attempting to elucidate the evolution of a group of plants, whether one is working with living or fossil forms, it is necessary to have some notion as to what we are looking for; it is, however, equally important that we should not be rigidly bound by a mental concept of what the primitive ancestral form *should* be and ignore everything that does not fit the preconceived concept.

Some bitter words have been exchanged in the past over the question of what constitutes a primitive angiosperm. The student of botany, whether beginning or advanced, who assumes that this problem is solved is certainly mistaken. Although it is not possible to consider the problem in all its aspects, I shall briefly discuss certain recent morphological studies of living plants in order to lend as much meaning as possible to the fossil evidence. In order to emphasize the magnitude of the task of understanding early angiosperm evolution, the following comments are offered from a recent study by Dr. H. H. Thomas, a leading student of angiosperm evolution:

In trying to distinguish a relatively primitive plant we look for one in which each character is primitive. In seeking a fossil ancestor we hope to find one which has all the characters of a flowering plant. But one of the earliest discoveries in genetics was that characters can be inherited separately, or rather in groups which are independent. We must therefore not place too much stress on the occurrence of a single character, such as the possession of free petals. It is likely that in the course of the evolution of a genus the different characters, floral and vegetative, will tend to keep in step to some extent, but we cannot rely on this unless supported by approved statistical correlation. . . . If plants and plant structures are to be studied from the evolutionary point of view it is not enough to arrange them in a sequence with reference to their forms. The causal aspect must be studied, and consideration given to what is known of the physiology of reproduction. (1957, p. 132–133)

It has been known for some time that several plants in the ranalian complex (the common buttercup and magnolia are representative), long considered as occupying a primitive position, lack vessels
in their wood. Vessel elements are dead cells at maturity and serve
to conduct water and dissolved materials; they are tube-shaped
and differ from tracheids in having one or more holes at each end.
They are a characteristic feature of the wood of both monocots
and dicots and are lacking in most pteridophytes and gymnosperms.
There are exceptions; they have been found in a few ferns, in the
lycopod genus *Selaginella*, in *Equisetum* and in the Gnetales, but
it is likely that they evolved independently in those groups.

The known number of species of vesselless dicots is now about
100 (in ten genera) and includes *Trochodendron* (Trochodendraceae),
Tetracentron (Tetracentraceae), *Drimys, Bubbia, Belliolum, Exo-
spemum,* and others in the Winteraceae. Even a few species of
vesselless plants in the ranalian group would support its supposed
primitive positions, but the occurence of a hundred presents more
than a suspicion. This information correlates, though not necessarily species by species, with a seemingly primitive carpel morphology. Although studies of the nature and evolution of the
carpel have long held a foremost place in botanical research, they
were greatly stimulated by A. C. Smith's discovery of a new family
with especially primitive floral characters (Degeneriaceae) on the
Fijian island of Vanua Levu. Although vessels are present in
Degeneria vitiensis the anthers and carpels are remarkable. The
former, lacking any differentiation into anther and filament, have
been described as "Broad microsporophylls." The single carpel of
the flower (Fig. 6-2D–F) is referred to as conduplicate, implying
that it is a leaflike structure folded lengthwise; it bears three main
veins and two rows of ovules which are attached some distance within the margin. The adjacent inner surfaces are covered with hairs
and it is not until fruit development sets in that these surfaces
actually become concrescent. The outwardly flaring "lips" are
stigmatic throughout the length of the carpel and the pollen tubes
develop as shown in Fig. 6-2F. Comparable conduplicate carpels
are known in certain genera of the Winteraceae and in the
Himantandraceae.

Returning for a moment to the matter of vesselless dicot woods,
we may add a brief comment on the fossil game *Homoxylon.** This

---

* Probably most of the species of *Homoxylon* should be transferred to *Sahnioxylon*
of Bose and Sah since *Homoxylon* was originally established by Hartig in 1848 for a
conifer. Bose and Sah (1954) discuss the problems of correctly classifying woods of
this type.

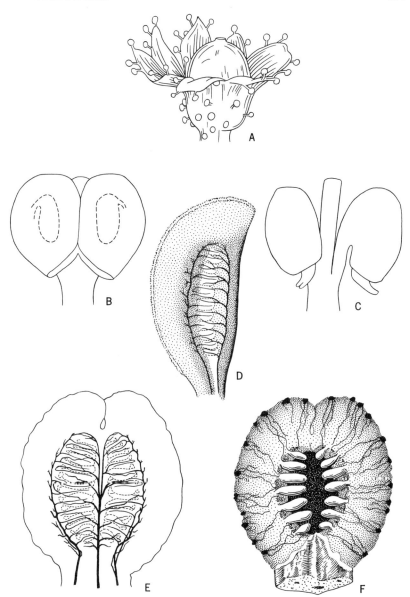

Fig. 6-2. A. *Lagenostoma lomaxi,* the lobed glandular cupule partially encloses a seed. B. Hypothetical stage in the evolution of a carpel, showing two partially fused cupulate organs. C. *Pilophorosperma;* two cupules, each containing a single seed. D. A primitive conduplicate carpel in side view. E. Same, unfolded showing vascular system. F. Same, showing placentation, distribution of glandular hairs (stippling), pollen grains and developing tubes. (A from Scott, 1923; B,C from Thomas, 1934; D,E,F from Bailey and Swamy, 1951.)

includes several species from Mesozoic horizons, some of which are considered to be angiosperms. Boureau has recently described two species from New Caledonia: *H. aviasii* from the basal Jurassic and *H. neocaledonicum* from the upper Triassic. They are woods with annual rings, they lack vessels, and the tracheids are scalariform with a tendency toward circular-bordered pitting in the latter part of the annual ring; the two are considered together here since they appear to be quite similar. Woods of this type, referred to *Homoxylon,* have been compared with certain living vesselless angiosperms, and there is the added interest in these two species in that they come from a region that is the center of primitive vesselless-living angiosperms. This type of wood is, however, quite close to that of some bennettitalean woods and so far as I am aware there is no positive assurance that any of the species assigned to *Homoxylon* are indubitably flowering plants.

At this point in our search for angiosperm ancestors I have chosen to consider a controversial and highly interesting group of plants, the Caytoniales, which were first described by H. H. Thomas from the rich mid-Jurassic plant bed at Cayton Bay in Yorkshire. It should be noted very clearly that the Caytoniales are not angiosperms but in the nature of their seed-bearing organs they come very close to being so. Nor is it implied that they are close ancestors of the modern ranalian plants considered above, yet the two groups do seem to offer some clues as to how angiospermy may have arisen. In so large and diversified a group (there are some 250,000 living species) it is possible that they originated along several distinct lines, but a discussion of this point is beyond the scope of the present account.

In most works on fossil plants the Caytoniales are assigned to the Pteridospermopsida or it is suggested that they are an offshoot of that group. The leaves, seed-bearing and male organs are assigned to the genera *Sagenopteris, Caytonia,* and *Caytonanthus* respectively. Several species of each are known and although not found in organic connection there is good reason to suppose that the association is significant. The leaves have been known for well over a century and are widely scattered geographically and stratigraphically, ranging from Upper Triassic to Lower Cretaceous. *Caytonia,* associated with *Sagenopteris,* has been found in Greenland, Sardinia, western Canada and probably Scania (Sweden), as well as in Yorkshire.

Caytonia sewardi Thomas consists of a central axis about 4 cm long with two rows of structures that will be referred to as fruits.

The fruit is about 4.5 mm in diameter, is recurved with a liplike
structure adjacent to the pedicel, and it enclosed a U-shaped row
of seeds. Although the fossils are preserved as compressions the
cellular organization of the seeds is known in some detail. The
fruits were originally thought to have been angiospermous, but a

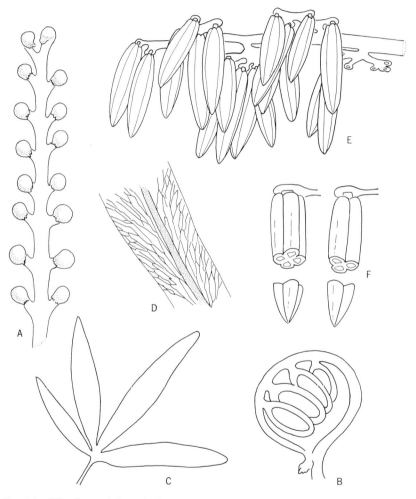

Fig. 6-3. The Caytoniales. A. *Caytonia nathorsti,* central axis bearing two rows of
cupules. B. A cupule or "fruit" in longitudinal section showing four seeds. C,D.
Sagenopteris phillipsi; C. a leaf with four leaflets; D. a portion of a leaflet showing
net venation. E,F. *Caytonanthus kochi,* from East Greenland; E. restoration of a portion of a microsporophyll; F. two anthers, one with four chambers and the other with
three. (A–D from Thomas, 1925; E,F from Harris 1937.)

later study by Harris revealed pollen grains within the micropyles; however, the opening must have been very small and it seems possible that the pollen reached the micropyles by means of a pollen drop mechanism. The axes are commonly found showing elliptic scars where the fruits were attached, the latter apparently being dropped by a dehiscence mechanism. Both seeds and fruits may be recovered in abundance by macerating the shale.

Caytonanthus arberi (Thomas) Harris, the microsporangiate organ, is known from central axes with short lateral branches which are subdivided and bear four-chambered "anthers." The pollen grains, unlike those of the angiosperms, are winged. Based on studies of well-preserved specimens from Scoresby Sound, Greenland (Fig. 6-3E), Harris notes that the anther is radially symmetrical, thus differing from the bilateral symmetry in extant flowering plants. The way in which the anthers are arranged on their branch system is also quite different from that of any modern flowering plant.

The leaves, *Sagenopteris phillipsi* (Brongiart), were known long before the reproductive parts were discovered. Although varying considerably in form they may be described as palmately compound with lanceolate to oval leaflets which are net-veined (Fig. 6-3C).

The known relationship of these plant organs goes beyond mere association which, however, is significant in view of the geographical distribution. In his original account of the Cayton Bay plants Thomas demonstrated a similarity between the epidermal cell structure of *Sagenopteris* and that of the fruit axes, and Harris has done the same for *Caytonia thomasi* and *Sagenopteris nilssoniana* from Greenland. But perhaps the most convincing evidence is the discovery of numerous seeds in the Yorkshire locality with pollen of the *Caytonanthus* type in the micropyles.

It has been pointed out many times that the Caytoniales do not compare closely with any living primitive flowering plants; however, they were very close to the angiospermous state and suggest one way in which the carpel may have evolved. The flowering plants are a varied assemblage and it is reasonable to suppose that they are polyphyletic. It may require several keys to unlock the mystery of their origins, and the problem centers around the carpel with its enclosed seed or seeds. It may therefore be well to add a few theoretical notes which are intended to serve as an introduction to the literature on the subject for students who may wish to pursue the matter in more detail.

Two concepts of carpel origin have attracted interest in recent years. Thomas has interpreted the apparent inverted *Caytonia* as having evolved from earlier pteridospermous plants. Beginning with the Upper Carboniferous *Lagenostoma,* in which the single seed was contained in a lobed cupule (Fig. 6-2A), one may regard the Triassic *Pilophorosperma* with its cupule lined with hairs as a later type, and in *Caytonia* the seeds are very nearly enclosed. According to this view the carpel found in primitive modern angiosperms, such as *Degeneria* and *Drimys,* is the result of a lateral fusion of two carpels of the *Pilophorosperma* (Fig. 6-2B, C) or *Caytonia* types.

In support of the concept of the carpel of the Ranales as a conduplicate or longitudinally folded structure Bailey and Swamy cite the following evidence: the three main vascular strands, the extensive stigmatic surface, the attachment of the ovules in a laminal position midway between the margin and center line, and their open character (being closed at the time of pollination only by the loose contact of the stigmatic hairs).

The fused cupule concept is admittedly speculative since the later stages are not known, whereas the concept of the carpel as an enfolded leaflike structure, although long favored by many botanists, finds no supporting evidence in the form of gymnospermous plants, living or fossil, which may be regarded as ancestral.

Continued studies of living plants will certainly shed light on this great problem, and I am not inclined to be pessimistic about possible contributions from the fossil record; many areas of the earth are still virtually unexplored paleobotanically. In concluding their discussion Bailey and Swamy note that:

> The most promising possibility of solving the riddle of the origin of the angiosperms appears to be an intensified search for paleobotanical evidence in parts of the earth's surface that have been inadequately explored, particularly Austro-Malayan and Indo-Malayan regions. (1951, p. 378)

Early Cretaceous Angiosperms

One of the most disturbing features of the paleobotanical record is the appearance, in approximately mid-Lower Cretaceous times, of angiosperm leaf impressions that are distinctly modern in aspect. Some of the more important floras of this age will be cited and a possible explanation offered for their sudden appearance. Nothing is known of the flowers of these plants, but the leaves do not suggest anything especially primitive. In general the floras are ones in which ferns and gymnosperms are the dominant elements.

In 1894 Saporta described several leaf fragments from rocks of Aptian age in Portugal some with parallel venation, others netted, which were interpreted as angiosperms. From a somewhat higher horizon in Portugal, considered to be Albian, a significant number and variety are recorded, including the following genera: *Salix* (willow), *Aralia, Braseniopsis, Myrica* (sweet gale), *Laurus* (laurel), *Viburnum, Eucalyptus, Magnolia* and *Sassafras.** The veracity of these generic identifications may be questioned, but there is no reason to doubt their angiospermous nature. Certain revisions of these floras have been given in a more recent study by Teixeira.

Seward has described an interesting assemblage of fossil plants from Cretaceous horizons in western Greenland which include the fern families Gleicheniaceae and Matoniaceae, as well as cycado-phytes, conifers, and *Ginkgo* foliage; associated with these are some questionable monocot leaf fragments and several undoubted dicots, including *Artocarpus* (breadfruit), *Quercus* (oak), *Meni-spermites* (moonseed), *Platanus* (sycamore), and others. The exact age of this flora is uncertain, but it is possibly derived from an early Lower Cretaceous horizon. Seward has commented on it as follows:

> My view is that the Greenland Cretaceous flora represents more fully than the floras of other countries the early stages in the transitional period from an older Jurassic-Wealden vegetation to the type of flora which continued into the Tertiary period, and still persists in regions remote from its original home. . . . The point I wish to make is that in the Cretaceous floras of Green-land Dicotyledons, which are surprisingly modern in the form of the leaves, occur in association with Ferns and Gymnosperms, which in other parts of the world are characteristic of floras distinguished by the absence of any recognizable examples of modern angiosperms. This fact lends support to the view that it was within the Arctic circle that the evolution of deciduous angiosperms progressed with greater rapidity and energy than in more south-ern latitudes. (1926. p. 155)

Several angiosperm leaves have been described from the Patuxent beds of Virginia and have been described under the names *Roger-sia, Proteaephyllum,* and *Ficophyllum.* These also appear as a very minor element in a predominantly pteridophyte-gymnosperm flora and at best one can say that they are leaves of dicots. In contrast, beds of Patapsco age are quite rich in plant fossils that are unquestionable angiosperms although the generic assignments

---

*See comment under "Notes" on page 187 concerning use of scientific and common names.

may in many cases be questioned. A listing of the species in these members of the Potomac group and of later Cretaceous horizons of the Atlantic coastal plain is given in a recent contribution by Dorf (1952).

In a comprehensive study of plants from many localities of Lower Cretaceous age in western Canada, Bell described the following genera from the Blairmore group which is considered to be of Albian age: *Populites* (poplar), *Ficus* (fig), *Trochodendroides, Cinnamomoides* (cinnamon), *Celastrophyllum, Sapindopsis, Fontainea* and *Araliaephyllum,* and a species of *Sapindopsis* from a somewhat lower horizon which may be upper Aptian.

Our knowledge of Lower Cretaceous angiosperms is not based wholly on leaves. Some years ago Marie Stopes described several dicot woods from Aptian horizons in England. Five genera were recognized (*Aptiana, Woburnia, Hythia, Sabulia, Cantia*), none of which can be said to display especially primitive characters. All have vessels, the first three have multiseriate rays, *Cantia* has uniseriate rays and vessels in which the pitting varies from round to oval scalariform, and in *Hythia* the vessel pitting ranges from round-bordered to scalariform.

Quite recently Samylina has reported a Lower Cretaceous flora (Aptian-Albian) from the Zyrianka river, a tributary of the Kolyma river, in eastern Siberia. Although the flora is predominantly one of ferns and gymnosperms, about 25% of the species are angiosperms. A total of 22 have been defined including the following: *Ranunculaecarpus* (Ranunculaceae?); *Sassafras* (Lauraceae); *Cercidophyllum* (Cercidophyllaceae); *Crataegites,* two species (Rosaceae); *Dalbergites* (Leguminosae); *Celastrophyllum* (Celastraceae?); *Zizyphoides* (Rhamnaceae); *Araliaecarpum* (Araliaceae); and several species are assigned to the indefinite *Dicotylophyllum.* Most of these are plants with small leaves, usually not exceeding 4 cm in length.

Summary Comments on Early Flowering Plants

An attempt has been made in the preceding pages to gather together the essence of what is known about early angiosperms and their origins, and it is quite clearly a very unsatisfactory story. We now have several tantalizing pieces of evidence which suggest that angiosperms existed far below the base of the Lower Cretaceous, of which *Sanmiguelia* is perhaps the most difficult to refute; then follows the much-discussed "sudden" appearance of angiosperms in modern aspect in the Lower Cretaceous. Several paleobotanists have suspected that the sudden appearance is an illusion and it has been suggested recently that the origins of the group should be sought in early Mesozoic or even Permian horizons. If *Sanmiguelia lewisi* is the authentic palm that it appears to be and the stratigraphic origin is correctly given as Triassic, then the Permian seems likely. But if this is the case why is so little of the earlier record preserved? In a recent discussion of the problem Axelrod notes that the early representatives of a group might be expected to be rare as fossils, and he suggests that the angiosperms originated in the diverse upland environments of the tropical zone during Permo-Triassic times. There is reason to believe that tropical latitudes of that age were appreciably broader than at present and in all probability included a very wide range of habitats. Unfortunately we find fewer and fewer upland sediments due to continued erosion as we go back in time.

In another useful review of the problem Takhtajan suggests that the rapid expansion of the angiosperms is due to the great plasticity of the group: "In no other group of plants do we observe such colossal differences as we see, for example, between magnolia and the cereal grasses, between the orchid and Haloxylon." (The last is a small xerophytic tree found in the arid steppes of Central Asia and the Middle East.) He also places considerable importance on vessels and insect pollination as factors which influenced successful evolution.

Directly interwoven with the problem of angiosperm origin as a whole is the question of a single or multiple derivation. That is, do the 250,000 extant species really constitute a single evolutionary line or several? Many botanists favor the former, but a possible multiple origin should not be discarded. Cheadle has shown that vessels originated independently in the monocots although he suggests that the group evolved from ancient (probably long extinct) members of the Nymphaeales which in turn were derived from the Magnoliales.

In such a stormy sea of controversy where the sailing is so hazardous it is cheering to encounter a fresh approach to the old problem and as an interesting example I conclude this section with a few notations on E. J. H. Corner's *Durian theory*. The durian (*Durio zibethinus* of the Bombacaceae) is a large forest tree native to the Malayan area and is characterized by slender twigs, simple leaves, and fruits that are huge, spiny, loculicidal capsules with one to five large seeds in each of its five cavities; the seeds are covered with a thick, white or yellow aril which may be regarded as a third integument. The fruits are a highly prized food of many forest animals from elephants to beetles.

The aril is often brilliant red and is found in some 45 families of flowering plants although the degree of its development varies from species in which the seed is completely enclosed to others in which it exists as an indistinct rudiment. It is present in many families of flowering plants that are not closely related, whereas in groups of related genera it is possible to trace apparent evolutionary sequences from the large, colorful, food-rich, arillate capsules to plants in which the fruits are dry capsules, often with winged seeds, or drupes, berries or nuts—and in general smaller. Such sequences are approximately correlated with increased power of dormancy (in many arillate species the seed must germinate immediately if they are to survive), and greater dissemination potentialities. The presence of the aril in numerous unrelated families in which these reduction series may be observed leads to the supposition that the arillate fruit is a primitive type.

The following structural correlations define a primitive angiosperm in terms of the durian theory; the large fruits must have been borne on stout branches; thus the ancestral form is postulated as a massive tree, possibly of cycadlike habit. The lack of dormancy power and inability to withstand desiccation suggests that the plants could not have survived outside of tropical rain forests. In summary it is argued by Corner that "the primitive angiosperm must have been a mesophytic, tropical, Cycad-like monocaulous tree with large pinnate leaves and peltate scales, probably monocarpic, and producing a terminal cluster of large arillate follicles." (1949, p. 414)

There are several attractive features to the theory: it fits in with other evidence suggesting a tropical origin of the angiosperms; it allows an early origin of plant types found in both monocot and dicot groups; it introduces the fruit as a neglected organ of plant evolutionary studies and one that can be correlated with paleobotanical studies more readily than the flower can. The general organization of the presumed primitive angiosperm (that is, in large compound leaves and sparsely branched, soft-wooded types) compares in some way with early seed plants such as the cycads and seed-ferns.

The theory is not offered as a cure-all or final answer to the problem of angiosperm origins but rather as a field of investigation which may lead to a better understanding of the problem as a whole. Any serious student of plant morphology would do well to read Corner's article.

Some Upper Cretaceous Angiosperm Floras

The difficulties of correctly identifying Cretaceous angiosperms have been noted; however, it seems worthwhile to introduce a few of the better known floras. Although all the names may not be valid, the abundance of genera and the families present in the Upper Cretaceous at least indicates the diversity of the flowering plants at that time.

In his account of the flora of the Amboy clays (Raritan formation) of New Jersey, Newberry listed 156 species, of which about four-fifths are dicotyledons; a few ferns are represented and several conifers. Later studies have revealed a richer assemblage of conifers but of particular interest at the moment is the diverse aggregation of dicots with 28 families recognized:

Juglandaceae	Magnoliaceae	Aceraceae	Cornaceae
Myricaceae	Menispermaceae	Rhamnaceae	Ericaceae
Fagaceae	Lauraceae	Vitaceae	Myrsinaceae
Ulmaceae	Rosaceae	Tiliaceae	Sapotaceae
Moraceae	Leguminosae	Passifloraceae	Ebenaceae
Proteaceae	Aquifoliaceae	Myrtaceae	Caprifoliaceae
Salicaceae	Celastraceae	Araliaceae	Asclepiadaceae

One of the largest fossil floras from any horizon is the lower Upper Cretaceous flora of the Dakota sandstone; the greater part of the collections were made by Charles H. Sternberg in the latter part of the past century in Ellsworth County, Kansas, although others have been made in Nebraska and Minnesota. Some 460 species were recognized by Lesquereux in his work which appeared in 1891. No comprehensive revision has been undertaken since that time and many of the indentifications may be questioned; there are, however, two points of lasting interest:

1. Aside from a few ferns, cycads, and conifers it is almost exclusively an angiospermous flora. Most of these are dicots with only a small sprinkling of questionable monocots.

2. It is a very diverse flora; the figure of 460 species may be in error, but the number is certainly high.

In a more recent study of a small flora of comparable age in southwestern Colorado, Brown has recognized *Ficus* (fig), *Nelum-*

bium, Mahonia (Oregon grape), *Sassafras, Platanus* (sycamore), *Celastrophyllum* (staff tree), *Juglans* (walnut), and *Sterculia.*
The Lower Medicine Bow flora from northwestern Colorado totals 64 recognizable forms of which 58 are assigned to species. They are mostly dicots, with a few monocots, the dominant elements being:

Ficus planicostata	*Dryophyllum subfalcatum*
Magnoliophyllum cordatum	*Rhamnus salicifolius*
Trochodendron nebrascensis	*Quercus viburnifolia*
Cinnamomum affine	*Viburnum marginatum*
Myrica torreyi	*Rhamnus cleburni*

In addition to these 10 dominants the flora includes a second species of *Cinnamomum* (cinnamon), several other species of *Ficus* (fig), two of *Magnolia* (magnolia), and two of *Sabalites* (sabal palm).

The region in Colorado where these fossils were obtained is presently a high, barren intermontane basin of the Rocky Mountain system with a mean temperature of 40 to 45° F; there is sagebrush-grassland on the ridges and a poplar-willow association along the flood plains of permanent streams. A few other plants are found but it is in general a very sparse vegetation. The fossil flora was much more diverse than the living one and the palm leaves (*Sabalites montana* and *S. eocenica*) alone leave no doubt as to the great difference in the general climatic relations of this region during the Upper Cretaceous. Dorf concludes that it was intermediate between warm temperate and subtropical and probably closer to the latter.

The Lance flora is another warm climate assemblage based on collections from several uppermost Cretaceous localities in east central Wyoming. Seventy recognizable forms have been found, with the following as the dominant elements:

Aracarites longifolia (araucaria)	*Dryophyllum subfalcatum*
Platanophyllum montanum (sycamore)	*Fraxinus leii* (ash)
Salix lancensis	*Viburnum marginatum*
Sequoia dakotensis	*Cornophyllum wardii*
Vitis stantoni (grape)	*Pistia corrugata*

The presence of palm leaves, several members of the Lauraceae, and a few other tropical or subtropical plants suggest a Cretaceous climate that was more uniformly warm and moist than the present one of that area. It may also be noted that the flora probably represents several different plant communities. Judging from the requirements of modern equivalents it is unlikely that conifers (*Araucarites longifolia* and *Sequoia dakotensis*), a ginkgo (*G. adiantoides*), palms (*Sabalites*), and several members of the laurel family would all have been growing in close association.

Even farther to the north Bell has described a rather large Upper Cretaceous flora from Vancouver Island with warm climate elements such as: a species of the fern genus *Anemia,* large palm leaves (*Geonomites*), *Artocarpus* and *Ficus* (Moraceae), *Cinnamomum* and *Laurus* (Lauraceae), *Ginkgo,* and several cycad leaves.

REFERENCES

The references for this Chapter appear at the end of Chapter 7.

7

the ANTHOPHYTA (Angiosperms)

Some Tertiary Floras of Europe and of Western United States

Any attempt to present a comprehensive summary of Tertiary floras is quite beyond the scope of this book. A great many have been described from various parts of the world during the last 150 years; the size of the floras varies from a few species to several hundred and the preservation ranges from excellent to very fragmentary. There are, however, some studies, or series of studies, that have been conducted in recent decades that are based on the best techniques available and which tell intriguing stories of changing climates and floral migrations.

Notes

1. In view of the large numbers of plant names that must of necessity be introduced in this chapter, the usual practice of citing the authority after the binomial has not been followed. In most cases the sources of information are evident and they are listed at the end of the chapter.

2. Since numerous Tertiary floras are considered here, it is necessary to introduce a rather large number of plant names and I am aware that many of them may be unfamiliar to the reader. The scientific names used in the original publications must serve as a basis, but whenever possible corresponding common names are also given; however, since many of the fossil species are extinct, it should be remembered that the common names may not be exactly equivalent to their usage with living plants. Moreover, with some species, genera, and families it is not possible to cite common names that are meaningful.

3. The problems of correctly identifying the foliage, seeds, and fruits of Cretaceous and Tertiary angiosperms are unique and considerable attention has been devoted to study techniques by the more competent modern workers. For discussions of this phase of the work the following references are especially recommended: Chaney and Axelrod, 1959; Dorf, 1938; Reid and Chandler, 1933; Szafer, 1946.

TERTIARY FLORAS OF ENGLAND AND WESTERN CONTINENTAL EUROPE

A series of fossil seed-fruit floras, ranging through the Tertiary, is known from western Europe which reveals a striking change in composition and climate. As a matter of reference the floras to be considered are listed below:

Flora	Age	Location
London Clay	Lower Eocene	Isle of Sheppey, southeastern England
Hordle	Upper Eocene	Hordle, Hampshire, southern England
Bembridge	Oligocene	Isle of Wight, southern England
Pont-de-Gail	Pliocene, basal	Pont-de-Gail, Cantal, France
Reuverian	Lower Pliocene	Limburg Province, Netherlands
Króscienko	Pliocene	Króscienko, southern Poland
Teglian	Upper Pliocene, base of	Limburg Province, Netherlands
Cromerian	Pliocene, top of (or Pleistocene?)	Cromer, Norfolk, England

The London Clay Flora

Early in the last century naturalists began to gather pyritized seeds and fruits from the beaches of the Island of Sheppey, near the mouth of the Thames River, and other localities in southeastern England. In 1840, after several years of study, J. S. Bowerbank produced a book dealing with the fossils and, although an excellent work from the standpoint of accuracy of description, sufficient modern comparative material was not available to render significant identifications possible. Bowerbank's interest in the flora was sufficient to have led him to establish "The London Clay

Club" to facilitate study of the fossils. Although one may assume that it never ranked among London's most popular Clubs, an interest in the seeds and fruits has continued to the present day and in 1933 Reid and Chandler's *The London Clay Flora* appeared, presenting in the finest tradition of British paleobotany a comprehensive study of this great and unique assemblage of fossil plants.

The London Clay flora, of early Eocene age, includes 314 species of seeds and fruits; of this number 234 have been identified, whereas the affinities of the remainder are considered doubtful. It is almost exclusively an angiosperm flora, there being but 7 conifers. Of the 100 genera, only 28 are still extant; thus it is family relationships that will primarily occupy our attention. The present-day distribution of the families which make up the London Clay flora are: 5 are entirely tropical (Nipaceae, Burseraceae, Icacinaceae, Bixaceae and Sapotaceae); 14 are almost exclusively tropical (Palmae, Olacaeae, Menispermaceae, Anonaceae, Lauraceae, Meliaceae, Anacardiaceae, Sapindaceae, Sapiaceae, Elaeocarpaceae, Sterculiaceae, Dilleniaceae, Myrsinaceae, and Apocynaceae); 21 families are equally tropical and extratropical and 5 are chiefly temperate. The Lauraceae and Icacinaceae, almost exclusively tropical families, are represented by 40 and 21 species respectively. Of modern floras the closest comparison lies with the forests of Indo-Malaya and particularly the Malay Islands.

A few genera may be mentioned because of their abundance or particular interest. The most frequently encountered fossil is the fruit of *Nipa burtini* which is mentioned as being "strewn in great abundance wherever the dark patches of pyrites nodules occur" at Sheppey. The modern *Nipa* is a brackish water plant found at sea level in the tropics of the Indo-Malay region. Also abundant are *Wetherellia variabilis* (Linaceae) and *Hightea elliptica* (Myrtaceae ?); the former is believed to be related to the modern *Hugonia* which is found in the tropics of Asia, Africa, Australia, and New Caledonia; the latter, represented in the London Clay collections by over 500 specimens, has proved difficult to identify and its position in this family is somewhat doubtful. Three members of the Icacinaceae are abundantly represented: *Iodes multireticulata, Stizocarya communis,* and *Icacinicarya platycarpa.* Fossils referable to modern temperate climate families are not absent from the flora, one of the very abundant species being *Petrophiloides richardsonii* of the Juglandaceae.

The London Clay flora presents many questions that are by no means alien to the paleobotany of other areas and ages. A few

that immediately arise are: Why are fruits and seeds present in the clays to the exclusion of leaves? How can one account for the relationship with a distant tropical flora in this northerly latitude? Is it possible that the fossils were transported by ocean currents perhaps hundreds or thousands of miles from their original point of origin? Is the implication of a tropical climate erroneous in that the plants concerned, although of tropical affinities, may in fact be genera that became acclimatized to a temperate climate? Mrs. Reid and Miss Chandler have presented one of the most comprehensive and penetrating studies of these recurrent problems and it seems opportune to draw upon their results. A serious student of fossil botany would do well to read the Introduction to *The London Clay Flora*.

As to the seed-fruit composition of the flora Reid and Chandler refer to H. N. Moseley's account of the Challenger Expedition pertaining to observations that were made some 70 miles northeast of Point D'Urville, New Guinea, where the Ambernoh River, the largest in New Guinea, empties into the sea.

We passed through long lines of drift-wood disposed in curves at right angles to the direction in which lay the river's mouth, . . . The logs had evidently not been very long in the water, being covered only by a few young Barnacles (*Balanus*) and Hydroids. . . . Various fruits of trees and other fragments were abundant, usually floating, confined in the midst of the small aggregations into which the floating timber was almost everywhere gathered. . . . Very small seeds were as abundant as large ones, the surface scum being full of them, so that they could be scooped up in quantities with a fine net. . . . I observed an entire absence of leaves, excepting those of the Palm, on the midribs of which some of the pinnae were still present. The leaves evidently dropt first to the bottom, whilst vegetation drift is floating from a shore. Thus, as the debris sinks in the sea water a deposit abounding in leaves, but with few fruits and little or no wood will be found near shore, whilst the wood and fruits will sink to the bottom farther off land. (1892, p. 373)

◀ **Fig. 7-1.** Seeds and fruits from the Eocene London Clay. A. *Minsterocarpum alatum,* worn fruit showing seeds, 3X. B. *Melicarya variabilis,* a nine loculed fruit, 3X. C. *Euphorbiotheca sheppeyensis,* longitudinal section of fruit, 3X. D. *Nipa burtini,* internal cast of seed, 1X. E. *Iodes multireticulata,* endocarp, 3X. F. *Anonaspermum ovale,* internal cast of seed showing branching and anastomosing ruminations, 3X. G. *Ampelopsis rotundata,* internal cast of seed, 7X. H. *Protocommiphora europaea,* a two-loculed endocarp, 5.5X. I. *Microtinomiscium foveolatum,* locule cast, 6.6X. J. *Wetherellia variabilis,* seed with remains of the testa, 8X. (From Reid and Chandler, 1933.)

Fig. 7-2. The beach at Sheppey; Eocene fruits and seeds come from the clay bank to the left.

Of the 73 genera that it has been possible to relate in some degree to modern ones 33 are lowland tropical forms, 29 may be lowland tropical, montane tropical, or extra tropical and 8 are montane tropical or extra tropical. *Nipa* is considered to be a key plant in that it is a lowland tropical form occupying brackish waters, whereas a significant number of other species may represent plants that grew at a higher altitude and were carried down to the point of deposition by a large river or were carried a much greater distance by ocean currents. It is known from the works of H. B. Guppy and others that seeds and fruits may be carried great distances, for example, from the West Indies to the shores of Britain and Scandinavia by the Gulf Stream. Several points militate against a long ocean voyage for the London Clay fruits. It is unlikely that they would be found in abundance along the shore at Sheppey if they had been carried for thousands of miles; very few, moreover, display any specialized structural features that would enable them to remain afloat for the long period of time necessary; finally, it has been shown that beach drift, even though it may include seeds and fruits that have come long distances, contains a predominance of local debris, and we have seen that the London Clay flora is strongly tropical in both numbers of specimens and species.

It was suggested by Bowerbank in 1840 that the fossils may have been transported by a river "which probably flowed from near the Equator towards the spot where the interesting remains are now so abundantly deposited." There seems to be little doubt that river transport is involved but not to the extent suggested. A direct route from the African tropics, which in itself would have involved a very considerable distance, was cut off by the Tethys Sea which occupied the present Mediterranean basin as well as much of south-central Europe, part of northern Africa, and much of the present Near East; this sea moreover communicated with the Indian Ocean. In this connection it is of some interest to mention a small flora described by Miss Chandler from the Danian (uppermost Cretaceous) of Egypt in which the fossils are closely related or identical with those of the London Clay.

An argument that is often used against plants as a reliable index of climate is the assumption that they may change in the course of time in their reaction to climate. That some plants are remarkably versatile in their ability to survive under a wide variety of climatic conditions is evident, and Reid and Chandler note clearly that conclusions as to climate must be drawn from a study of the bulk of the fossil flora concerned. One composed of few species and few specimens might well lead to erroneous concepts, but when dealing with several hundreds of species and thousands of specimens this objection is hardly tenable.

A study of later Tertiary floras shows that the bulk of the London Clay plants did *not* survive under the cooler climatic conditions that developed even in late Eocene times. Reid and Chandler also present evidence from Pleistocene floras to indicate that, under changing climatic conditions, plants are much more apt to migrate to a favorable locale than to adapt in their current position.

Paleoclimatological problems may be complicated and not infrequently meet with bitter argument. It is not implied that the authors of the London Clay flora have given a final and faultless answer, but their study is extraordinarily well documented.

Some Later Tertiary Floras of Europe

The Hordle flora of Hampshire, England, is one of dominantly southern Asiatic relationships, of the south China and Burma areas, appreciably to the north of the Malay Islands. There are also secondary affinities with the flora of Japan, north China, and the Himalayas and, as compared with the London Clay flora, there is a noticeable increase in northern elements with European affinities. In this category may be cited genera such as *Salvinia* (Salviniaceae), *Pinus* (Pinaceae), *Potamogeton,* and *Limnocarpus* of the Potamogetonaceae, *Stratiotes* (Hydrocharitaceae), *Nuphar* (Nymphaeaceae), *Corydalis* (Papaveraceae), *Rubus* (Rosaceae), and *Omphalodes* (Boraginaceae). Of the dominant lowland tropical plants of the London Clay only three of these, *Nipa* (Nipaceae), *Iodes* (Icacinaceae), and *Tetrastigma* (Vitaceae) survived until the end of the Eocene in Europe.

The Bembridge flora from the Oligocene of the Isle of Wright is considered as being a warm temperate one. Among the genera which particularly suggest such a climate are *Acrostichum* (Leather fern, based on a leaf fragment), which is found today in southern Florida and the Keys in fresh water marshes, brackish swamps, and mangrove swamps; an *Acanthus* (Acanthaceae) which is allied with modern species found in the mangrove swamps of eastern Asia and Australia; *Phyllanthera vectensis* (Asclepiadaceae) is compared with the modern *P. perakensis,* a liane of the Malay Peninsula; several palms (*Sabal, Palmophyllum,* and *Palaeothrinax*) are represented by leaf fragments. The flora includes plants of more northerly affinities, indicating a change from the climate of London Clay times; a few of these are *Pinus, Typha* (Cat-tail), *Engelhardtia* (Juglandaceae), *Papaver* (Papaveraceae), and *Abelia* (Caprifoliaceae).

The Pliocene floras, such as the Pont-de-Gail, Reuverian, Krościenko, Teglian, and Cromerian, show a progressive shift from an eastern Asiatic affinity to one with western Europe. The Pont-de-Gail flora still retains some of the Indo-Malayan elements; although having a dominant relationship with eastern Asia, the west European relationships of the Reuverian and Krościenko floras are becoming pronounced; in the Teglian, the East and West relationships are about equal, whereas the Cromerian is definitely a west European flora. It is not possible to consider all of these in detail and I have, therefore, chosen to discuss two, the Reuverian and Krościenko floras, which are large and quite well known.

The Reuverian flora of Limburg is an especially large one; nearly 300 species are known, of which somewhat more than one-half have been identified with considerable certainty. So far as I am aware this is one of the largest of Tertiary floras; in fact, the authors (Reid and Reid) note that it was probably much richer in species of trees and shrubs (both absolutely and proportionally) than the modern wooded areas of that region. The highest concentration of modern equivalents is found in the mountains of western China; in some cases the fossils are so similar that they cannot be distinguished from the Chinese ones, notably: *Gnetum scandens, Stewartia pseudo-camellia, Magnolia kobus, Zelcova keaki, Pyrularia edulis,* and *Prunus maximoviczii.* Somewhat fewer genera are represented by closely allied species in Japan, eastern Tibet, and the Malay Peninsula, and fewer still have relationships with the flora of Europe.

Not only is the alliance marked by the number of species belonging to Chinese and Japanese genera which are not represented in Europe, but by the curious fact that when the genus lives in both Europe and China, it is often the Chinese or Japanese species that most resembles the Reuverian plant. Thus in the genus *Pterocarya* (wingnut), *P. limburgensis* is more closely allied to *P. hupehensis* than to *P. caucasica.* In *Styrax* (snowball) the alliance is with *S. japonica* and *S. obassia.* In the genus *Betula* (birch) we have *B. digitata,* an ally of the Chinese *B. ulmifolia* and belonging to a section of the genus no longer found in Europe. In *Cornus* (dogwood) we have either *C. controversa* or a closely allied species. The only *Clematis* (virgin's bower) is the Chinese *C. grata,* not the nearly related *C. vitalba.* But perhaps the most remarkable of these cases is in the genus *Eupatorium;* for not only is the alliance of the Reuverian plant closer to *E. japonicum,* but it is closest to an unnamed variety of *E. japonicum,* collected by Pere Faurie from a single mountain in Japan. . . . The alliance is shown almost entirely by species whose northernmost geographical range now lies much farther south than Limburg; but they are mountain plants in China, and are not found anywhere on the plains. In other words, though now living in southern latitudes they are temperate forms, and belong to the moist and temperate forest-belt found only on the Chinese mountains, and in the similar moist regions of the Himalaya and Japan. (1915, p. 16)

Some of the Reuverian plants that are still found in Europe, and in fact ranging considerably to the north of Limburg, are: *Picea excelsa* (Norway spruce), *Quercus robur* (English oak), *Carpinus betulus* (European hornbeam), *Corylus avellana* (European hazel), *Prunus spinosa* (blackthorn), *Ilex aquifolium* (English holly), *Vitis vinifera* (wine grape). In view of their ability to withstand considerable cold the authors suggest they could have survived through the rigors of the Pleistocene by moving to the south of France or the Caucasus.

Another large Pliocene flora, with about 40 species in common with the Reuverian, has been described by Szafer from Krościenko in southern Poland. The fossil seeds and fruits were obtained from an examination of about 100 cubic meters of the sediments in which they occur; it is a study that will probably take a foremost place in the classics of paleobotanical literature.

About 18% of the species of the flora are most closely related with living modern European ones, 37% with eastern Asiatic ones, and 19% with eastern American ones. Szafer concludes that two distinct elevations are represented in the flora: a lower one in which the following prevailed: *Carpinus* (hornbeam), *Pterocarya* (wingnut), *Alnus* (alder), *Liriodendron* (tulip tree), *Ceanothus, Vitis* (grape) and *Styrax* (snowball), and an upper elevation in which the following were predominant: *Picea* (spruce, 3 species), *Tsuga* (hemlock, 2 species), *Abies* (fir), and *Pinus* (pine).

Taken in total the Krościenko flora is regarded as indicating a climate with a mean temperature some 9° C higher than at present, with less contrast in summer-winter extremes, and a rainfall about twice that in the region today.

In summary, we have in this sequence from early Eocene London Clay times to the modern British flora, extending over a period of about 60 million years, a change from a lowland Malayan flora which changed rather sharply at the end of the Eocene to a flora still of eastern Asiatic affinities but more cosmopolitan—a gradual shift through the Pliocene to a dominantly west European flora. The change included a pronounced decrease in the percentage of woody genera and a progressive increase in genera of wide range. A reversal of this later trend occurs from the Cromerian to the present and is accounted for by the extinction of many genera during the Pleistocene.

LATE CRETACEOUS AND TERTIARY FLORAS
OF WESTERN NORTH AMERICA

During the past few decades a series of intensive studies have been undertaken dealing with the Tertiary floras of the Pacific Coast states. These investigations have been carried on for the most part by R. W. Chaney, D. I. Axelrod, and several others and much that follows is taken from their published accounts. Since the Tertiary floral succession has not been studied elsewhere in as great detail as in this area, and the story is a particularly interesting one from the standpoint of shifting floras and climates, it has seemed to me more effective to consider it in some detail at the expense of neglecting Tertiary floras elsewhere in the world. In order to render the account more complete, consideration is also given to some of the late Cretaceous floras of the region. There are now hundreds of publications dealing with the fossil plants of this age and area, many of which are rather weighty volumes, and in attempting to extract the essence of this voluminous literature I am aware of the shortcoming of the following account.

Upper Cretaceous Floras

The Upper Cretaceous floras of the western states flourished under conditions of more equitable and abundant rainfall and milder temperatures than prevail today; warm temperate to subtropical forests grew over much of the present desert and high mountain terrain.

The following plants have been found in the Medicine Bow formation in southwestern Wyoming and adjacent Colorado: *Cercidophyllum* (Katsura tree), *Cinnamomum, Dryophyllum* (evergreen oak), *Ficus* (fig), *Laurus, Lindera, Magnolia, Platanophyllum, Rhamnus* (buckthorn), *Sabalites, Viburnum,* and *Metasequoia.* Such an assemblage suggests rather heavy rain throughout the year and mild temperatures without much frost, in striking contrast to the low rainfall and great variation in temperature that prevails in that area now.

In the somewhat older floras of the Frontier formation and Aspen shales of southwestern Wyoming we encounter several ferns including *Anemia, Asplenium, Cladophlebis, Microtaenia,* and *Tempskya,* with such angiosperms as *Cinnamomum, Ficus, Laurus, Liquidambar* (sweet gum), *Prunus, Quercus* (oak), *Salix* (willow), and *Sassafras.* The distribution of the tree-fern *Tempskya* through the northwestern states has been mentioned previously and in itself is an important key to the Upper Cretaceous climate.

The Fruitland and Kirtland formations of the San Juan Basin of northwest New Mexico have yielded the following ferns: *Anemia, Asplenium, Onoclea,* and angiosperms: *Cornus, Ficus, Laurus, Quercus, Sabal,* and *Salix.* The late Upper Cretaceous Vermejo flora of southern Colorado and adjacent New Mexico includes the following: *Artocarpus, Cissites, Credneria, Ficus, Laurophyllum, Magnolia, Platanus,* and *Sabalites.*

Looking far to the northward the Upper Cretaceous of Alaska has yielded a flora that has a considerable assemblage of plants that do not occur there now. Over 200 species have been described by Hollick which include 15 genera of gymnosperms and 73 genera of angiosperms. *Ginkgo* was common, being represented by several species; five species of the cycadophyte *Nilssonia* have been described; *Sequoia* and *Metasequoia* are present, and recently Arnold has recorded a new genus *Parataxodium* which is thought may be the ancestral stock from which *Metasequoia* and *Taxodium* (bald cypress) were derived. Among the numerous dicots are : *Juglans* (walnut), *Populus* (poplar), *Hickoria* (hickory), *Betula* (birch), *Quercus* (oak), *Ficus* (fig), *Aristolochia* (birthwort), *Nymphaeites, Castaliites, Menispermites, Magnolia, Laurus, Platanus* (sycamore), *Credneria, Sorbus, Cassia, Acer* (maple), *Rhamnus, Vitis* (grape), *Aralia, Hedera* (ivy), and *Viburnum.*

The Cretaceous floras are not as well known as the Tertiary ones in western North America and the problems of their identification are much greater. Yet allowing for numerous mistakes it seems evident that they reflect a warmer climate through the New Mexico to Alaska area, with a heavier and more equitable rainfall throughout the year than prevails at present.

Tertiary Floras

In a recent review of certain Tertiary floras of North America Axelrod notes that over 200 are now known which include, in total, some 800 genera and several thousand species. The succession of these floras that have been excavated from Eocene to Pliocene horizons in the Rocky Mountain and Pacific Coast states tells a story of cooling climates accompanied by an increase in winter-summer extremes and a lowering of the annual rainfall to the point of advanced desert conditions in the intermontane areas. In attempting to summarize the vast amount of evidence now available I have been particularly concerned with the difficulty of understanding the vegetation of this great area at any one time during the Tertiary, to say nothing of grasping the whole sequence of changes during the past 60 or 70 million years. The present tremendous variation in habitats, and consequently plant communities, varies not only in going from north to south but, in many parts of the region, may be altered drastically in a few miles. Because of the several mountain ranges that have arisen during the Tertiary it is likely that such abrupt floristic shifts are more pronounced now than at any other time since the late Cretaceous; nevertheless there have been innumerable local floristic differences.

Eocene

During the middle Eocene a vast lake extended across southwestern Wyoming and adjacent parts of Colorado and Utah. The sediments that accumulated in this body of water, partly as a result of volcanic action, are especially famous for the beautifully preserved fossil fish that are found a few miles west of Kemmerer, Wyoming. Plants are also abundant at certain horizons and in a review of the flora Brown has recognized 135 species (exclusive of thallophytes) of macrofossils; in addition, many species have been established on fossil pollen.

The Green River flora includes: a fern (*Lygodium kaulfussi*); two cycads based on pollen (*Cycadipites, Dioonipites*); three pines known from pollen and a three-needle pine known from foliage; two species of *Potamogeton,* an aquatic monocot—one from pollen and another from a fruiting head; two palms have been recorded from pollen and three from leaves. A partial list of the dicots known from leaves, pollen, or both includes: *Salix* (willow), *Juglans* (walnut), *Hickoria* (hickory), *Tilia* (linden), *Alnus* (alder), *Betula* (birch), *Carpinus* (hornbeam), *Rhus* (sumac), *Schmalzia, Sapindus* (soapberry), *Koelreuteria* (golden rain-tree), *Liriodendron* (tulip tree), *Oreodaphne, Liquidambar* (sweet gum), and *Acer* (maple).

The Green River flora is one of the few that have been studied thus far from both pollen and macrofossils (leaf impressions); this advantage of checking one against the other offers a better chance for correct identifications and will undoubtedly be pursued to that end in future investigations. A few discrepancies appear in the flora that cause one to wonder; for example, three species of oaks have been identified from leaves but no oak pollen has been found. In view of the abundance with which the oaks discharge pollen one's first reaction is to question the identification of the leaves. Brown also cautions that in dual studies of this sort it is desirable to obtain both pollen and macrofossils from the same horizon or, if separated, this must be taken into account. In conclusion he suggests a warm temperate climate that was sometimes well watered, sometimes dry, and also offers the following interesting comment on the altitude:

> The ecologic group of species comprising *Hickoria, Juglans, Liquidambar, Ailanthus,* and *Tilia,* if judged by its modern equivalent, suggests that the Green River flora, with the exception of some conifers, grew at an altitude above the sea level of probably not more than 3000 feet and very likely at a considerably lower level. The Green River formation, now from 5000 to 10,000 feet above the sea, must therefore have been uplifted several thousand feet since the deposition of the sediments that contain the fossil flora. (1934, p. 51)

The late Eocene Goshen flora of west-central Oregon includes, among 45 species, the following as dominant elements: *Quercus*

(Fagaceae), *Ficus* (Moraceae), *Aristolochia* (Aristolochiaceae), *Drimys* and *Magnolia* (Magnoliaceae), *Anona* (Anonaceae), *Chrysobalanus* (Rosaceae), *Inga* (Leguminoseae), *Ilex* (Aquifoliaceae), *Cupania* and *Allophyllus* (Sapindaceae), *Meliosma* (Sabiaceae), *Nectandra* and *Ocotea* (Lauraceae), *Tetracera* (Dilleniaceae), *Calyptranthes* (Myrtaceae), *Diospyros* (Ebenaceae), *Cordia* (Boraginaceae).

The present distribution of the genera of the Goshen flora is far to the south of Oregon. What seem to be the closest modern equivalents are found in the tropical rain forest of the Pacific slope of Panama and the temperate rain forest of Costa Rica. On the basis of this comparison it is postulated that the late Eocene climate of the region had a mean annual temperature of 65 to 70° F and a rainfall of 70 inches or more.

The Chalk Bluffs flora, of Middle Eocene age, is known from several localities in north central California and presents problems of interpretation that seem worth considering. The most abundant species in the flora are:

Thouinopsis myricaefolia	*Mallotus riparius*
Cercidophyllum elongatum	*Chaetoptelea pseudo-fulva*
Persea pseudo-carolinensis	*Gordonia efregia*
Platanophyllum whitneyi	*Liquidambar californicum*
Platanus appendiculata	*Rhus mixta*

Chaetoptelea, known from both foliage and fruit, is one of the most characteristic fossils of the flora and is compared closely with *C. mexicana* which is found at subtropical elevations in southeastern Mexico and Central America. *Cercidophyllum,* also with foliage and fruits preserved, is abundant in the early Tertiary floras of North America; the modern *C. japonicum* (Katsura tree) is a large forest tree of southwestern China and is found north to Hokkaido in Japan. *Persea pseudo-carolinensis* is compared with the swamp red bay of southeastern United States which is often associated with *Platanus, Magnolia, Gordonia,* and *Liquidambar.* The characteristic fruits, seeds, and leaves of *Liquidambar* are well represented. The leaves are of both the three-lobed and five-lobed forms, the former being compared with *Liquidambar* from southern Mexico.

The climate in which the flora existed was probably similar to that of the Goshen plants, but there are some obvious inconsistencies which cannot be overlooked. For example, *Artocarpus* (breadfruit), *Rhamnidium,* and *Tabernaemontana* which are tropical genera are associated with temperate climate *Carya* (hickory),

Fig. 7-3. Some plants from the Middle Eocene Chalk Bluffs flora of Nevada County, California. A, *Laurophyllum litseaefolia* MacGinitie; B, *Platanophyllum whitneyi* (Lesquereux) MacGinitie; C, *Mallotus riparius* MacGinitie. (From MacGinitie, 1941.)

Acer (maple), and *Fraxinus* (ash). This occurrence of climatically divergent elements in a fossil flora is not an uncommon problem and has been discussed at some length by MacGinitie. He points out that there are several possible explanations: the identifications may be in error; the requirements of the modern equivalent plants may have changed or we may have an inadequate understanding of their present distribution and climatic requirements; the elements of the fossil flora may include upland plants as well as lowland ones.

The remarkable succession of fossil forests in the Yellowstone Park area has been mentioned in Chapter 1 where our attention was focused on the mode of origin of the deposits which include silicified stumps and logs as well as leaf-bearing beds. Knowlton's original study of the plants appeared over 60 years ago and their age has been regarded as Miocene; Professor Dorf has recently initiated an intensive restudy of the fossils and believes the age to be Eocene. This seems much more acceptable in view of the plant assemblage and what we now know about western Tertiary floras in general. As to the composition of the flora he notes:

> The buried forests were apparently a mixture of warm temperate forms, such as *Sequoia, Pinus, Platanus, Magnolia, Juglans, Cercidophyllum,* and *Castanea* (chestnut), and of subtropical forms, such as *Persea, Ficus, Columbia,* and *Laurus*. The assemblage is interpreted to indicate a warm lowland region with annual rainfall of 50 to 60 inches. (1959, p. 95)

The abundance of dicot genera is in striking contrast to the modern forests. Today 84% of the Park is forested and three-quarters of it is lodgepole pine; a large portion of the remaining quarter is forested with other species of pine, as well as fir, spruce, and Douglas trees. The only deciduous trees that are at all abundant are *Populus tremuloides* (quaking aspen), *P. angustifolia* (narrow-leaved cottonwood), *Betula glandulosa* (birch), and *Amelanchier alnifolia* (shadbush). To this writer one of the most fabulous features of the fossil forests is the presence of great silicified *Sequoia* stumps in excess of 14 feet in diameter.

Summer at present in the Yellowstone country is pleasant but fleeting. I have not had the pleasure of being there in mid-winter, but visits in late May, when a light snowfall was experienced each night, and to the Idaho country just to the south in September, when the weather behaved comparably, gives one the impression of a sub-Arctic climate! It was certainly milder in western Wyoming during the Eocene.

Oligocene

The Florissant beds of the high Rockies to the west of Colorado Springs consist of lake sediments resulting from volcanic action. They are noted for the abundance of leaf impressions, insects, and associated silicified stumps and they attracted attention as early as 1874; numerous publications have appeared from the studies of Lesquereux, Knowlton, Kirchner, Cockerell, and others. The present account is taken from the recent revision of the flora by MacGinitie.

The flora, probably of early Oligocene age, is a diverse one of well over 100 species. A few mosses are present, as well as *Equisetum,* and several conifers including *Abies* (fir), *Pinus* (pine), *Picea* (spruce), *Torreya, Chamaecyparis* (cypress), and *Sequoia.* There are a few monocots, with Typha (cat-tail) being quite abundant. The most abundant fossil is *Fagopsis longifolia* (Betulaceae) which is thought to have occupied a dominant place along the ravine banks comparable to that of the birches and alders today. The elm family is represented by *Celtis* (hackberry), *Ulmus* (elm), and *Zelkova;* the last, an Asiatic genus, is one of the dominant plants in the flora. Others occurring in abundance are a poplar related to the western black cottonwood (*Populus trichocarpa*); seven genera of the Rosaceae including *Amelanchier* (shadbush), *Cercocarpus* (mountain mahogany), *Crataegus* (hawthorn), *Malus* (apple), and *Prunus; Ptelea* (Rutaceae) is recognized by its characteristic winged fruits; several members of the flora represent genera now confined to Asia such as *Ailanthus* (Simarubaceae) and *Koelreuteria* (Sapindaceae) —the beautiful golden rain tree (*Koelreuteria paniculata*) is cultivated as far north as Missouri today.

Many of the Florissant species have their closest living counterparts in the modern flora of southwestern United States and parts of Mexico. In brief, the environs of the Florissant Oligocene flora is thought to have been one with an absolute minimum temperature of not below 20° F and an average annual temperature of not less than 65°, with an annual rainfall of about 20 inches. It was a mesophytic forest rich in species growing along stream and lake shores at an elevation of almost 3000 feet. The living forest is composed largely of yellow pine and some Douglas trees, Engelmann spruce, Colorado blue spruce, and limber pine; the only arborescent dicot of importance is aspen, whereas along the streams are found alder, birch, several willows, wild roses, chokecherry, shrubby potentilla, serviceberry, and a few others.

In his consideration of the climatic relations of the Florissant fossils MacGinitie offers a comment that is worth noting as one of very general application:

> There is another aspect of the Tertiary environment which has a bearing on the presence of apparently contradictory groups of plants in the general ecological picture of a fossil flora. All evidence points to higher average temperatures in western or southern states in pre-Pliocene times than obtain today. This appears to have been the result, not of higher maxima, but of higher *minima* of temperature. The extreme cold waves of winter, which bring freezing weather today into the Gulf littoral and below zero temperatures to the central states, must have been greatly mitigated in Middle and Lower Tertiary time. (1953, pp. 39–40)

He demonstrates the importance of temperature *minima* by a comparison of the California coast from Monterey north to Fort Bragg, with northeastern Kansas and north central Missouri. Both regions have an annual average temperature of 55° F, but the average January temperature at Columbia, Missouri, is 31° and that of the California area 48°; the absolute minimum in the former is about −26° and in the latter about 21°. The much higher minimum temperature permits the cultivation of some subtropical trees along the California coastal strip.

By Upper Oligocene times the tropical elements in the latitude of Oregon had largely disappeared, for the Bridge Creek flora of the John Day Basin in the eastern part of the state presents a distinctly cooler climatic assemblage. The dominant elements are *Quercus consimilis, Metasequoia occidentalis, Alnus carpinoides,* and *Umbellularia.* A comparison has been drawn between this Oligocene flora and the modern redwood forest, other genera in common being *Pteridium* (bracken fern), *Equisetum* (horsetail rush), and *Asarum,* whereas elements which occur today near the redwood forest but on more open slopes and valleys are *Philadelphus* (mock orange), *Crataegus* (hawthorn), and *Fraxinus* (ash).

In discussing certain European floras it was noted that the mid-Tertiary brought in not only temperate climate genera but ones that had a wider range, and a comparable trend appears in the western American floras. For example, we also find in the Bridge Creek flora *Platanus* (sycamore), *Juglans* (walnut), and *Celtis* (hackberry) which are found today in other parts of the west than the redwood forest as well as some that are not represented in western North America at present, as: *Carpinus* (hornbeam), *Fagus* (beech), *Ulmus* (elm), and *Tilia* (linden).

Fig. 7-4. Plants of the Bridge Creek flora. Leaf and seed catkin of *Alnus carpinoides*. (From specimens in the U. S. National Museum.)

Miocene

Two distinct Miocene floras are recognized by Axelrod:

1. The Arcto-Tertiary flora of the northern Great Basin and Columbia Plateau which is typified by temperate hardwood deciduous and conifer forests living under climatic conditions of 35 to 50 inches of rain distributed quite evenly throughout the year.

2. The Madro-Tertiary flora, occupying the area from southern California east into Mexico, composed of semiarid, live-oak woodland chaparral, and thorn forest with only 15 to 25 inches of rain distributed biseasonally; the winters were mild and the summers hot.

Vast outpourings of lava took place in Miocene times east of the Cascade range, resulting in the area known as the Columbia Plateau; some concept of the magnitude of this volcanic action may be gained

Fig. 7-5. Plants of the Bridge Creek flora.   A. *Metasequoia occidentalis.*   B,C. Leaf and fruit of *Ostrya oregoniana.*   D. *Quercus* sp.   E. *Zelkova hesperia.*   (From specimens in the U. S. National Museum.)

from the fact that the total accumulation of lava is estimated at about 100,000 cubic miles. New lakes and swamps developed as a result of the damming of streams and valleys, and numerous fossil localities were formed as plant remains, and ash from local volcanoes, accumulated in these newly formed bodies of water. The geological action of the time must have been violent in the extreme; this and other aspects of Tertiary paleobotany are vividly described in Chaney's *The Ancient Forests of Oregon.*

The Mascall flora which has been excavated from sediments in east central Oregon between Dayville and Mount Vernon includes nearly 70 species of which the following 10 (78% of the flora) are most abundant:

Mascall flora plants	*Most closely related living species*
Taxodium dubium	T. distichum (bald cypress) of southeastern United States
Quercus pseudo-lyrata	Q. borealis (red oak) of eastern United States, and Q. kelloggii (Californian black oak) found in California to central Oregon on the west side of the mountain ranges
Carya bendirei	(Hickories are found today in eastern North America and extend into Texas)
Platanus dissecta	Resembles P. racemosa of western United States
Quercus merriami	Q. rubra of eastern United States
Acer bolanderi	A. grandidentatum (maple) of the Rocky Mountains and A. leucoderme (maple) of southeastern United States
Metasequoia occidentalis	M. glyptostroboides, the dawn redwood
Ginkgo adiantoides	
Acer negundoides	A. negundo, the box elder, which ranges across the U. S., and A. henryi of Asia
Ulmus speciosa	U. fulva (the slippery elm) of eastern North America

Thirty-four of the Mascall species (more than half the total) have modern equivalents in eastern North America; most of the trees in the fossil flora were deciduous including the two most numerous gymnosperms, *Taxodium* and *Metasequoia*. The climate in which these plants lived is postulated as being comparable with that of Ohio or of Szechuan, China, at middle altitudes.

The Miocene Tehachapi flora far to the south is composed of a very different assemblage of plants. It has been obtained from a thick, white ash bed in Kern County, California, and is approximately 800 miles south of the Columbia Plateau. The closest relationship with a living flora lies in northern Mexico where 65% of the Tehachapi fossils have modern equivalents. The dominant elements, with their Sonoran (Mexico) relatives, are:

Tehachapi Miocene flora	*Modern equivalents; Sonora, Mexico*
Quercus convexa	*Q. oblongifolia*
Q. browni	*Q. chrysolepis* (Californian live oak)
Sabal miocenica	*S. uresana*
Dodonaea californica	*D. viscosa*
Amorpha oblongifolia	
Robinia californica	*R. neo-mexicana*
Populus prefremonti	*P. wislizeni*
Prunus prefasciculata	*P. fasciculata* (Desert almond)

The fossil leaves represent two distinct vegetational units, an arid, subtropical lowland group including the palm *Sabal* (both leaves and petrified stems are known), *Bursera* (Burseraceae), *Colubrina* (Rhamnaceae, buckthorn family), *Euphorbia* (Euphorbiaceae, spurge family), *Ficus* (Moraceae), and *Pithecolobium* (Leguminosae, pea family), whereas the remainder are from upland floras. This floristic comparison with northern Mexico and southwest United States indicates a Miocene climate of hot summers ranging up to 105° F and possibly as low as 25° in the winter; at the lower elevations rainfall was in the vicinity of 12 inches, increasing to about 25 in the adjacent uplands.

Extensive uplift of the Cascade-Sierra Nevada mountain range in early Pliocene times resulted in still greater aridity of the Great Basin area. This allowed a more northerly extension of the xerophytic elements of the north Mexican vegetation noted above. The Weiser flora from southwestern Idaho is made up of 32 genera in which oak, maple, pine, juniper, ash, sycamore, *Arbutus* (Ericaceae, heath family), *Pseudotsuga* (Douglas tree), and *Castanopsis* (Fagaceae, beech family) are dominant elements. The inferred climate is one with warm, dry summers and a rainfall of 20 to 30 inches. Although more temperate and xeric than the earlier Tertiary floras it is not as strongly so as later Pliocene vegetation. The immediate vicinity from which the Weiser flora was excavated is an arid region (less than 13 inches rain) of semidesert shrubs in which black sage and salt-bush are dominant.

By later Pliocene times the vegetation had approached a compo-

Fig. 7-6. Late Tertiary plants from Nevada. A, *Populus pliotremuloides* Axelrod; B, *Populus alexanderi* Dorf, both Mio-Pliocene. C, *Sassafras* sp.; D, *Sequoiadendron chaneyi* Axelrod; E, *Quercus hannibali* Dorf; F, *Alnus smithiana* Axelrod, all late Miocene. (From Axelrod, 1956.)

sition close to that of the present time. For the northern area, for example, Chaney notes:

The Pliocene floras of Oregon give us, for the first time in the records of past vegetation, a sense of modern trees—the sorts of trees which are still living in western America. While the gap in time is still several million years, it seems clear that during this epoch the climate and topography of Oregon were much as we know them today. The Pliocene is a record of an immediate yesterday, not of remote ages. (1956, p. 43)

REFERENCES

Arnold, Chester A. 1955. A new Cretaceous conifer from northern Alaska. *Amer. Journ. Bot.,* 42: 522–528.

Axelrod, Daniel I. 1939. A Miocene flora from the Western Border of the Mohave Desert. *Carnegie Inst. Washington Pub.,* 516: 1–129.

————. 1940. Late Tertiary floras of the Great Basin and Border Areas. *Bull. Torrey Bot. Club,* 67: 477–487.

————. 1950. Evolution of desert vegetation in western North America. *Carnegie Inst. Washington Pub.,* 590.

————. 1952. A theory of angiosperm evolution. *Evolution,* 6: 29–60.

————. 1956. Mio-Pliocene floras from West-Central Nevada. *Univ. California, Pub. Geol. Sci.,* 33: 1–322.

————. 1958. Evolution of the Madro-Tertiary Geoflora. *Bot. Rev.,* 24: 433–509.

Bailey, Irving W. 1944. The development of vessels in angiosperms and its significance in morphological research. *Amer. Jour. Bot.,* 31: 421–428.

———— and Nast, C. G. 1943. The comparative morphology of the Winteraceae II. Carpels. *Journ. Arnold Arb.,* 24: 472–481.

———— and Smith, A. C. 1943. The family Himantandraceae. *Journ. Arnold Arb.,* 24: 190–206.

———— and Smith, A. C. 1942. Degeneriaceae a new family of flowering plants from Fiji. *Journ. Arnold Arb.,* 23: 356–375.

———— and Swamy, B. G. L. 1948. *Amborella trichopoda* Beill., a new morphological type of vesselless dicotyledon. *Journ. Arnold Arb.,* 29: 245–254.

————. 1951. The conduplicate carpel of dicotyledons and its initial trends of specialization. *Amer. Journ. Bot.,* 38: 373–379.

Bell, W. A. 1956. Lower Cretaceous floras of western Canada. *Geol. Survey Canada Mem.,* 285: 1–331.

————. 1957. Flora of the Upper Cretaceous Nanaimo group of Vancouver Island, British Columbia. *Geol. Surv. Canada Mem.,* 293: 1–84.

Berry, Edward W. (et al.) 1911. The Lower Cretaceous deposits of Maryland. *Maryland Geol. Survey,* 1911. Pp. 1–622.

————. 1919. Upper Cretaceous floras of the eastern Gulf region in Tennessee, Mississippi, Alabama, and Georgia. U. S. Geol. Survey Prof. Paper, 112: 1–177.

Bose, M. N., and S. C. D. Sah. 1954. On *Sahnioxylon rajmahalense* Sahni, and *S. andrewsii,* a new species of *Sahnioxylon* from Amrapara in the Rajmahal Hills, Bihar. *The Palaeobotanist,* 3: 1–8.

Boureau, Edouard. 1954. Decouverte du genre *"Homoxylon* Sahni dan les terrains secondaires de la Nouvelle-Caledonie. *Mem. Mus. National Hist. Nat.,* Paris ser. C, 3: 129–143.

Brown, Roland W. 1934. The recognizable species of the Green River flora. U. S. Geol. Survey Prof. Paper, 185-C: 45–68.

—————. 1950. Cretaceous plants from southwestern Colorado. U. S. Geol. Survey Prof. Paper, 221-D: 45–66.

—————. 1956. Palmlike plants from the Dolores formation (Triassic) of southwestern Colorado. U. S. Geol. Survey Prof. Paper, 274-H: 205–209.

Chandler, Marjorie E. J. 1954. Some Upper Cretaceous and Eocene fruits from Egypt. *Bull. Brit. Mus. Nat. Hist., Geology,* 2: 149–187.

Chaney, Ralph W. 1925. A comparative study of the Bridge Creek flora and the modern redwood forest. *Carnegie Inst. Washington Pub.,* 349: 1–22.

—————. 1925. The Mascall flora—its distribution and climatic relation. *Carnegie Inst. Washington Pub.,* 349: 23–48.

—————. 1947. Tertiary centers and migration routes. *Ecological Monographs,* 17: 139–148.

—————. 1956. The ancient forests of Oregon. Condon Lecture, Oregon State System Higher Education, Eugene. 56 pp.

————— and Axelrod, D. I. 1959. Miocene floras of the Columbia Plateau. *Carnegie Inst. Washington Pub.,* 617: 1–229.

————— and Sanborn, E. I. 1933. The Goshen flora of west central Oregon. *Carnegie Inst. Washington Pub.,* 439: 1–103.

Cheadle, Vernon I. 1953. Independent origin of vessels in the Monocotyledons and Dicotyledons. *Phytomorphology,* 3: 23–44.

Corner, E. J. H. 1949. The Durian theory or the origin of the modern tree. *Ann. Bot.,* 13: 367–414.

Dorf, Erling. 1936. A Late Tertiary flora from southwestern Idaho. *Carnegie Inst. Washington Pub.,* 476: 75–124.

—————. 1938. Upper Cretaceous floras of the Rocky Mountain Region. I. Stratigraphy and paleontology of the Fox Hills and Lower Medicine Bow formations of southern Wyoming and northwestern Colorado. *Carnegie Inst. Washington Pub.,* 508: 1–78.

—————. 1942. Upper Cretaceous floras of the Rocky Mountain Region II. Flora of the Lance formation at its type locality, Niobrara County, Wyoming. *Carnegie Inst. Washington Pub.,* 508: 79–159.

—————. 1952. Critical analysis of Cretaceous stratigraphy and paleobotany of the Atlantic coastal plain. *Bull. Amer. Assoc. Petrol. Geol.,* 36: 2161–2184.

—————. 1959. The fossil forests of Yellowstone Park, Wyoming. *Internat. Bot. Congress, Montreal,* Abstracts II: 95.

Erdtman, Gunnar. 1948. Did dicotyledonous plants exist in early Jurassic times? *Geol. Stockholm Förhand.* March–April: 265–271.

Fontaine, William M. 1889. The Potomac or Younger Mesozoic flora. *Mon. U. S. Geol. Survey,* pp. 1–377.

Guppy, H. B. 1917. *Plants, Seeds and Currents in the West Indies and Azores.* William and Norgate, London. 531 pp.

Harris, Thomas M. 1932. The fossil flora of Scoresby Sound, East Greenland. Part 2. Description of seed plants incertae sedis, etc. *Medd. om Grönland,* 85 (3): 1–112.

—————. 1933. A new member of the Caytoniales. New Phyt. 32: 97–114.

————. 1937. The fossil flora of Scoresby Sound, East Greenland. pt. 5. *Meddelel. om Grönland*, 112 (2): 1–112.

————. 1940. On *Caytonia* Thomas. *Ann. Bot.*, n.s. 4: 713–734.

————. 1941. *Caytonanthus*, the microsporophyll of *Caytonia*. *Ann. Bot.*, n.s. 5: 47–58.

Hollick, Arthur and Martin, G. C. 1930. The Upper Cretaceous floras of Alaska. U. S. Geol. Survey Prof. Paper, 159: 1–116.

Hughes, N. F. and Couper, R. A. 1958. Palynology of the Brora coal of the Scottish Middle Jurassic. *Nature*, 181: 1482–1483.

Knowlton, Frank H. 1916. Flora of the Fruitland and Kirtland formations. U. S. Geol. Survey Prof. Paper, 98-S: 327–344.

————. 1899. Fossil flora of the Yellowstone National Park. *U. S. Geol. Survey Mon.*, 32, pt. 2: 651–882.

Kryshtofovich, African N. 1918. On the Cretaceous flora of Russian Sakhalin. *Journ. College Sci. Imper. Univ. Tokyo*, 40 (8): 1–73.

Kuhn, Oskar. 1955. Das erste Dicotylenblatt aus dem Jura. *Orion*, 10 *Jahrgang*, No. 19–20: 802–803.

Lesquereux, Leo. 1891. (Ed. post. by F. H. Knowlton). The flora of the Dakota group. *U. S. Geol. Survey Mon.*, 17: 1–400.

Lignier, O. 1908. Végétaux fossiles de Normandie. V. Nouvelles recherches sur le *Propalmophyllum liasinum* Lignier. *Mem. Soc. Linn. Normandie*, Caen, 23: 1–14.

MacGinitie, Harry D. 1941. A Middle Eocene flora from the central Sierra Nevada. *Carnegie Inst. Washington pub.*, 534: 1–178.

————. 1953. Fossil plants of the Florissant beds, Colorado. *Carnegie Inst. Washington Pub.*, 599.

Moseley, H. N. 1892. *Notes by a naturalist on the "Challenges,"* 2nd ed. London. 540 pp.

Newberry, J. S. 1895. (Ed. post. by A. Hollick). The flora of the Amboy clays. *U. S. Geol. Survey Mon.*, 26: 1–260

Reid, Clement, and Reid, Eleanor M. 1915. The Pliocene floras of the Dutch-Prussian border. *Meded. van de Rijksopsporing van Delfstoffen*, No. 6, Gravenhage, pp. 1–177.

Reid, Eleanor M. 1921. A comparative review of Pliocene floras, based on the study of fossil seeds. *Quart. Journ. Geol. Soc. London*, 76: 145–159.

———— and Chandler, M. E. J. 1926. Catalogue of Cainozoic plants in the Department of Geology. Vol. I. The Bembridge flora. *Brit. Mus. Nat. Hist.*, pp. 1–206.

————. 1933. The London Clay flora. *Brit. Mus. Nat. Hist.*, pp. 1–561.

Samylina, V. A.. 1959. New occurrences of angiosperms from the Lower Cretaceous of the Kolyma basin. *Acad. Nauk USSR (Botany)*, 44 (4): 483–491.

Saporta, Gaston. 1894. Flore fossile du Portugal. *Travaux Géol. du Portugal, Lisbon*, pp. 1–286.

Scott, Dukinfield H. 1923. *Studies in Fossil Botany.* Adam and Charles Black, Ltd., London. 446 pp.

Seward, Albert C. 1904. The Jurassic flora. II. Liassic and Oolitic floras of England. *Cat. Mes. Plants Brit. Mus.*, pp. 1–183.

————. 1926. The Cretaceous plant-bearing rocks of western Greenland. *Phil. Trans. Roy. Soc. London*, 215B: 57–175.

Simpson, J. B. 1937. Fossil pollen in Scottish Jurassic coal. *Nature* 139: 673.

Stopes, Marie C. 1912. Petrifactions of the earliest European angiosperms. *Phil. Trans. Roy. Soc. London*, 203B: 75–100.

————. 1915. The Cretaceous flora. Part II. Lower Greensand (Aptian) plants of Britain. *Cat. Mes. Plants Brit. Mus.,* pp. 1–360.

Szafer, Wladyslaw. 1946. The Pliocene flora of Kroscienko in Poland. Polsk. *Akad. Umiet., Krakow,* 72: 91–162.

Swamy, B. G. L. 1949. Further contributions to the morphology of the Degneriaceae. *Journ. Arnold Arb.,* 30: 10–38.

Takhtajan, Armen L. 1954. Origins of angiospermous plants. Soviet Science press, 1954. *Amer. Institute Biol. Sci.,* transl. 1958, pp. 1–69.

————. 1957. On the origin of temperate flora of Eurasia. *Acad. Nauk. U.S.S.R.,* 42: 1635–1653.

Teixeira, Carlos. 1948. Flora Mesozóica Portuguesa. *Servicos Geol. de Portugal.,* pt. I. Pp. 1–118.

Thomas, H. Hamshaw. 1925. The Caytoniales, a new group of angiospermous plants from the Jurassic rocks of Yorkshire. *Phil. Trans. Roy. Soc. London,* 213B: 299–363.

————. 1931. The early evolution of the angiosperms. Ann. Bot. 45: 647–672.

————. 1934. The nature and origin of the stigma. *New Phytologist,* 33: 173–198.

————. 1957. Plant morphology and the evolution of the flowering plants. *Proc. Linn. Soc. London,* 168 Session, 1955–56, pts. 1 and 2: 125–133.

————. 1958. Palaeobotany and the evolution of the flowering plants. *Proc. Linn. Soc. London,* 169 Session, 1956–57, pts. 1 and 2: 134–143.

————. 1958. Fossil plants and evolution. *Journ. Linn. Soc. London, Botany,* 56: 123–135.

Traverse, Alfred. 1955. Pollen analysis of the Brandon lignite of Vermont. *Bureau Mines, Rept. Invest.,* 5151: 1–107.

8

the LYCOPODOPHYTA

Introduction

The lycopods have held a prominent position in paleobotany since the earliest days of the science. This interest has centered largely in the Carboniferous members of the group because of the abundance of fossils, the diversity of form that they exhibit, and their arborescent habit. Some of them vie in size with all except the largest of modern forest trees and the fossil record indicates that in places they grew in more or less pure stands. As a result, they enter into our economy today as an important contributor of the raw material that became coal.

The lycopods were, however, well established by late Silurian times and possibly earlier. The record indicates that several lines of development evolved during Middle and Upper Devonian times and the group was established as one of the dominant elements in the vast, low-lying Upper Carboniferous swamps of North America and Europe. The later record of the lycopods is relatively scanty; the arborescent forms disappeared with the close of the Paleozoic.

The living members of the group are represented chiefly by the genera *Lycopodium* and *Selaginella,* plants that rarely attain a height of more than a foot or two. They are typical microphyllous plants with a dichotomous branching system or exhibit a combination of monopodial and dichotomous branching.

The spores are produced in sporangia which are borne singly at the base of a scalelike sporophyll; numerous sporophylls are usually aggregated into small but distinct cones. In some species of *Lycopodium* the sporangia are borne on ordinary leaves, and groups of these fertile leaves alternate with sterile ones along the length of the stem. They are homosporous, that is, the spores in any one species are all of essentially the same size.

Selaginella is heterosporous; there is a striking difference between the small microspores (which develop into the male gametophytes) and the megaspores (which develop into the female gameto-

214

phytes). An individual sporangium contains only one type, but both microsporangia and megasporangia may be found in the same cone. It is suggested that the student review the life cycle and general morphology of these plants before studying the fossil forms.

As usual there are conspicuous gaps in our knowledge of racial development in the group and there are a few particularly vexing fossils which suggest that it was a more diverse assemblage than has been supposed until quite recently. A wholly satisfactory classification for the lycopods is thus not possible; the following outline is intended more as a guide to the better known fossils than as a distinct system of classification.

1. Early lycopods of the Silurian to mid-Devonian
2. Lepidodendraceae
3. Sigillariaceae
4. Some herbaceous lycopods, Selaginellaceae
5. Pleuromeiaceae

EARLY LYCOPODS OF THE SILURIAN-DEVONIAN

It is possible that *Aldanophyton antiquissimum* (see Chapter 2) from the Middle Cambrian of Siberia is a lycopod, but since the spore-bearing organs are not known it can be accepted as only a suggestion that the group existed at that early date.

The oldest fossil that we can assign with confidence to the lycopods is *Baragwanathia longifolia* Lang and Cookson (Fig. 8-1) from the Silurian of Australia. The plant probably exceeded somewhat in size our largest species of *Lycopodium;* fragments of the dichotomously branching stems were found up to 28 cm long and although most specimens are 1 to 2 cm in diameter, they are known to have reached 6 cm. These stems bore slender leaves 1 mm wide and 4 cm long and were apparently rather lax in texture. Reniform sporangia containing spores are present in certain areas of the stem associated with ordinary leaves; the general organization thus appears to be quite like that of the living *Lycopodium lucidulum* although it has not been established for certain as to whether the sporangia in *Baragwanathia* were actually attached to the leaf on its upper surface near the base or directly to the stem. The woody tissue of the stem consists of a primary stele, with annular tracheids, similar to that of *Asteroxylon* but somewhat more complex, there being at least 12 rays of xylem.

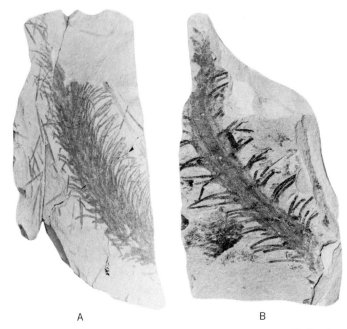

A   B

Fig. 8-1. *Baragwanathia longifolia.* A. Portion of a leafy shoot. B. Specimen showing sporangia. (A photograph courtesy Isabel Cookson; B from Lang and Cookson.)

It is a little disturbing to find in the Silurian a plant that compares so closely with modern lycopodiums; one may hesitate in claiming that species like *L. lucidulum* are direct descendants of this ancient Australian plant but it seems possible. It was a plant of respectable size and insofar as our knowledge goes indicates that the lycopods were among the forerunners of early land vascular vegetation.

Drepanophycus is a genus that deserves particular attention as an early to mid-Devonian lycopod. Based on German specimens, it is depicted by Kräusel and Weyland (Fig. 8-2) as a plant of *Lycopodium*-like habit with a creeping stem which produced upright dichotomously forking shoots attaining a height of about 45 cm; the stems were clothed with rather stout spiny appendages, some of which had a single sporangium on the upper surface.

It was rather widely distributed as specimens have been reported from the Gaspé in Canada, Norway, Germany, Scotland, Wales, Belgium, and China. Thus far only specimens from German and Welsh deposits have been found fertile, but there seems to be little doubt that most of those from other localities are referable to

Drepanophycus. Some of the Belgian fossils from the Emsien (upper part of the Lower Devonian) attained a diameter of about 2.5 cm and it seems likely that these may have exceeded the estimated height of 45 cm. The spines were penetrated by a single, slender vascular strand and have a distinctive broad, expanded base on the larger branches. A single sporangium is borne on the upper surface of the leaf and occupies a position about midway between the proximal and distal ends; there is a suggestion from Welsh specimens that it may occasionally have been borne at or near the tip. The Lower Devonian *Sugambrophyton pilgeri* Schmidt was similar to *Drepanophycus,* the branches of the dichotomously forking shoot system attaining a breadth of 2.6 cm. The leaves, however, are distinctive in that the ones on the basal part of the plant are simple, whereas those on the upper parts are smaller and dichotomize two or three times.

Protolepidodendron (Fig. 8-3) is best known from the Middle Devonian of Germany and is characterized by leaves that fork near the tip. Radially elongate sporangia are occasionally found on the upper surface of the leaves. The shoots reached a diameter of

Fig. 8-2. Restoration of *Drepanophycus spinaeformis* from the Lower Devonian of Germany. (From Kräusel and Weyland, 1935.)

about 2 cm and were supplied with a triangular protostele which is not typically lycopodiaceous. *Protolepidodendron* has also been reported from the district of Chanyi in China (probably Middle Devonian); the stems are up to 6 mm broad and bear leaves 3.5 mm long. In describing these Halle notes that distinct leaf scars are not present, suggesting that the plants were herbaceous and did not shed their foliage as did the larger Carboniferous lycopods.

Although the plants considered above are mid-Devonian or older it seems expedient to consider one Upper Devonian lycopod, *Colpodexylon trifurcatum* Banks, at this point. It was a plant with dichotomously forking stems up to 2.5 cm in diameter, but fragments up to 2 feet long suggest that it may have been taller than the earlier Devonian lycopods. The leaves were three-forked and persistent on the stem; the stele is composed of primary, scalariform tracheids and varies in cross-sectional shape from only slightly to quite deeply lobed.

LEPIDODENDRACEAE

The arborescent lycopods of the Carboniferous were among the dominant elements in the earth's vegetation of this age and they present unique aspects of morphology and growth that are unpar-

Fig. 8-3. Restoration of *Protolepidodendron scharyanum.* (From Kräusel and Weyland, 1935.)

alleled among plants of the present day. The fossil remains are especially abundant as compressions in the roof shales of Upper Carboniferous coal mines and as petrifactions in coal balls. Perhaps the most commonly encountered genus is *Lepidodendron* which includes many scores of species. As is always the case with larger plants our information comes from scattered remains of roots, stems, leaves, and reproductive organs. Although all these organs have never been correlated for a single species, our understanding of *Lepidodendron* as a genus is remarkably comprehensive and it is in many ways representative of the arborescent lycopods. The following description is a composite one in that fossil remains representing several species and certain closely allied genera are included.

Mature trees (Fig. 8-4) are known to have exceeded 100 feet in height and 3 feet in diameter. The branching pattern was dichotomous although the equality of the two branches at each division varied considerably; the root system consisted of four major branches which likewise dichotomize several times, the ultimate subdivisions bearing numerous spirally arranged rootlets. Linear grasslike leaves, which varied tremendously in length in different species (and possibly on a single plant), were borne toward the terminal portions of the ultimate branchlets and, upon falling from the stem, left a characteristic scar which retained its identity even on the oldest parts of the trunk. The spore-bearing cones are basically similar to those of the living lycopods, but usually are much larger.

Stem Anatomy

Unless otherwise indicated the following description is based on *Lepidodendron scleroticum* Pannell, a species that is abundant in certain Illinois coal balls. It has been selected for initial study because it is known from numerous specimens which represent various branch orders of the plant. The smallest twigs (Fig. 8-5) measure about 4 mm; in the center is a minute, solid protostele and this is surrounded by phloem, a rather uniform cortex of parenchyma cells, and outermost are several conspicuous leaf bases. The larger branches display a more complex anatomy and reveal typical aspects of the plant. Figure 8-6 is a representative sector of a branch about 4 cm in diameter; the central pith, which is partially destroyed in this specimen, is enclosed by the primary wood (Fig. 8-7) which is in turn surrounded by a band of secondary

Fig. 8-4. A. Restoration of an arborescent lycopod of the *Lepidophloios* type. B. A young plant prior to the first dichotomy of the trunk, based in part on *Lepidophloios pachydermatikos.* (B from Murdy and Andrews.)

wood (Fig. 8-8). The latter consists of scalariform tracheids and rays which are one cell broad and vary from one to a score or more cells high. In the nature of the wood we find the first of several characters that contrast with the structure of modern forest trees;

it composes a small portion of the total volume of the stem. The large thin-walled tracheids were probably efficient water conductors, but as a supporting tissue they were quite inadequate. The cambium and secondary phloem lie immediately outside the secondary wood, but these tissues are rarely preserved.

The cortex in *L. scleroticum* (Fig. 8-6) consists of two rather well-defined zones. The inner part is a groundwork of roundish thin-walled cells and scattered through it are numerous clusters or "nests" of cells, with somewhat thicker walls, filled with a dark material which renders them quite conspicuous. The nests, a distinctive feature of this particular species, are more or less isodiametric and are composed of at most a few dozen cells. The outer cortex also consists of a groundwork of thin-walled parenchyma and dispersed through it is an anastomosing network of longitudinally elongate fiber strands.

Outside the cortex is an extensive development of tissue that has been called periderm or cork. This is a unique feature of the arborescent lycopods, both in its structure and relative abundance, and will be described in some detail. Rather early in the growth

Fig. 8-5. *Lepidodendron scleroticum*, a small protostelic twig, 20X.

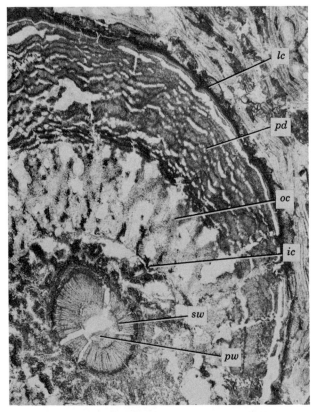

Fig. 8-6. *Lepidodendron scleroticum.* Portion of a branch about 4 cm in diameter; *pw,* primary wood; *sw,* secondary wood; *ic,* inner cortex; *oc,* outer cortex; *pd,* periderm; *lc,* leaf cushions. (From Pannell, 1942.)

of the stem a periderm cambium was initiated in the outermost part of the cortex; this is quite distinct from the wood cambium and it continued to divide, probably throughout the life of the plant, resulting in the formation of an extensive periderm. In *L. scleroticum* the periderm cells are radially aligned, six to ten times longer than broad, and with sharply tapered ends. The conspicuous holes scattered through this tissue as it appears in Fig. 8-6 may represent areas of less resistant cells that served a secretory function.

The periderm was more complex in some other species; *Lepidodendron johnsonii* Arnold, from the Pennsylvanian of Chaffee County, Colorado, presents some especially interesting features. It consists chiefly of the usual radially aligned fibrous cells and in this case they are twenty to thirty times as long as they are broad. In their equal length and regular arrangement they resemble the storied structure found in the wood of certain dicots. In the inner region the fibrous cells are interrupted by radial rows of tangentially elongate ones. In tangential section they appear oval-shaped and some are divided into two cells by a transverse wall. It has been suggested that they may have afforded elasticity to the stem, permitting movement within the tissue during growth, or in response to bending of the trunk and branches. A second interesting feature of this Colorado species is the presence of periodic tangential rows of glands. These are of complex organization; each gland extends vertically for an undetermined distance and is composed of thin-walled, brick-shaped cells, within which are oval masses of smaller ones with dark contents presumably of a waxy or gummy nature.

Although the periderm usually seems to have been initiated in the extreme periphery of the cortex, this is not always the case; it may start 15 or 20 cells deep and, in a related genus, *Lepidophloios,*

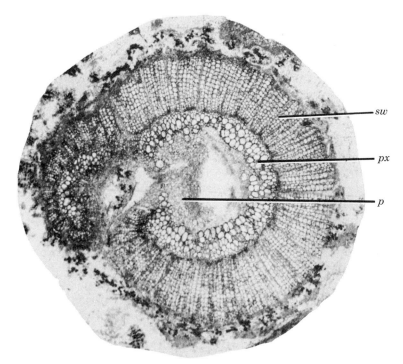

Fig. 8-7. *Lepidodendron scleroticum;* this specimen shows an early stage of an unequal dichotomy, 14X; *p*, pith; *px*, primary wood; *sw*, secondary wood.

Fig. 8-8. *Lepidodendron scleroticum,* portion of the stele of a large branch showing extensive development of secondary wood, 5X. (From Pannell, 1942.)

even deeper. Periodicity of growth in the periderm is suggested in some of the lycopods; in *Sigillaria scutellata* it is banded due to variation in cell size, lending a superficial appearance to the structure of the wood of a modern conifer. In a Lower Carboniferous lycopod from Scotland, *Levicaulis arranensis* Beck, there is evidence that the periderm developed from a series of cambia or meristematic regions rather than a single one. As to magnitude of development, Professor Baxter has in his collections at the University of Kansas, a slab of petrified periderm from an Iowa locality with the extraordinary radial dimension of 40 cm; by the most conservative estimates this implies a trunk close to 4 feet in diameter.

It is thus evident that the periderm was a complex tissue in the arborescent lepidodendrons and their immediate relatives and it made up a considerable part of the trunk. As a result its function has been the subject of some controversy. Several investigators have concluded that it should not be compared structurally or functionally with the cork of modern trees. In his recent study of this tissue in *Levicaulis,* Beck suggests that it be termed secondary cortex rather than periderm and he also points out that it may well have been a living tissue. There can be no doubt that it was the chief supporting element of the arborescent lycopods, and it is also apparent that further study is needed before we will fully understand it.

Leaves

Terminal branchlets of *Lepidodendron* and other genera are occasionally preserved with the foliage attached (Fig. 8-9), and isolated leaves are common in petrifactions associated with the stems. The generic name *Lepidophyllum* has long been used for isolated specimens of foliage, but is invalid since it was originally used for a living South American flowering plant. Miss Snigirevskaya has recently proposed the name *Lepidophylloides,* one that is appropriate and correct.

In the roof shales immediately overlying the coal from which *Lepidodendron scleroticum* is found leaves occur which are about 3 to 4 mm wide and nearly 30 cm long. The anatomical structure

Fig. 8-9. Terminal leafy branchlets of a *Lepidodendron* from the Upper Carboniferous of Holland, about 0.5X. (From a specimen in the Swedish Natural History Museum.)

of these lycopod leaves is of particular interest and several species
have been described from the American Pennsylvanian. The type
shown in Fig. 8-10C is constantly associated in coal balls with *L. scleroticum,* there being no reasonable doubt as to the identity of the
two organs.

An elongate group of tracheids composes the single central strand,
which is surrounded by thin-walled cells which are interpreted as
phloem. There is next a more or less continuous sheath referred
to as transfusion tissue; these are somewhat elongate cells with
transverse end walls and reticulate wall thickenings. Referring to
the lamina of the leaf, it is distinctive by virtue of the abundance
of hypodermal tissue; that is, immediately beneath the epidermis
there are several rows of thick-walled, elongate fibrous cells arranged

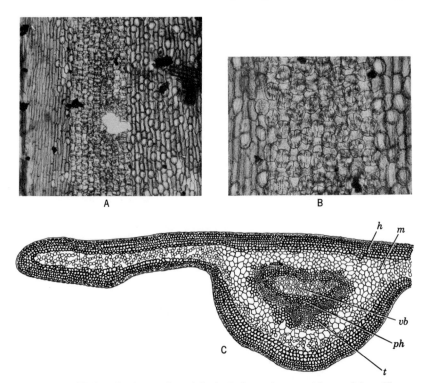

Fig. 8-10. A. Underside of a portion of the leaf of an arborescent lycopod from Kansas
showing one of the stomatal bands. B. Stomatal band at a higher magnification.
C. Cross section of a leaf (one side of lamina omitted) found associated with *Lepidodendron scleroticum* (mesophyll not actually preserved); *h,* fibrous hypodermal tissue;
m, mesophyll; *vb,* tracheids; *ph,* phloem; *t,* transfusion cells. (A,B from a preparation
by Gilbert A. Leisman.)

in distinct rows. Between the upper and lower hypodermal layers there was a loosely compacted tissue which presumably was photosynthetic. Such hypodermal tissue is generally characteristic of plants found growing under extreme xerophytic conditions. The stomates were confined to two longitudinal bands on the under side. These show exceptionally well in maceration preparations of the epidermis.

The length of the leaves borne by the arborescent lycopods varies greatly and there is some evidence to suggest that there may be some variation correlated with growth stages of an individual plant. In the collections of the British Museum of Natural History there are several specimens attributed to *Lepidodendron lycopodioides* with leaves that measure 13, 15, and 55 mm long respectively; a specimen of *L. sternbergii* with leaves 6 cm long and 3.5 mm broad; a specimen of *L. acutum* with leaves 20 mm long and about 2.5 mm broad. Zeiller has given some exceptionally interesting data for lycopods from the Valencienne coal basin of France: in *L. lycopodioides* the leaves were found to be 1 cm long on the small branchlets and 3 to 5 cm long on the larger branches; in *L. obovatum* the leaves on small branchlets are 4 to 5 cm long while those on larger ones attain a length of 60 to 80 cm; and in *L. dichotomum* branch specimens bear leaves 2 to 6 cm long, while on fossils that presumably represented the trunk, the leaves reach a length of 40 cm. A possible explanation for this variation within a species will be given in a later paragraph dealing with the mode of growth of these plants.

A *Lepidodendron* specimen has been found in the West Virginia coal fields with leaves 76 cm long and 5 mm wide and there are reports in the literature of lycopod leaves up to 1 meter in length which I am inclined to believe are not exaggerated.

It was noted in the description of the stem anatomy that the periderm cambium usually lies near the periphery of the stem immediately within the leaf bases. This is the opposite of the relationship that we find in modern (dicot) trees and shrubs in which the periderm cambium is innermost and produces cork cells toward the outside. The result is that in *Lepidodendron* and related genera the external pattern of leaf cushions remains intact and distinct throughout the life of the plant (Fig. 8-11). A great many species have been established on compressions or impressions of the exterior of the shoot system; some of these are certainly invalid and we still have inadequate data on the variations within a single plant.

A well-preserved leaf cushion shows the area where the leaf was actually attached and within this are three scars, a central one rep-

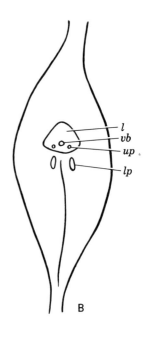

Fig. 8-11. A. Exterior of a large *Lepidodendron* branch or trunk showing character-istic leaf cushion pattern, about 0.5X. B. Diagram of a single leaf cushion; *l*, leaf scar; *vb*, vascular trace scar; *up*, upper parichnos scars; *lp*, lower parichnos scars.

resenting the vascular strand, and two lateral *parichnos* scars. The latter represent channels of thin-walled, rather loosely arranged parenchyma cells which were continuous from the mesophyll of the leaf through the cushion to the cortical tissues of the stem. In some lepidodendrons there is another pair of larger scars just below the leaf scar; these appear to represent the outer surface of channels of tissue very similar to the parichnos and, like the latter, probably served an aerating function.

Cones

A great many species of cones have been described which were borne on the lepidodendrons and related plants; most of these have been given the name *Lepidostrobus*. A description of *Lepidostrobus diversus* Felix from the middle Pennsylvanian of Indiana will illus-trate most of the distinctive features of the lepidostrobi. They

(Fig. 8-12) are slender, attaining a maximum diameter of 1 cm and a length of a little over 11 cm. Many closely imbricated, spirally arranged appendages (sporophylls) are borne on the central axis. Each sporophyll consists of a pedicel attached at about 90° to the

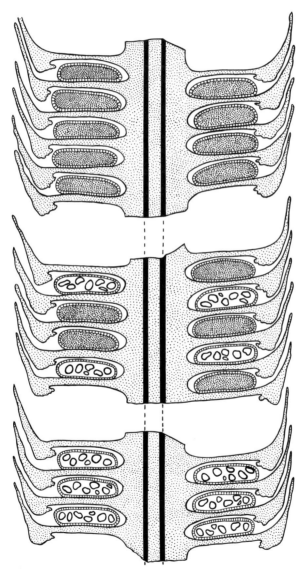

Fig. 8-12. Longitudinal section showing representative portions of the cone of *Lepidostrobus diversus;* distal part of cone is microsporangiate, basal part is megasporangiate, and central region is mixed. (Adapted from Felix, 1954.)

cone axis; it is rather broadly flared toward the distal end and turns up sharply to form the lamina which is long enough to overlap one or two sporophylls above. Also at the junction of pedicel and lamina there is a downward projection known as the heel; although not present in all lycopod cones it is a conspicuous feature of many and probably served to enclose the sporangium more effectively during its maturation process. A single vascular strand, which originates from the cone stele, passes out through the pedicel and up into the lamina.

A single radially elongated sporangium is attached to the upper surface of the pedicel. In the upper portion of the cone each sporangium contains large numbers (at least several thousands) of microspores measuring about 20 μ in diameter; the sporangia in the basal part of the cone each contain 16 megaspores which are over 700 μ in diameter. The central region was transitional with mega- and microsporangia irregularly mixed.

Immediately distal to the sporangium, there is in many of the heterosporous lycopods (Fig. 8-20A), including the living *Selaginella,* a small tongue-shaped structure called a ligule that is sunken in a ligular pit. This is also associated with the vegetative leaves and its former presence is sometimes shown in branch compression specimens as a minute dot just above the leaf scar. The ligule, function and origin of which are unknown, is considered to be characteristic of the heterosporous lycopods and lacking in the homosporous ones; however, its presence has not been demonstrated in *L. diversus* and it is my feeling that the attention that has been devoted to it is quite out of proportion to its importance.

One of the problems encountered in studying cones is illustrated by this species; if, instead of having a complete one, only basal and terminal fragments were available several interpretations might be possible: the two might be taken as parts of a single cone; they might be interpreted as representing a single species in which mega- and microsporangia were borne in separate cones; the microsporangiate part might be interpreted as the cone of a homosporous species and unrelated to the other. The evidence that we have to date indicates that a great many, if not most, of the arborescent lycopods of the Carboniferous bore bisporangiate cones that were generally similar in their organization to *Lepidostrobus diversus.*

There was, however, a great deal of size variation and some attained a magnitude of slender baseball bats; where extensive forests of the plants were established the showers of spores must have been quite impressive at times! The size that some attained is indicated by the following measurements, based on cones attributed to *Lepidostrobus* that are preserved in the collections of the British Museum and the Geological Survey in London; all are from the Carboniferous of Britain:

A specimen 13 mm in diameter × 15.5 cm long; complete.

A specimen 25 mm in diameter × 32.5 cm long; diameter very uniform and broken at one end, thus not complete.

A specimen 30 mm in diameter (with an axis 5 mm in diameter) × 16 cm long; not complete.

A specimen 50 mm in diameter (with an axis 10 mm in diameter) × 19 cm long; a fragment only.

A specimen 22 mm in diameter with an axis 4 mm in diameter; length 28 cm and probably complete.

Němejc has described lycopod cone compressions from the Carboniferous of Bohemia under the name *Sporangiostrobus* which attained a diameter of 6 cm. Macerations revealed microspores 50 µ in diameter and the megaspores measured 2.5 mm. The affinities of these cones have not been established and it is not necessarily implied that they belong in the Lepidodendraceae; they are noted here to indicate the dimensions attained in the lycopods in both cone and megaspore size.

Among the numerous species of heterosporous lycopod cones in the Carboniferous there is a distinct evolutionary sequence involving: reduction in the number of megaspores per sporangium; increase in size of the megaspores that remain; enclosure of the megasporangium by the sporophyll to form a seed. Our knowledge is not sufficiently complete so that we can cite an unbroken evolutionary line, but the general tendency is illustrated by the following examples:

Lepidostrobus noei Mathews from the lower Mississippian of Kentucky is a cone in which each megasporangium contains several hundred spores, each about 350 µ in diameter, whereas the microspores, produced in huge quantities, are 50 µ in diameter.

L. diversus, described above, has 16 megaspores per sporangium.

L. foliaceus Maslen from the Pennsylvanian is reported to have not more than four megaspores per sporangium.

L. braidwoodensis Arnold is based on a compression specimen from Illinois; maceration of the sporangia revealed, in each one, a single large megaspore about 2 mm in diameter. Three much smaller undeveloped spores are found at the apex of the large one; thus, of a single tetrad within each megasporangium, only one spore matured.

Lepidocarpon is a genus established by D. H. Scott in 1901 for megasporangiate cones that are similar in their basic organization to *Lepidostrobus* but in which the sporangium is almost completely enclosed by the sporophyll. In the course of development of the cone the sporophyll grew laterally up around the sporangium which is a radially elongate sac but differs from *Lepidostrobus* in that it is broad at the base and tapers upward to an acute apex. In a section tangential (Fig. 8-13) to the axis of the cone the unique morphology of *Lepidocarpon* is most clearly revealed; the sporophyll encloses the sporangium except for a radially elongate slit or "micro-

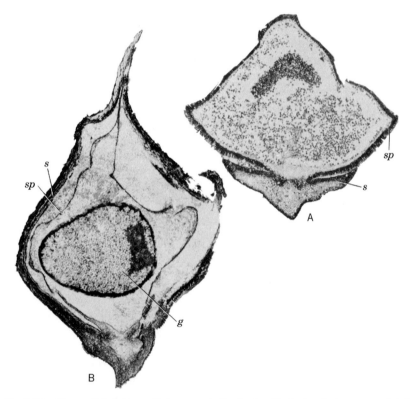

Fig. 8-13. Tangential sections of: A. sporophyll of a *Lepidostrobus* (microsporangiate) cone, 20X; B. a seed of *Lepidocarpon magnificum,* 9X; *s,* sporophyll; *sp,* sporangium wall; *g,* gametophyte. B is a composite photograph, the gametophyte having been taken from another specimen.

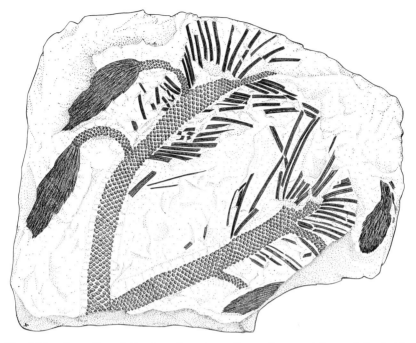

Fig. 8-14. A portion of the branch system of an arborescent lycopod (probably *Lepidophloios* sp.) from West Virginia; the cones on the short side branches at the left are approximately 10 cm long.

pyle" at the apex. A single megaspore matured in each sporangium and specimens have been found in English and American coal balls with the gametophyte preserved.

So far as I am aware Scott's original specimen is the only one on record with the seeds attached to the cone axis; they are abundant enough as isolated specimens and apparently were shed from the cone axis by an abscission mechanism. As to the corresponding microsporangia, the available evidence suggest that they were borne on separate cones. Large microsporangiate cones have been found associated with an Illinois species, *Lepidocarpon magnificum* Andrews and Pannell; they are 5 cm in diameter and in excess of 16 cm long. Spores, identical with those in the microsporangia, were also found in the seeds. It is thus likely that many of the purely microsporangiate cones found in the Upper Carboniferous belong to *Lepidocarpon*.

In summary, *Lepidocarpon* is a seed in the sense that it is an integumented megasporangium and, at least with respect to this organ, it represents the peak of evolution attained by the lycopods. However, it is quite apparent that the "integument" here is not homologous with the integument of the seed plants considered in later chapters. There is no reason to doubt that *Lepidocarpon* is terminal in a line of lycopod evolution which became extinct toward the close of the Paleozoic.

The arborescent lycopod stem compressions, some of which are definitely referable to *Lepidophloios* (Fig. 8-15), occasionally have conspicuous round scars (Fig. 8-16) the significance of which has been much debated. Some investigators have held that they represent the scars of vegetative branches that were sloughed off by an abscission mechanism, possibly similar to the living cottonwoods, whereas others have been of the opinion that they represent specialized cone-bearing branches. Unfortunately, distinct generic names have been established for such specimens; *Halonia* being applied to ones in which the scars are arranged in a spiral pattern and *Ulodendron* for those in which the scars are arranged on two opposite sides of the branch.

There are actually very few specimens of Carboniferous lycopods known with the cones attached and I feel that any generalized statement concerning the halonial or ulodendroid scars is premature; the

Fig. 8-15. *Lepidophloios scoticus* from the Lower Carboniferous of Scotland, about 0.5X. (Kidston collection, Geological Survey Great Britain.)

Fig. 8-16. *Lepidophloios scoticus.* The regularly spaced round scars probably represent points where small cone branches were attached. (British Crown copyright, by permission of the Controller of her Britannic Majesty's Stationery Office.)

following comments will be confined to a few specimens in which the evidence seems clear-cut. Some years ago I obtained a fine specimen of the terminal part of a *Lepidophloios* branch system from the West Virginia coal fields. The dichotomizing branches are close to 2 cm in diameter and bear leaves at their tips that were probably several decimeters long. Pendant side branches about 9 mm in diameter are borne, about 8 cm apart, and each one is terminated by a cone, the largest being 3.5 cm in diameter and 15 cm long. Petrified *Lepidophloios* stems found in Kansas coal balls shed some light on the anatomy of these fertile branches. Small terete vascular strands depart from the stele of the main branch (apparently a very unequal dichotomy of the latter) and pass out to the periphery where they terminate in the ulodendroid scars. It is therefore established that in some *Lepidophloios* species cones were borne on relatively small special branches and these were probably shed by an abscission mechanism after discharge of the spores.

Root System

The underground organs of *Lepidodendron* and closely allied plants are assigned to the genus *Stigmaria;* they will be referred to as a "root" system and although there is no doubt that they functioned as such, in their gross morphology and anatomy they are quite unlike the roots of other plants.

Stump casts are not uncommon in Upper Carboniferous rocks and perhaps the finest example is an exhibit in Victoria Park, Glasgow (Fig. 8-17). In the course of developing the park a path was cut through what had once been a quarry and several stumps were discovered; subsequent excavation was carried out to expose more and a building erected for protection against the weather. The stumps clearly reveal a sequence of events beginning with a lycopod forest that flourished in the region in Carboniferous times; it was a low-lying area and subject to flooding, and probably as a result of land subsidence several feet of mud accumulated, bringing about the death of the trees. The remaining stumps eventually decayed but not until the surrounding matrix had consolidated somewhat; then the cavities were filled with mud, resulting in casts of the original stems and roots. More sediments accumulated later and this was followed by an uplift that raised the area to become dry land. Finally, through the curiosity of man the remnants of the long buried forest have been revealed.

In most cases the root system originated at the base of the trunk as four massive primary roots, the largest known being 80 cm in diameter; these radiate out horizontally or dipped down slightly into the ground attaining a length of 14 meters. They dichotomized at least twice, but the divisions took place within a few feet of the trunk. This stigmarian rootstock in turn bore appendages (rootlets) which were arranged in a very regular (quincuncial) pattern (Fig. 8-18) which may be interpreted as a spiral in which the root-

lets of successive spirals form nearly longitudinal rows. The rootlets
occasionally dichotomized and reached a length of 37 cm.

The main rootstock of *Stigmaria ficoides* contained a stele (Fig.
8-19) of radially aligned tracheids. A few of the innermost cells are
spiral tracheids, whereas the rest are scalariform; thus the wood is
almost exclusively secondary. It may be added that several species
of *Stigmaria* have been recognized and in one or two of them some
primary centripetal wood (as in the stems) was present. The wood
was followed by a broad cortical region which is usually decayed
and finally there is, in larger specimens, a conspicuous band of periderm. The vascular trace to the rootlets departs from the inner-

Fig. 8-17. Stump casts of arborescent lycopods excavated from Carboniferous rocks
in Victoria Park, Glasgow.

Fig. 8-18. *Stigmaria ficoides,* from the Coal Measures of Newcastle, England; portion of a rootstock with rootlets. Length of longest rootlet shown is 21 cm. (British Museum Natural History.)

most part of the root stele and passes horizontally through a broad ray in the wood.

Close to their point of attachment to the rootstock the rootlets appear in cross section as follows: there is a broad outer cortex followed by a rather clearly delimited middle one; the cells of the latter are smaller, less regular in shape, and there are some air spaces between them. There is next an inner cortex enclosing the vascular strand which is monarch, that is, there is one protoxylem point, and occasionally a few rows of secondary wood cells are present.

Within a few millimeters of their departure from the rootstock (Fig. 8-19B) the rootlets lose their middle cortex; a conspicuous gap is thus evident between the outer and inner cortical layers. Occasionally a *connective* may be observed which is a remnant of the middle cortex and joins the inner cortex, with the vascular strand, to the outer cortex.

The stigmarian root system presents several unusual features which are difficult of interpretation. The stele differs from that of the stem in the lack of any appreciable amount of centripetal primary wood (it would be interesting to find a petrified specimen

showing the transition between the base of the trunk and the rootstock). The regular arrangement of the rootlets is quite unlike the
irregular distribution of branch rootlets in most plants. A feature
of special physiological interest is the lack of root hairs; in view of
the abundance and excellent preservation of these fossils in English
coal balls it seems certain that specimens with root hairs would have
been found if they did exist. A rough comparison with the root sys-

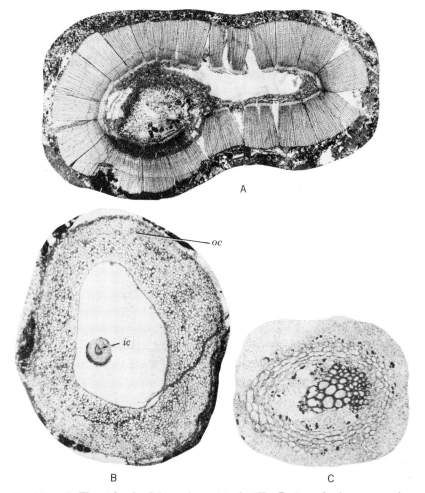

Fig. 8-19. A. The stele of a *Stigmaria* rootstock, 2X. B. A rootlet in cross section,
7X; *ic,* inner cortex; *oc,* outer cortex; the vascular strand is the minute cluster of cells
within the inner cortex. C. Inner cortex and vascular strand highly enlarged; a few
secondary tracheids are evident.

tem of modern seed plants of comparable size suggests that the stigmarian root system had a very low absorption area; this, coupled with the small amount of wood, seems to imply a conducting system of low efficiency and explains at least in part the xerophytic nature of the leaves which are constructed to lower drastically the transpiration rate.

The lycopods had reached arborescent dimensions by Upper Devonian times although none are known that were as large as the Carboniferous forest giants. In searching through museum collections for a representative or two that might bridge the gap between the small plants of the early to mid-Devonian and the great forest trees of the Carboniferous it seemed to me that *Bothrodendron kiltorkense* in some way meets the need. This plant was first found in the Upper Devonian of Kiltorcan Hill, County Kilkenny, Ireland, and has also been reported from Bear Island. In its gross habit it seems to have been similar to the later lepidodendrons; the main stem is known to have attained a diameter of 1 foot and rose from a stigmarian rootstock; equal dichotomies characterized the branching pattern of the aerial shoot system and the plant attained a height of 25 feet. The branches seem to have been exceptionally slender, to a point that the plant probably presented a delicate, almost filmy appearance in life. The leaf scars are small, nearly round, and in many specimens they are arranged in a perfectly whorled pattern, whereas in others it appears to be a very low angle spiral. The leaves measure 2 to 15 cm long and 1 mm wide.

The cones are imperfectly known from compressions but present a few features that are interesting enough to make one wish for better preserved material. They were borne terminally and consisted of whorls of sporophylls, each with an elongate sporangium on the upper surface, whereas the distal portion of the sporophyll, according to Johnson, attained a length of 20 cm. There is a rather remarkable specimen in the Museum in Stockholm labeled *Lepidostrobus bailyanus* from Kiltorcan which is probably a cone of *B. kiltorkense;* its most striking feature is the long slender sporophyll which reaches a length of 9 cm. There is thus a resemblance in size between the sporophylls and ordinary leaves and, in life, there probably was no conspicuous distinction between the sterile and fertile branch tips. Johnson was able to observe 10 to 20 megaspores in some of the sporangia and, in others, smaller bodies which are thought to have been microspores.

Some years ago Watson described a cone from the Upper Carboniferous of England that was believed to have been borne on the stems of *Bothrodendron mundum;* since the cone was not actually found attached it was given the name *Bothrostrobus.* It is a heterosporous cone about 8 mm in diameter (Fig. 8-20); the sporangia are nearly globose, being slightly taller than broad and are further characterized by large ligules; not more than four megaspores were present in each sporangium.

Mode of Growth in the Arborescent Lycopods

In the foregoing discussion I have tried to point out some of the more important features of the lycopods as well as some of the unsolved problems; one of the most intriguing of these is the mode of growth of these strange trees.

Restorations of plants assigned to *Lepidodendron* and *Lepidophloios* show them as trees of great height with a columnar trunk that dichotomized to form a dense crown above. Scott, in his *Studies in Fossil Botany,* cites a trunk found in the Lancashire coal fields that was 114 feet long from the base up to the first branching. This is probably an exceptionally tall specimen but it emphasizes a distinctive feature of the plants, that is, the great height that was reached prior to the initiation of branching. Some light is shed on this by a recent discovery of *Lepidophloios* stems from Kansas

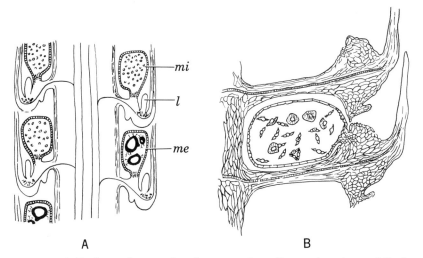

A   B

Fig. 8-20.   A. *Bothrostrobus mundus,* diagrammatic median section of cone; *l,* ligule; *me,* megasporangium; *mi,* microsporangium. B. *Spencerites insignis,* portion of a cone showing one sporophyll with distally attached sporangium. (A from Watson, 1907; B from Berridge, 1905.)

coal balls which are approximately 12 cm in diameter and composed almost wholly of primary tissues. There is no secondary wood and only a narrow band of periderm 3 to 5 mm thick which was formed deep within the cortex. It is evident that these stem or branch specimens grew by means of a rather large apical meristem and, assuming a mode of growth comparable with that in the modern cycads and tree ferns, a very tall unbranched trunk could have been formed. But in the case of the lycopods an obvious problem correlated with this mode of growth is the matter of the photosynthetic tissue available to the plant. It is difficult to understand how a trunk could have grown to a height of many feet, actually well over 100 in some cases, with only the small leaves a few centimeters long that seem to have been borne by many of these plants (Fig. 8-9). If, however, the trees in this initial stage of growth were clothed with leaves up to a meter in length it is more readily understood.

There is evidence cited above from Zeiller's Valencienne specimens that some of the lycopods may have started with such long leaves and they became progressively smaller as the branch orders decreased in size.

In summary, it has been suggested by Andrews and Murdy that in at least some of the lycopods the trunk attained a fairly large diameter at an early stage in growth and developed as a uniform columnar trunk up to 100 feet or more (Fig. 8-4B). The apical meristem then divided to form two branches which may have been equal or unequal, but in either case successive branch orders became progressively smaller, reaching a minimum diameter, at which time growth ceased. There was probably considerable variation in different species, particularly in relation to the amount of secondary growth.

SIGILLARIACEAE

The sigillarias were Carboniferous lycopods with certain basic points of similarity to the lepidodendrons; the two diverge, however, in several ways, including features of the leaf cushion pattern and cone structure.

The arrangement of the leaf cushions in *Sigillaria* (Fig. 8-21) appears to be a spiral one as in *Lepidodendron,* but the development of distinct longitudinal ridges results in an alignment of the cushions in vertical rows. The gross outline of the leaf cushion is hexagonal, round, or oval.

A   B

Fig. 8-21. A. *Sigillaria mammillaris* Brongniart, portion of a trunk, about 0.3X. B. *Sigillaria coriacea* Kidston, fragment of stem showing details of leaf scars. (British Museum Natural History.)

Sigillaria was certainly not uncommon, for stem impressions are widely distributed geographically and stratigraphically through the Upper Carboniferous; great trunk casts are exposed in the cliffs along the Nova Scotia coast at Joggins and I had occasion several summers ago to visit an open-pit coal mine in northern Pennsylvania where at one point we found vast quantities of sigillarian stem impressions of various branch orders. Thus, although the plants are abundant as compressions or casts, petrified stems are conspicuous by their scarcity. The great English paleobotanist Williamson remarked on this rather forcefully in 1872 and his impression holds good today:

> Considering the abundance of Sigillariae in the coal-measures, it is marvelous that *indisputable* specimens displaying their internal organization should be so rare; but such is the case. After years of search I have only met with three specimens of the Sigillarian character of which there can be no doubt.

The lack of petrified stems is perhaps, in part, apparent rather than real; so far as I am aware there are no reliable anatomical criteria for distinguishing sigillarian stems from those of the lepidodendrons when the characteristic leaf bases are not preserved.

Adding to the uncertainty of stem identification is the problem of correlating these organs with isolated petrified cones. The most informative specimens that are considered referable to *Sigillaria* are those of *Mazocarpon oedipternum* Schopf from the Calhoun coal-ball horizon (upper Pennsylvanian) of Illinois. It is a heterosporous species with the megaspores and microspores borne in separate cones.

The sporophylls of the megasporangiate cones (Fig. 8-22) were arranged in a low spiral or verticillate order and they are distinguished in this species by a conspicuous heel. There are eight megaspores in each sporangium, being arranged in a peripheral position and surrounded by parenchymatous tissue. Several megaspores have been found with fully developed female gametophytes, each with a single archegonium. Large numbers of microspores were produced in each sporangium of the microsporangiate cones; the unique parenchymatous tissue of the megasporangia is not present here.

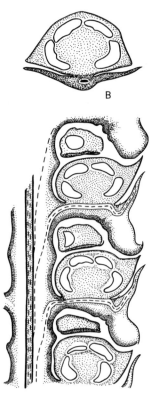

B

A

Fig. 8-22. *Mazocarpon oedipternum.* A. Radial longitudinal section through a portion of megasporangiate cone; B. a single megasporophyll in tangential view, about 4X. (From Schopf, 1941.)

Another genus of cones, *Sigillariostrobus,* known from compression specimens is also referred to the Sigillariaceae. These vary in different species from 15 to 25 cm long and from 1.5 to 2.5 cm in diameter. They were borne singly at the end of a pedicel or specialized branch. One of the reasons for regarding these cones as sigillarian is the discovery by Zeiller, in 1884, of specimens in which the pedicel displayed a leaf cushion structure comparable with that of the stems.

They are heterosporous with mega- and microsporangia borne in separate cones. In three Polish species described by Bochenski it was found that 12 megaspores were present in each megasporangium and the megaspores were large, exceeding 2 mm in diameter. Thousands of microspores were present in each microsporangium. The sporangia were apparently shed from the cone axis as they approached maturity as many specimens have been described which retain their sporangia only in the basal region. The megasporangiate cones sometimes present a very distinctive appearance as compression fossils in that the general outline of the cone may be retained perfectly, but apparently the only structures preserved are the large megaspores. Thus, at first glance the cones give the appearance of being simply a large mass of megaspores (Fig. 8-23). This unique mode of preservation has been explained as a result of the reduction, to a thin, clear film, of the parenchymatous central cushion around which the megaspores are clustered. Thus the more resistant megaspores are held together in very nearly their original position, whereas the remains of the parenchymatous tissue are not visible except by special treatment.

HERBACEOUS LYCOPODS

Not all of the Carboniferous lycopods were large trees. Fossil specimens characterized by frequent branching, leaves only a few millimeters long, and minute cones are known from Paleozoic, as well as Mesozoic rocks, that were small, low-growing herbaceous plants. Some of these were so similar in appearance to the extant *Selaginella* that they have been referred to that genus; others have been described under the names *Selaginellites* and *Lycopodites.* Owing to considerable variation in the quality of preservation, the fact that some specimens are fertile and others not, and to the divergent views of different investigators, the nomenclatural his-

Fig. 8-23. Compression specimens of part of a megasporangiate *Sigillariostrobus* from the Upper Carboniferous of Pennsylvania, about 2X.

tory of the fossil herbaceous lycopods offers a state of confusion that is impressive even for paleontological tangles. To the best of this writer's understanding the following is a brief summary of the usage of the two names cited above:

Selaginellites was instituted by Zeiller in 1906 for small, apparently herbaceous plants bearing cones that are clearly heterosporous.

Lycopodites has been employed by most authors for plants of similar diminutive size, with minute leaves resembling those of the modern *Lycopodium,* and which presumably were homosporous. So far as I am aware, however, no fossil specimens referred to *Lycopodites* have actually been proven to be homosporous. Thus in practice it has been used simply for presumably herbaceous lycopods that have not been clearly established as representing homo- or heterosporous plants.

In the following paragraphs a few of the better known fossils are introduced.

Selaginellites crassicinctus Hoskins and Abbott, from the Des Moines series (Pennsylvanian) of Kansas, is based on a cone with an over-all diameter of 5 mm. The sporophylls are about 3 mm long and bear megasporangia toward the basal part of the cone and both megasporangia and microsporangia in the upper region. The megasporangia usually contain four spores which have a characteristic equatorial flange and a maximum diameter of about 800 μ; large numbers of microspores, each about 80 μ in diameter, are contained in the microsporangia. A small ligule not more than 225 μ long is present.

Selaginellites canonbiensis Chaloner is a cone from the Upper Carboniferous of Scotland and is 3.5 mm in diameter; the lower portion is megasporangiate and the upper microsporangiate. There were four or possibly more megaspores, each 630 μ in diameter, in a megasporangium. The microspores measured about 43 μ in diameter.

Selaginellites polaris Lundblad from the Triassic of East Greenland is known from rather large cones measuring about 7 mm in diameter; the megasporangia contained numerous spores each 412 μ in diameter, whereas the microspores average about 40 μ.

As noted above, where the comparison has seemed sufficiently close to justify it, some authors have assigned these heterosporous herbaceous lycopods to the genus *Selaginella.* Miss Lundblad has described specimens showing both cones and vegetative parts from the Rhaetic of Scania, Sweden as *Selaginella hallei.* The foliage is dimorphic, the leaves on the lower side of the stems being 1 to 2 mm long, and those on the upper side are closely adpressed and about half that length. The cone is only 5 mm long and 1 mm in diameter and apparently very similar in construction to that of a living *Selaginella.* The megasporangia contained four spores, each about 426 μ in diameter and the microspores are 50 μ in diameter.

Selaginella amesiana Darrah is a Pennsylvanian specimen that is remarkable for the preservation of the female gametophytes. It is a partially petrified cone found in a nodule from the famous Mazon Creek area of northern Illinois. The cone is 3.5 mm in diameter and the portion preserved includes many megasporangia; these contain four megaspores each and in many of them the female gametophyte is well preserved. The nuclei are clearly defined and in some cells structures can be discerned which appear to be chromosomes.

Thus the evidence seems to suggest rather strongly that the modern *Selaginella* and possibly *Lycopodium* are plants of very ancient lineage. The similarity between some of the fossil species and the modern selaginellas implies close affinity, so close in fact that it seems reasonable to regard the latter as stemming directly from Carboniferous forms such as have been described.

PLEUROMEIACEAE

Pleuromeia sternbergi (Münster) Corda represents a curious line of Lower Triassic lycopods possibly related to the living *Isoetes*. The stem (Fig. 8-24) is unbranched and attained a height of 2 meters and a maximum diameter of 9 cm. The leaves, which are not as densely arranged on the stem as in the Carboniferous lycopods, are 11 cm long and about 1.5 cm broad except for the broadly flaring basal portion. A ligule was present and two vascular bundles traverse the length of the leaf. As the trunk increased in height the leaves were shed from the basal region.

The base of the plant flared out into a four-lobed root system with the four points turned upward; numerous rootlets are borne over the outer surface.

The stem is terminated at the top by a massive spike of sporophylls which may be termed a cone. Although complete specimens have not been found, the fragments bear either microspores or megaspores; thus the plant was heterosporous and apparently dioecious. The megasporangia are large, being about 2 cm broad and contained megaspores 500 to 700 μ in diameter; the microsporangia are about 1.2 cm broad and the microspores measure 15 to 25 μ in diameter.

According to Seward the original specimen on which Münster's description was based came from a split stone taken from the tower of Magdeburg Cathedral, an unusual paleobotanical collecting site!

The sporangia are reported as borne on the under (abaxial) side of the sporophylls, although this has been questioned by certain

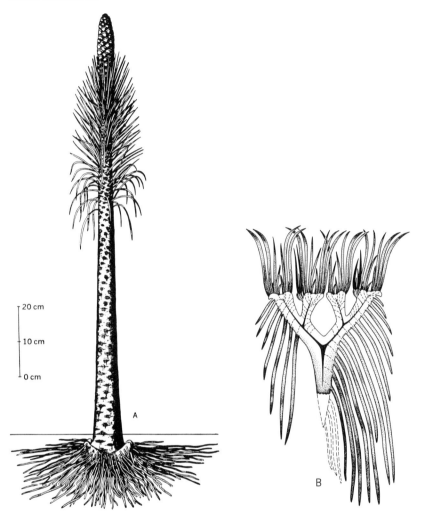

Fig. 8-24. A. Restoration of *Pleuromeia sternbergi*. B. Diagrammatic sketch of *Stylites gemmifera*. (A from Hirmer, 1933; B from Rauh and Falk, 1959.)

authors. The abaxial position has been explained as possibly resulting from an ancestral type in which sporangia were peltately arranged, those on the upper (adaxial) surface having been lost.

Some support for this theory may be found in the Upper Carboniferous *Spencerites insignis* (Williamson) Scott; this seems properly regarded as a lycopod cone but it is distinctive in certain characters. The axis contains a typical lycopod stele and the sporophylls are arranged in alternating verticils of ten to a whorl.

Each sporophyll is constructed of a rather slender pedicel and an upturned lamina (Fig. 8-20B). The distal part of the pedicel expands into upper and lower fleshy lobes with a single sporangium attached to the upper one. This presents a contrast to other lycopod cones in which the sporangium is attached either near the proximal end of the pedicel or along the greater length of it. No ligule has been observed. The spores have a characteristic equatorial wing and measure about 280 μ in diameter, a size that is problematical in that most microspores are much smaller than this and megaspores are usually somewhat larger. It seems most likely that the portion known is simply the megasporangiate portion of the plant. The systematic position of *Spencerites* is uncertain and it was introduced here because of the pads or lobes of the pedicel; although only the upper one bears a sporangium, the symmetry suggests that formerly one might have been borne on the under side as well.

The theory of an ancestral type with peltate sporangiophores from which forms like *Pleuromeia* and *Spencerites* might have evolved is not, however, supported by any pre-Carboniferous evidence.

SUMMARY

The lycopods are an ancient line of plants in which the distinctive characters were acquired in the Silurian, or possibly pre-Silurian times. The record seems to indicate that the group gained evolutionary momentum in the mid-Devonian and was already quite diverse by the close of that geologic period. Arborescent forms had then made their appearance and the forest trees of the Carboniferous were huge by any standard. There is sufficient evidence to indicate that small herbs have existed throughout the racial life span of the group; *Selaginella*-like plants are known from the Carboniferous on up through the Mesozoic to the present and there seems to be no reason to doubt that this genus, as we know it today, has existed with little change for some 250 million years. It also seems to me more than likely that *Lycopodium* may stem back directly to the Silurian *Baragwanathia,* but the fossil record of plants that can be closely compared with *Lycopodium* is a scanty one.

Several lines of arborescent lycopods flourished in the Carboniferous and these are probably derived from a common stock in the

Devonian. Some evidence for this is found in *Archaeosigillaria primaeva,* described by David White in 1907, from the Upper Devonian of New York. When originally excavated the specimen measured some 5 meters in length and consisted of a rather bulbous base 38.5 cm in diameter and shortly above this the trunk tapered gradually to a diameter of 12 cm at the upper end. Its chief point of interest lies in the nature of the leaf cushions which, in the lower third of the trunk are sigillarian in form, whereas in the central part they become more lepidodendroid and distinctly so in the upper portion. It thus appears to combine features of *Sigillaria* and *Lepidodendron* as these genera are known in typical Carboniferous specimens. The leaves were narrow, not over 3 cm long and apparently persistent in their attachment since specimens are found as low as 70 cm from the base.

An abundance of stem remains in coal ball petrifactions attests to the dominance of the arborescent lycopods in the Upper Carboniferous; at one locality in Illinois approximately two-thirds of the plant material is *Lepidodendron* and in the southeastern Kansas coal fields the petrifactions contain large quantities of steles and periderm. An unusual bit of evidence bearing on the abundance of the lycopods, as well as a distinctive type of plant preservation, is recorded by Scott (1920, p. 182) in a description of the Russian paper-coal:

At Tovarkovo, in the province of Toula, in Central Russia, beds of a peculiar kind of coal, called leaf-coal or paper-coal, have long been known. The seams are about 8 inches thick, and lie near the surface of the ground, only covered by sand. The so-called coal has all the appearance of a bed of excessively thin dead leaves, intermixed only with structureless organic matter of the nature of humic acid. The leafy films, on investigation, have proved to consist entirely of layers of cuticles, belonging to the stems of ancient plants, from which all other tissues had rotted away, ages before. The cuticles are perfectly fresh and pliable, and not in any way fossilized, although geologists are agreed that the bed (which covers an area of many square miles) belongs to the Carboniferous Limestone horizon, in the lower part of the Carboniferous formation. The cuticle is in some cases complete, corresponding to the whole circumference of the stem which it once enclosed; it is perforated by numerous small round holes, regularly disposed, and corresponding in arrangement with the leaf-traces of a *Bothrodendron.*

As to the acquisition of the seed habit, *Lepidocarpon* was not the only one to achieve this level of development. A similar organ has been described for *Miadesmia membranacea* Bertrand although this was probably a plant of much smaller habit; the seed here displayed a characteristic fringed integument and a more attenuate micropyle. It seems to me that *Miadesmia* and *Lepidocarpon* are sufficiently distinct as to suggest that they evolved independently as seeds from separate lines of lycopods. There is of course no reason to suppose that these plants were at all closely related to the "true" gymnospermous groups such as the pteridosperms, cordaites, and cycadophytes; the seed integument here is certainly not homologous with that of the two lycopod seeds.

The arborescent representatives seem to have declined rapidly toward the end of the Paleozoic. This decline, however, may not have been as sudden as the macrofossil record has indicated. For example, Harris has described 13 new species of *Triletes* from the Rhaetic rocks of East Greenland; this is a genus of spores generally attributed to the lycopods and the Greenland megaspores range in size from 250 μ to 1400 μ.

Pleuromeia has been the subject of considerable speculation as to its position in the lycopod clan. Some interesting information bearing on this problem has come from the discovery of a new genus of living lycopods, *Stylites,* assigned to the Isoetaceae. Before dealing with this directly it seems appropriate to add a few notations on the fossil history of *Isoetes* (quillwort).

The quillworts are among the most distinctive of all pteridophytes; there are several scores of species found in aquatic environments or places that are moist for part of the year. The axis of the plant is extremely short, bearing abundant roots on the under side, and a dense rosette of long, strap-shaped leaves above. Other than those formed in the early part of the growing season, most of the leaves are fertile, bearing a large sporangium on the upper surface near the base. Both mega- and microsporangia occur on the same plant although the two may be developed in different seasons, according to Pfeiffer.

Isoetes is a genus of some antiquity; R. W. Brown has reviewed the American specimens and recognizes two species:

I. horridus (Dawson) Brown from the Paleocene of Wyoming seems identical in habit with modern species; the dense rosette of sporophylls and casts of the sporangia and spores leave no doubt as to the identity of the fossil. Other specimens attributed to it have been found in Lower and Upper Cretaceous horizons.

Another species, *I. serratus* Brown, although not as well preserved, was found in the Frontier formation (Upper Cretaceous) of southwestern Wyoming. The implication is thus quite clear that *Isoetes* had evolved to essentially its present morphology by mid-Mesozoic times.

Returning to *Pleuromeia*, one is tempted to interpret it as a diminutive descendent of the arborescent lycopods of the Carboniferous, one in which branching of the subterranean system is much reduced. *Isoetes* might correspondingly be regarded as a lycopod in which the entire shoot and root system is reduced to a point where there is very little of either left. Fitting into this picture is a Lower Cretaceous plant *Nathorstiana arborea* Richter (see Mägdefrau, 1932) which has been described as an elongated *Isoetes*. It is a small plant with an erect stem a few centimeters long which is unbranched.

A few years ago a new genus of living isoetaceous plants was discovered which adds a significant fragment to this particular picture puzzle of lycopod evolution. Two species have been described, *Stylites andicola* Amstutz and *S. gemmifera* Rauh, both of which were found growing around the boggy margins of a lake in central Peru at an elevation of 4750 meters. Although *Isoetes* rarely branches, the plants of *Stylites gemmifera* dichotomize at least three times (Fig. 8-24B). Rosettes of leaves, apparently very like those of *Isoetes*, are found at the apex of each branch whereas the "caudex" bears fleshy roots along one side only. Thus, although *Pleuromeia-Nathorstiana-Stylites-Isoetes* do not present an unbroken sequence they present significant links in what appears to have been a line of evolution in which the axis became progressively reduced to the extreme state found in *Isoetes*.

The lycopods are an ancient group of plants including a diverse assemblage in respect to size and general morphology; growth mechanics was unique and is still poorly understood in the arborescent forms. Like most of the other major divisions of plants, we are only beginning to unravel their past history.

REFERENCES

Amstutz, Erika. 1957. *Stylites,* a new genus of Isoetaceae. *Ann. Missouri Bot. Gard.,* 44: 121–123.

Andrews, Henry N., Jr., and Murdy, William H. 1958. *Lepidophloios*—and ontogeny in arborescent lycopods. *Amer. Journ. Bot.,* 45: 552–560.

———— and Pannell, E. 1942. Contributions to our knowledge of American Carboniferous floras II. *Lepidocarpon. Ann. Missouri Bot. Gard.,* 29: 19–28.

Arnold, Chester A. 1938. Note on a lepidophyte strobilus containing large spores, from Braidwood, Illinois. *Amer. Mid. Nat.* 20: 709–712.

————. 1940. *Lepidodendron johnsonii,* sp. nov., from the Lower Pennsylvanian of Central Colorado. *Univ. Michigan, Contrib. Mus. Paleont.,* 6: 21–52.

Banks, Harlan P. 1944. A new Devonian lycopod genus from southeastern New York. *Amer. Journ. Bot.,* 31: 649–659.

Beck, Charles B. 1958. *"Levicaulis arranensis,"* gen. et sp. nov., a lycopsid axis from the Lower Carboniferous of Scotland. *Trans. Roy. Soc. Edinburgh,* 63: 445–456.

Benson, Margaret. 1908. *Miadesmia membranacea,* Bertrand; a new Palaeozoic lycopod with a seed-like structure. *Phil. Trans. Roy. Soc. London,* 199B: 409–425.

Berridge, E. M. 1905. On two new specimens of *Spencerites insignis. Ann. Bot.,* 19: 273–279.

Bochenski, Tadeusz A. 1939. On the structure of Sigillarian cones and the mode of their association with their stems. *Polish Acad. Sci.,* 7: 1–16.

Brown, Roland W. 1939. Some American fossil plants belonging to the Isoetales. *Journ. Washington Acad. Sci.,* 29: 261–269.

Chaloner, William G. 1958. A Carboniferous *Selaginellites* with *Densosporites* microspores. *Palaeontology,* 1: 245–253.

Darrah, William C. 1938. A remarkable fossil *Selaginella* with preserved female gametophytes. *Harvard Univ. Bot. Mus. Leaf.,* 6: 113–135.

Dijkstra, S. J. 1956. Lower Carboniferous megaspores. *Mededelingen Geol. Stichting (Heerlen),* n.s., No. 10: 1–18.

———— and Piérart, P. 1957. Lower Carboniferous megaspores from the Moscow Basin. *Mededelingen Geol. Stichting (Heerlen),* n.s. No. 11: 5–19.

Felix, Charles J. 1952. A study of the arborescent lycopods of southeastern Kansas. *Ann. Missouri Bot. Gard.,* 39: 263–288.

————. 1954. Some American arborescent lycopod fructifications. *Ann. Missouri Bot. Gard.,* 41: 351–394.

Graham, Roy. 1935. Pennsylvanian flora of Illinois as revealed in coal balls. II. *Bot. Gaz.,* 97: 156–168.

Hirmer, Max. 1933. Rekonstruktion von *Pleuromeia sternbergi* Corda, nebst bemerkungen zur Morphologie der Lycopodiales. *Palaeontographica,* 78B: 47–56.

Hoskins, John H. and Abott, Maxine L. 1956. *Selaginellites crassicinctus,* a new species from the Desmoinsian series of Kansas. *Amer. Journ. Bot.,* 43: 36–46.

Johnson, Thomas. 1913. On *Bothrodendron (Cyclostigma) Kiltorkense* Haughton sp. *Sci. Proceed. Roy. Dublin Soc.,* n.s. 13: 500–528.

Kisch, Mabel H. 1913. The physiological anatomy of the periderm of fossil lycopodiales. *Ann. Bot.,* 27: 281–320.

Kräusel, Richard, and Hermann Weyland. 1935*a.* Neue Pflanzenfunds in Rheinischen Unterdevon. *Palaeontographica,* 80B: 171–190.

————. 1935*b.* Pflanzenreste aus dem Devon. IV. *Protolepidodendron* Krejci. *Senckenbergiana,* 14: 391–403.

Lang, William H., and Isabel C. Cookson. 1935. On a flora, including vascular land plants, associated with *Monograptus,* in rocks of Silurian age, from Victoria, Australia. *Phil. Trans. Roy. Soc. London,* 224B: 421–449.

Lundblad, Britta. 1948. A Selaginelloid strobilus from East Greenland (Triassic). *Meddeleser Dansk Geol. Foren.,* 11: 351–363.

————. 1950. On a fossil *Selaginella* from the Rhaetic of Hyllinge, Scania. *Svensk Bot. Tidskrift,* 44: 477–486.

Macgregor, Murray and Walton, John. 1948. *The Story of the Fossil Grove.* Published by City of Glasgow Public Parks, 32 pp.

Mägdefrau, Karl. 1931. Zur Morphologie und phylogenetischen Bedeutung der fossilen Pflanzengattung *Pleuromeia. Beih. Bot. Centralbl.,* 48: 119–140.

Mathews, G. B. 1940. New Lepidostrobi from central United States. *Bot. Gaz.* 102: 26–49.

Nathorst, Alfred G. 1902. Zur Oberdevonischen Flora der Bären- Insel. Kongl. *Svenska Vetensk.-Akad. Handl.*, 36 (3): 1–60.

Pannell, Eloise. 1942. Contributions to our knowledge of American Carboniferous floras. IV. A new species of *Lepidodendron*. *Ann. Missouri Bot. Gard.*, 29: 245–274.

Pfeiffer, Norma E. 1922. Monograph of the Isoetaceae. *Ann. Missouri Bot. Gard.*, 9: 79–232.

Rauh, Werner and Falk, Heinz. 1959. *Stylites* E. Amstutz, eine neue Isoëtacee aus den Hochanden Perus. *Sitzungsberichte Heidelberger Akad. Wissen. Jr.*, 1959, pp. 1–83.

Schmidt, Wolfgang. 1955. Pflanzen-reste aus der Tonachiefer-Gruppe (Unteres Siegen) des Siegerlandes. *Sugambrophyton pilgeri*, n.g., n. sp., eine Protolepidodendracee aus den Hamberg-Schichten. *Palaeontographica*, 97B: 1–22.

Schopf, James M. 1941. Contributions to Pennsylvanian paleobotany *Mazocarpon oedipternum*, sp. nov. and Sigillarian relationships. *Illinois State Geol. Surv., Rept. Investig.*, No. 57: 1–40.

Scott, Dukenfield H. 1898. On *Spenserites*, a new genus of lycopodiaceous cones from the Coal Measures, founded on the *Lepidodendron spenceri* of Williamson. *Phil. Trans. Roy. Soc. London*, 189B: 83–106.

————. 1901. On the structure and affinities of fossil plants from the Palaeozoic Rocks. IV. The seed-like fructifications of *Lepidocarpon*, a genus of lycopodiaceous cones from the Carboniferous formation. *Phil. Trans. Roy. Soc. London*, 194B: 291–333.

————. 1920. *Studies in Fossil Botany*. A. and C. Black, Ltd., London, 434 pp.

Snigirevskaya, Natalie S. 1958. An anatomical study of the leaves (phylloids) of some lycopsids in the Donets basin coal balls. *Acad. Nauk USSR* (*Botany*), 43: 106–112.

Watson, D. M. S. 1908. The cone of *Bothrodendron mundum*. *Mem. Manchester Lit. Phil. Soc.*, 52(1): 1–15.

White, David. 1907. A remarkable fossil trunk from the middle Devonic of New York. *N. Y. State Museum Bull.*, 107: 327–340.

Williamson, William C. 1887. A monograph on the morphology of *Stigmaria ficoides*. *Palaeontographical Soc. London*, 1886 vol.: 1–62.

Zeiller, René. 1888. *Études des Gites Minéraux de la France; Bassin Houiller de Valenciennes*. Paris.

9

the ARTHROPHYTA

Although problems of classification are not lacking, most of the plants that have been assigned to this division stand apart quite clearly from all other groups. The ecological requirements of the arthrophytes and general evolutionary pattern have followed rather closely those of the lycopods as far as geologic time is concerned. Several early and mid-Devonian fossils of considerable morphological interest have been referred to the group although it will be evident that their exact relationships to later forms is still uncertain; but by late Devonian times the die was cast and a burst of evolution in the Carboniferous resulted in some of the most unique plants that have ever lived. It was a diverse and plentiful assemblage, yet certain basic characters hold them together as a unit. Apparently decline set in rapidly at the close of the Paleozoic and a single genus, *Equisetum*, survives today. A few botanists have devoted serious consideration to possible relationships with other major groups and it has been suggested that they may have been progenitors of the seed plants; such speculations are certainly in error.

The terms *arthrophyte, sphenopsid,* and *articulate* have been used as common names; the last seems particularly apt because of the characteristic jointed nature of the stems and whorled arrangement of the leaves and sporangia-bearing organs which correlate with a comparably unique internal organization. Plants in other groups are jointed or articulated but are not readily confused with *the* articulates.

The student who is not well acquainted with *Equisetum*, the lone generic survivor of this great race of plants, would do well to familiarize himself with its numerous distinctive features. Briefly, the various species range from 1 to 4 or 5 feet high (a few tropical ones have been reported much higher, but their stems are at most 2 or 3 cm in diameter) and the shoot system consists of underground rhizomes which may produce a great profusion of upright shoots; in some species the latter are sparsely branched and in others they are regularly and lavishly so. Two types of upright shoots may be present, fertile ones that carry on but little photosynthesis and bear a cone at the apex, and sterile shoots. Other species bear only one type of shoot which is both fertile and photosynthetic. In all cases the leaves are inconspicuous scalelike structures borne in a sheathing whorl at the node. Internally, through the internodal region there is a large central pith canal surrounded by a ring of only a few tracheids which are associated with minute protoxylary canals; in this character they present a striking divergence from the larger Carboniferous forms in which several inches of secondary wood may be present. Another series of canals is present in the cortex.

The cone of *Equisetum*, which is terminal on the shoot system, consists of whorls of stalks which flare out into a disc, making an umbrellalike organ, and around the rim of the disc there are several pendant sporangia.

Classification

There are a few problems in particular that seem to me to have a direct bearing on any attempt at classifying the articulates. The earlier Devonian fossils are in part fragmentary and in other cases not as distinctly "articulate" as the Carboniferous plants; they are not sufficiently well known to enable us to understand their interrelationships or affinities with the later members, and they are therefore treated under the heading "Certain Early Articulates." A surprising number of new cone types have turned up recently, indicating that the Sphenophyllales and Equisetales are more complex assemblages than has been supposed. Two especially complex sporangiate organs, *Eviostachya* and *Cheirostrobus,* have been found in the Devonian and Lower Carboniferous respectively; aside from their intrinsic interest they suggest a far more diverse Devonian assemblage than is presently known and I have chosen to consider them at the close of the chapter simply as *"Problematical Articulates."*

Certain early articulates
Equisetales
   Calamitaceae
   Equisetaceae
Sphenophyllales
Some problematical articulates

SOME EARLY ARTICULATES

Rather than follow a strict chronological order it will perhaps lend greater clarity if the two best known Devonian articulates, *Calamophyton* and *Hyenia,* are described first.

Some especially fine specimens of *Calamophyton* (*C. bicephalum* Leclercq and Andrews) have been discovered in Middle Devonian rocks near the village of Goé in eastern Belgium. The plants (Fig. 9-1) attained a height of at least 2 feet and consisted of a "main axis" which tended to divide in a more or less digitate fashion, as many as seven branches being formed at very nearly the same point. These in turn branched monopodially or with a mixture of monopodial and dichotomous forking. It is not known how the structure referred to as the main axis was borne and most of the specimens are predominantly sterile or fertile.

The ultimate branches bear sterile appendages ("leaves") that are about 10 mm long (Fig. 9-2A). They were slender, dichotomous structures, probably terete in cross section, which divided two to four times. Successive divisions were at right angles to each other so that the appendage as a whole was clearly three-dimensional. They exhibit a weak tendency toward a whorled arrangement.

The ultimate fertile appendages are quite complicated for so ancient a plant. Each one (Fig. 9-2B) is also about 10 mm long; the primary stalk divided into two equal branches or "heads," from which the specific name *bicephalum* is taken. Each of these gave rise to three short side branches which terminate in a pair of slender sporangia.

There is sufficient similarity between the sterile and fertile appendages so that they may be regarded as homologous organs. They were probably rather rigid in life, especially the sterile ones since their three-dimensional branching pattern is retained in fossilization, that is, they dip down into the rock matrix and much tedious work with fine steel needles was required to expose them.

Hyenia is probably closely related to *Calamophyton* and has been found associated with it; in fact, the only clear distinction between the two lies in the general habit. In *Hyenia elegans* Leclercq a stout rhizome (Fig. 9-3B) bore numerous upright shoots that were rarely branched. *Hyenia vogtii* Høeg from the Middle Devonian of Spitsbergen offers an interesting problem in branch morphology. The plant is known from axes up to 9 mm thick which presumably correspond to the upright leafy shoots of other hyenias; however,

Fig. 9-1. *Calamophyton bicephalum,* a specimen from the Middle Devonian of eastern Belgium, about 0.5X. (From Leclercq and Andrews, 1960.)

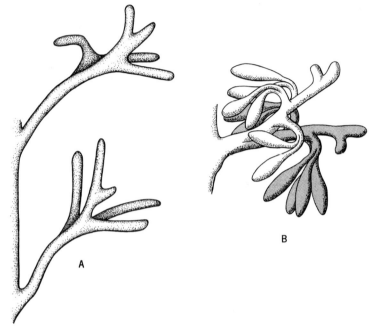

Fig. 9-2. *Calamophyton bicephalum.* A. Two sterile appendages (leaves) showing their three-dimensional form, 5X. B. A fertile appendage; the sporangiophore divides into two parts, each of which bears three short branches that terminate in paired sporangia, 4X. (From Leclercq and Andrews, 1960.)

these axes produced lateral branches which resemble the main one but are somewhat more slender. Leaves 1 to 1.5 cm long are arranged in fairly regular verticils on both primary and secondary branches. Of special note is the position that the branches occupy. According to Professor Høeg who described these fossils:

> In our new species, branching is frequent and regular; it is of considerable interest to note that the branches are strictly lateral, and that they take the place of leaves, thus showing that they are homologous with those organs. We have a parallel case of development within another group of primitive plants, still living, the bulbils of *Lycopodium selago* and other bulbilliferous species of *Lycopodium* being inserted in the place of leaves . . .

A few Lower Devonian fossils have been reported which lack the quality of preservation and completeness of the Middle Devonian plants but may be regarded tentatively as early articulates.

Protohyenia janovii Anan'ev from western Siberia (Fig. 9-3A) is known from short unbranched shoot fragments, some of which bear sterile appendages that forked several times and probably served as leaves. Other shoots bore appendages of essentially the same form, but they were terminated by oval sporangia which measure about 3 × 2 mm. It is thought that these shoots were attached to a rhizomatous stem like that of *Hyenia,* but this has not been confirmed. There is an unmistakable similarity here between the sterile and fertile appendages and it is not difficult to accept the supposition that the latter might have evolved into the more complex organs of the Middle Devonian fossils.

EQUISETALES

Calamitaceae

This family includes the arborescent articulates, commonly known as the calamites. The larger members of the group were

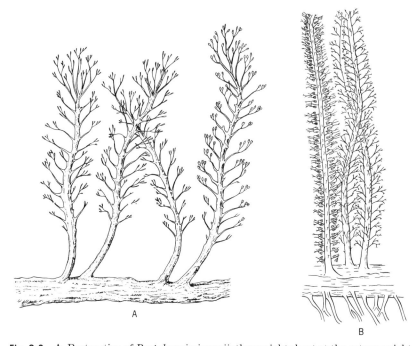

Fig. 9-3. A. Restoration of *Protohyenia janovii;* the upright shoot at the extreme right is fertile, the side branches terminating in sporangia. B. Restoration of a portion of a plant of *Hyenia elegans;* the left-hand shoot is fertile, whereas the other two are sterile, about ⅙X. (A from Ananiev, 1957; B from Leclercq, 1940.)

trees that probably attained a height of 50 feet or more; consequently we are confronted with the usual assortment of vegetative and reproductive organs. Although there are several generic names and numerous species under each category (leaf, stem, cone, etc.), they belong for the most part to a rather clearly defined group of plants. Thus the objective in the following pages will be to convey an understanding of a "generalized" calamite, following the approach used with the arborescent lycopods.

Stem Structure

In general habit the calamites were jointed, profusely branched plants; if one can imagine the vegetative shoots of the common *Equisetum arvense* enlarged about thirty or forty times some concept may be gained of the way these plants appeared in the Carboniferous forests.

Several genera have been recognized on the basis of differences in the structure of the secondary wood; with small twig specimens, in which this tissue is present only in the initial stages of development, such generic distinctions cannot be made.

Figure 9-4A shows a small twig from an Illinois coal ball. A pith is present and bordering its periphery is a ring of small protoxylary canals, so-called because the first formed wood cells are found along the lateral and outer sides of each canal. There is very little primary wood in any of the calamite stems, cambial activity having been initiated very early. The phloem, formed on the outside of the wood, is rarely preserved. There is next a broad cortical region of essentially uniform, isodiametric cells.

Somewhat larger branch specimens bring in two characteristic features of the shoot system, a pith cavity and strongly developed secondary wood. The pith cavity is not the result of decay but is rather a natural canal due to the failure of the apical meristem to form this tissue other than as a thin peripheral band. Three distinct genera have been established on the basis of marked differences in the secondary wood;

Arthropitys is the most frequently encountered wood type being represented by numerous specimens and species; delimitations of the latter are difficult, thus our discussion here will be at the generic level. Figure 9-5A shows a sector of an *Arthropitys* stem that was probably about 12 cm in over-all diameter, only the peripheral pith and wood being shown. The xylem is composed of *wood sectors* (*ws*) which radiate out from the protoxylary canals as a focal point,

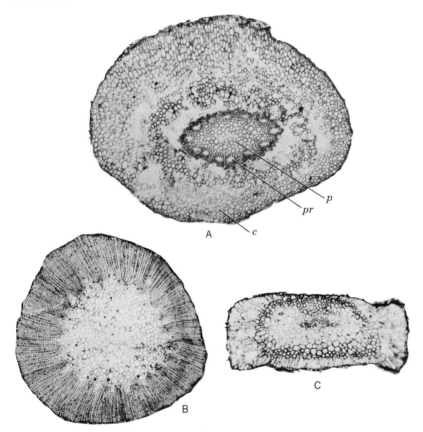

Fig. 9-4. A. A small *Calamites* twig, 30X; *p*, pith; *pr*, protoxylary canal; *c*, cortex.
B. A calamite root, 5X. C. A leaf, probably referable to *Asterophyllites*.

and large *interfascicular rays* (*ir*) which lie between them. The
relationship between the two varies greatly, even among different
specimens that apparently were derived from a single species. The
rays may be broadly wedge-shaped masses of tissue that quickly
lose their identity in the radial direction or they may extend undi-
minished in breadth to the outer periphery of the wood.

The wood sectors are composed of tracheids and small wood rays.
The tracheids, although apparently constant within a species, display pitting on their radial walls that is scalariform, circular-
bordered, or more or less intermediate. The wood rays are generally
one to three cells wide and vary greatly in height.

In *Calamodendron* the wood sectors are flanked on either side
by bands of vertically aligned fibrous cells; thus the radial sequence

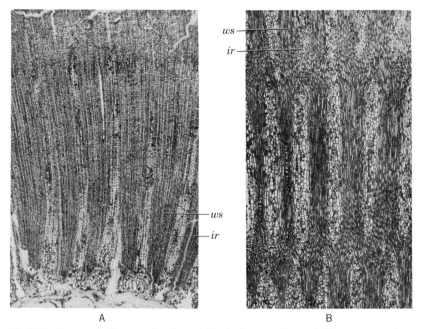

A   B

Fig. 9-5. The wood of a calamite stem of the *Arthropitys* type. A. Transverse section; B. tangential longitudinal section showing part of three nodes; *ir,* inter-fascicular ray; *ws,* wood sector. (Both about 5X.)

is wood sector, fiber band, interfascicular ray, fiber band, wood sector, etc.

In the third genus, *Arthroxylon,* the cells of the interfascicular rays are vertically elongated almost to the point of being fibrous. The wood sector is similar to that of the other genera although in *A. williamsonii* Reed the ray cells in this region are also vertically elongate; thus in a tangential section it is quite difficult to differentiate between interfascicular ray and wood sector.

Very little is known about the tissues outside the xylem other than in the small twigs. Many years ago W. C. Williamson described an English *Arthropitys* specimen in which the wood attained a radial thickness of 5 cm and this was partially surrounded by a fibrous periderm of radially aligned cells 5 cm thick which is similar in general aspect to the periderm of the lycopods. A short time ago I had occasion to examine a slide of this specimen preserved in the British Museum and it seemed to me rather likely that the periderm is correctly associated with the wood, but since the two are not in organic connection one cannot be sure.

We have recently discovered *Arthropitys* specimens in considerable abundance in an Illinois coal ball locality which reveal two novel features. Some are exceedingly large, the radial dimension of the secondary wood exceeding 12 cm; so far as I am aware this is appreciably larger than any previously known petrified articulate stem and suggests an over-all diameter in life of at least 35 cm (allowing for a large pith canal and broad cortex).

Some of these large stelar fragments retain a cortex several centimeters thick of apparently roundish parenchymatous cells, there being *no* evidence of periderm tissue.

It is thus evident that additional information is needed concerning the cortical anatomy of these unique plants and it will almost certainly be forthcoming in the near future.

Pith casts of the calamitean stems (Fig. 9-6) are among the most frequently encountered elements of the roof shale floras of Carboniferous coal mines. When the trunks and branches fell into the swamps, the large pith canal was soon filled with mud which hard-

Fig. 9-6. A pith cast of *Calamites cruciatus;* the round scars represent points where branches were attached, slightly reduced. (Swedish Natural History Museum.)

ened to form a cast; occasionally they are found in their original cylindrical form, but more often they were flattened as the load of sediments above increased. As the stem tissue around this mud core decayed, the primary rays tended to break down more rapidly than the wood sectors; thus the casts display a ridge and furrow appearance, the furrow corresponding to the apex of the wood sector and the ridge to the primary ray.

The generic name *Calamites* was established for the casts, a great many of which probably represent the *Arthropitys* type stem since the latter were more abundant than *Calamodendron* or *Arthroxylon*. The pith casts do not have characters that allow assignment to the respective three "wood genera."

The pith casts may reach a considerable size; Seward cites a specimen from the Radstock Coal Measures in England that is 27 cm in width. Allowing for crushing this indicates a pith cavity of perhaps 20 cm in diameter and would imply a stem with an over-all diameter greatly in excess of the petrified ones mentioned above.

Roots

The anatomy of the roots (Fig. 9-4B) differs from that of the stems in several ways. They lack the jointed organization, a solid and well-preserved pith is often present even in quite large specimens, and the protoxylary canals are absent. The cortical tissue is rarely preserved, but in specimens in which it is still intact, large lacunae are present.

Foliage

The calamitean stems bore foliage (Fig. 9-7) that is referable to the genera *Asterophyllites* and *Annularia*. A useful monographic study of the American species has been made recently by Abbott.

In *Asterophyllites* (there are four American species) the number of leaves in a whorl varies from 4 to 40 with the different species and they show a strong tendency to be cupped upward. With the exception of the very small-leafed *A. charaeformis* (Sternberg) Goeppert, in which they are only 2 to 3 mm long, the leaves are long and essentially linear with a width-length ratio of 1:12 up to an extreme of 1:100 in *A. longifolius* (Sternberg) Brongniart.

There are ten American species of *Annularia;* the number of leaves per whorl varies from 8 to 32 and they tend to stand out at right angles to the stem, forming a perfect rosette in compression

A   B

Fig. 9-7. Foliage of the calamites. A. *Asterophyllites* sp.; B. *Annularia* sp.; both from the Upper Carboniferous of Missouri, about natural size. (B from Andrews, 1947.)

fossils. The individual leaves may be lanceolate but are more often ovate to spatulate; in both genera they are uninerved.

Most of our knowledge of the anatomy of the leaves (Fig. 9-4C) comes from a study made some years ago by Thomas on specimens from the Halifax Hard Bed in the Lower Coal Measures of Lancashire; most of them are probably *Asterophyllites* leaves. The epidermis has a rather heavy cuticle and within this is a single row of radially aligned cells which constituted the mesophyll; they are separated by conspicuous air chambers as are the spongy mesophyll cells of many modern plants.

The vascular bundle consisted of a single central strand of tracheids surrounded by a narrow zone of thin-walled cells which probably functioned as the phloem. Peripheral to this is a conspicuous layer of elongate cells which are partially filled with dark contents; this has been termed the melasmatic tissue and is considered to be a food reservoir region. Directly above the bundle is a patch of fiber cells which served to add rigidity to the leaves.

Cones

The diverse nature of the Carboniferous articulates is most forcefully illustrated by the variety in the morphology of their spore-bearing organs, and in many cases the preservation is far better than the paleobotanist dares hope for.

The calamitean cones vary considerably in size, and the way in which they were borne differs from *Equisetum* where we usually find a single one terminating the shoot. In some of the calamites the cones were borne individually at the nodes, others were arranged in terminal groups or infructescences, and still others on specialized branches. C. E. Weiss has given some excellent illustrations showing the general habit of the fertile shoots, Fig. 9-8 being taken from his account; the serious student would do well to consult this excellent work.

Calamostachys binneyana Carruthers from the European Coal Measures is probably the most common and one of the best known of the calamitean cones (Fig. 9-9A). It attained a length of 3.4 cm and a maximum diameter of 7.5 mm, consisting of equidistant alternating verticils of bracts and sporangiophores. There are about 12 bracts in a whorl, which are fused to form a disc at the base, whereas the upturned distal ends are separate. The fertile appendages are arranged about six in a whorl and each one consists of a sporangiophore which terminates in a peltate or cruciate head bearing four sporangia. To indulge in a fine point, but one that is quite apparent in well-preserved specimens, the bracts in successive whorls alternate; it may be noted in the figure that one whorl shows the upturned lamina, whereas in the one above or below, it appears to be missing. All of the many specimens of this cone that have been found in English coal balls indicate without doubt that it was homosporous. In 1909 Thomas reported a *Calamostachys* cone on a twig with foliage of the *Asterophyllites* type.

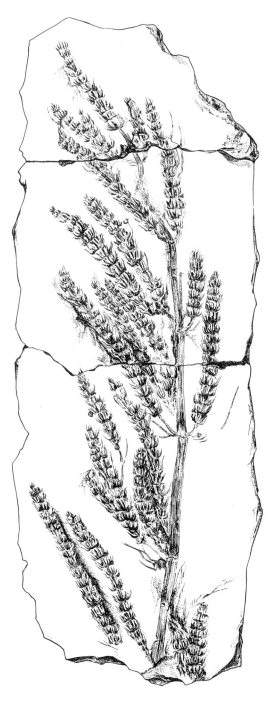

Fig. 9-8. *Calamostachys lud-
wigi,* a branch bearing numerous cones. (From C. E.
Weiss, 1884.)

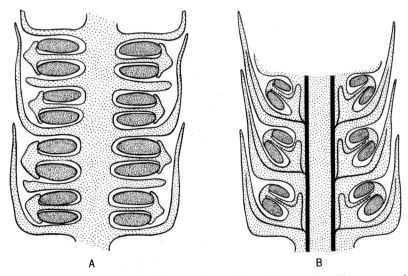

A   B

Fig. 9-9. Diagrammatic longitudinal sections of, A. *Calamostachys binneyana* and B. *Palaeostachya andrewsii.* (B from Baxter, modified.)

Calamostachys casheana Williamson, from the Coal Measures of England, is quite similar to *C. binneyana* in its general organization but it is heterosporous. Some of the sporangia contain numerous small spores, whereas in others the number is much lower and the spores three to four times as large in diameter. Figure 9-10 is drawn from a specimen in the Williamson collection (British Museum), the portion present being beautifully preserved. It is unfortunate that this heterosporous species should be as rare as *C. binneyana* is abundant; however, additional information has recently turned up in the form of a spectacular cone, *C. americana* Arnold, from the Pennsylvanian of Illinois. It is 4 cm in diameter and at least 12 cm long. The axis bore whorls of bracts (40 to 45 each) which were united in the basal region to form a horizontally expanded disc. Successive whorls are spaced about 4 mm apart and midway between them is a whorl of sporangiophores; the latter are about 1 cm long and the distal head bore four radially elongate sporangia. Aside from its size the most significant feature of the cone lies in its spores; some sporangia contain microspores which range from 85 to 114 μ in diameter, whereas others contain megaspores from 150 to 260 μ in diameter. Some sporangia are reported to contain megaspores at one end and microspores at the other; so far as I am aware such an association within a single sporangium has never been observed in any other plant. Although the specimens on which the original description is based are very well preserved, the sporangium walls are quite delicate and not always intact. It is therefore possible that the apparent occurrence of two kinds of spores in a single sporangium may be the result of some slight distortion and breaking of the tightly compressed sporangia.

Several interesting cones have been assigned to the genus *Palaeostachya* which is distinguished by the alignment of the sporangiophores at an angle of about 45° to the axis and their attachment immediately above the bracts. Whether *Calamostachys* and

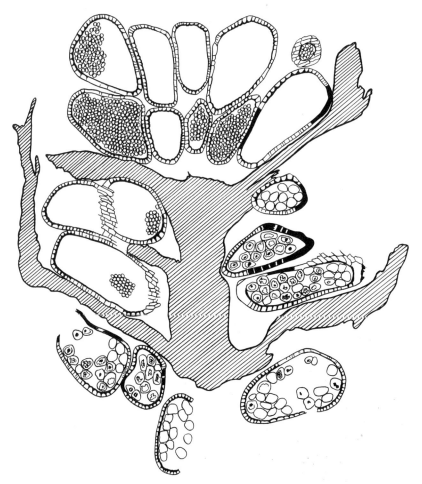

Fig. 9-10. Oblique longitudinal section of *Calamostachys casheana*. (From a slide in the Williamson Collection, British Museum Natural History.)

Palaeostachya should be regarded as two distinct genera, or whether there are really more than two in the assemblage, is debatable.

P. andrewsii Baxter from the Des Moines series of Iowa (Fig. 9-9B) is a cone with 24 bracts in a whorl and these are free down to their point of attachment to the axis; there are usually 12 sporangiophores per whorl, each one terminating in a peltate head with four sporangia. The spores are all the same size but quite large, being 270 to 320 μ in diameter. Although heterospory is not known for certain in the genus, it seems probable that we are dealing with a megasporangiate fragment of a heterosporous cone (the species is based on a specimen from which both apical and basal parts are missing) or possibly one in which microspores and megaspores were produced in separate cones.

P. decacnema Delevoryas from the Des Moines series of Kansas is a somewhat smaller cone with a diameter of about 8 mm and spores that are 45 to 50 μ in diameter, whereas the sporangiophores are attached slightly higher, relative to the associated whorl of bracts.

A much discussed feature of *Palaeostachya* is the course followed by the vascular strand from the cone axis to the sporangiophore. In the British *P. vera* Seward the trace is described as passing up to about the middle of the node and then descending through the cortex before it diverges out into the sporangiophore. A similar course has been observed in *P. andrewsii,* but in *P. decacnema* the trace goes directly to the sporangiophore. It has been suggested that the looped character of the strand, where it does occur in *Palaeostachya,* indicates a "phyletic slide" downward of the sporangiophore from the position that it occupies in *Calamostachys.*

The cones of *Mazostachys pendulata* Kosanke (Fig. 9-11C) are semipetrified fossils which were found in an ironstone concretion in northern Illinois. They are 4 mm in diameter, about 2.6 cm long, and the 12 bracts at each node are fused in the basal portion. Six sporangiophores are attached just below the node and each bears two pendant sporangia. The spores are about 60 μ in diameter. *Annularia sphenophylloides* foliage was found in the same concretion associated with the cones.

Two other cones may be mentioned briefly since they seem to present some similarity to *Mazostachys;* these were described by C. E. Weiss in the books cited in the References at the end of the chapter. *Cingularia typica* Weiss (Fig. 9-11B) had strap-shaped sporangiophores which are attached immediately beneath the whorl of bracts; each sporangiophore is divided at the distal end with two

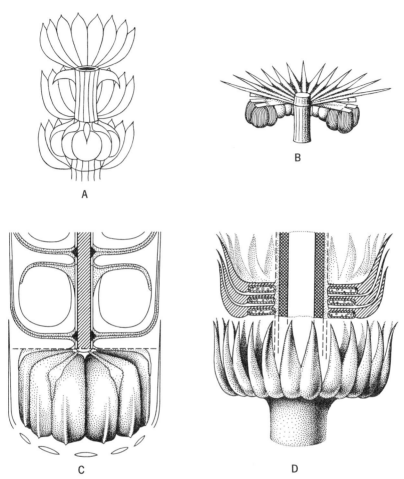

Fig. 9-11. Representative portions of the cones of. A. *Stachannularia tuberculata;* B. *Cingularia typica;* C. *Mazostachys pendulata; Kallostachys scottii.* (A, B from C. E. Weiss, 1876; C from Kosanke; D from Brush and Barghoorn, 1955.)

sporangia borne on the under side of each bifurcation. The sporangia are nearly spherical and about 5 mm in diameter. In *Stachannularia tuberculata* (Sternberg) Weiss the general organization of the bracts and sporangiophores is similar to the above, but here each sporangiophore bears but one pendant sporangium.

In contrast to the cones considered thus far, in *Kallostachys scottii* Brush and Barghoorn (Fig. 9-11D) there is only one kind of appendage which may appropriately be termed a sporophyll. The cone is about 2 cm in diameter and was probably homosporous.

Each whorl consists of 12 units which are fused in their basal portion and the ascending distal lobes are bifurcated. In Fig. 9-11 two whorls are included, one showing the external appearance and the other a longitudinal section. On the under side of the basal part of the sporophyll there are two radially elongate sporangia; since these are borne side by side and closely appressed, only one appears in longitudinal section.

The cone axis has a large central canal or cavity; exarch primary xylem is quite well developed, occasional protoxylem canals can be observed, and some secondary xylem is present. Cortical canals are also present.

The whorled arrangement of the appendages as well as the protoxylary and cortical canals point to the articulates, but the extensive development of exarch primary wood suggests lycopod anatomy. The cone is perhaps best considered tentatively as representing a distinct line of arthrophytes.

Equisetaceae

By comparison with the Carboniferous record the fossil history of the articulates from early Mesozoic times to the present is quite scanty. Numerous specimens, chiefly impressions or casts, of stems have been referred to *Equisetites;* many suggest plants closely related to the modern *Equisetum,* but others are too incomplete or poorly preserved to allow a dependable generic comparison.

Stem impressions of *Equisetites* are distinguished from the calamitean plants partly by the fusion of the leaves into a sheath, and if the assumption that they are closely allied to *Equisetum* is correct, they attained an appreciably greater size than the living species. The stem casts of the Triassic *Equisetites platyodon* Brongniart are about 6 cm in diameter and the stems of *E. columnaris* Brongniart from the Jurassic of Yorkshire reached a comparable size; this is insignificant by calamitean standards, yet appreciably larger than any surviving Equisetums.

A few of the *Equisetites* species have been preserved with cones intact. *E. woodsi* Jones and de Jersey from the Jurassic of Queensland was a plant with stems about 7 mm in diameter and a leaf sheath 11 mm long, of which the free teeth compose the distal 5 mm. Associated cones, which are believed to have been borne on these stems, are rather large, measuring 2.4 cm long and 1 cm in diameter. Several hundred sporangiophores were borne on the cone axis; the head of the appendage was dome-shaped, 1.5 mm in diameter, and three rows of 50 to 60 sporangia hang down from the under side; this arrangement of the sporangia and the high number is quite remarkable. There are a few intriguing suggestions that plants closely akin to *Equisetum* did live in the late Paleozoic. In his flora of Autun and Epinac (France) Renault described a cone under the name of *Bornia radiata* which has been variously referred to *Asterocalamites* and *Archaeocalamites,* but its morphology is more interesting than its taxonomic history. The cones are 5 mm in diameter and 13 to 15 mm long, but since they are only fragments the original length is not known. No sterile appendages are present as in the calamite cones, but in his description Renault notes that the fertile appendages are occasionally interrupted by whorls of leaves; there are eight to ten sporangiophores to a whorl and each apparently was slightly forked at the distal end and bore four sporangia.

In the modern *Equisetum* the outer wall of the spore splits in a distinctive fashion, resulting in four straplike appendages known as elaters which apparently add buoyancy to the spores and allow them to be more readily air-borne. The gametophytes of *Equisetum,* which develop directly from the spores, are usually unisexual; thus, since several spores usually become entangled, another asset of the elaters may be the development of several gametophytes in close proximity to one another. But whatever nature's intended use of the elaters may be they present a unique diagnostic character. It is therefore of interest to note that elater-bearing spores (*Elaterites triferens* Wilson) have been recorded from Des Moines age (Pennsylvanian) rocks of Iowa. Each spore of *Elaterites* bears three elaters. Somewhat comparable spores (*Equisetosporites chinleana* Daugherty) have been found in the Upper Triassic of Arizona which have two elaters.

There is, therefore, enough evidence to indicate that *Equisetum* is a very ancient type reaching back into the early Mesozoic or perhaps earlier; the precise relationship of this modern relict to other articulates is, however, still obscure.

SPHENOPHYLLALES

This is a distinct line of articulates which appears first in the late Devonian and persisted to at least the early Triassic. The stems and leaves are referred to the genus *Sphenophyllum;* the latter are wedge-shaped and arranged in regular verticils, whereas the wood of the stems presents unique features. Of particular interest is a diverse assemblage of cones, generally referred to the genus *Bowmanites*.

The sphenophylls were small plants probably forming a low-growing herbaceous undergrowth in the Carboniferous forests. The largest specimen that I am aware of is one reported by Seward, in the collections of the Austrian Geological Survey; it consists of a stem 85 cm long and 4 mm in diameter which bears a branch 61 cm long. Thus, a meter in height is probably close to a maximum and their slender stems were rarely over 1 cm in diameter.

Stem Anatomy

The wood of *Sphenophyllum* stems (Fig. 9-12) is unlike that present in any other plant group. The center is occupied by a triangular, primary stele of tracheids only and with the small protoxylem cells located at the apices. A considerable amount of secondary wood was formed in the older stems, the tracheids opposite the protoxylary points being conspicuously smaller than those in between.

A common Pennsylvanian species, *S. plurifoliatum* Williamson and Scott, is distinguished by rays in the secondary wood which consist of groups of vertically aligned cells located between the truncated angles of the tracheids. These groups are connected, by narrow horizontal cells, into a unified system. The tracheids of both primary and secondary wood have closely aggregated, bordered pits.

S. insigne Williamson, a rather common British Lower Carboniferous species, has rays of the normal type, that is, they are uniform throughout their radial extent, and the tracheids are scalariform. This species also has protoxylary canals, but these have been reported as occasionally occurring in *S. plurifoliatum.*

A new species (*S. constrictum* Phillips) has recently been described from Indiana and Kansas coal balls in which the cortex and thick, fleshy leaves (which are once forked) are composed of large, thin-walled cells which lends a succulent aspect to the plant.

Sphenophyllum plants seem to have branched rather sparsely but in various ways. In his study of numerous stems from Illinois and Iowa, Baxter described a well-preserved specimen in which three branches were found departing simultaneously from the angles of the primary wood and apparently in between pairs of leaves rather than in their axils; another specimen revealed an apparent dichotomy of the stem.

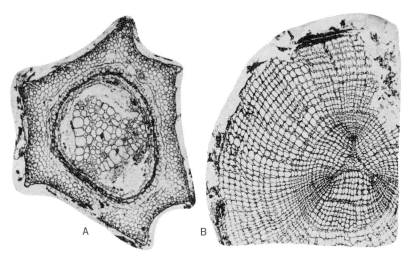

Fig. 9-12. *Sphenophyllum* sp. A. A young stem in which secondary wood had just started to develop, 15X. B. Part of a larger stem showing the distinctive central triangular primary wood and the abundant development of secondary wood, 10X.

Leaves

The leaves are arranged in whorls of six, eight, or nine and are wedge-shaped in their over-all outline (Fig. 9-13). The whorls are flattened in much the same fashion as with *Annularia;* thus, in life they must have stood out at a 90° angle from the stem. In her study of the American fossils Abbott recognized 18 species which vary a good deal in size and the degree of dissection; brief notations on a few of these suggest the range of leaf form in the genus:

S. emarginatum Brongniart. Regularly wedge-shaped, up to 16 mm long and 10 mm wide with a nearly entire (i.e., not dissected) distal margin.

S. cornutum Lesquereux. Up to 2 cm long; divided to about half the length of the leaf into eight or more equal lobes. As in other species a single vein enters the base of the leaf, dichotomizes several times, and a single strand penetrates each lobe.

S. tenerrimum Ettingshausen. Attain a length of 9 mm long and may be dissected in the middle to the base of the leaf, thus dividing it into two parts.

S. thoni Mahr. The distal margin is rounded and finely divided (fimbriate).

Fig. 9-13. *Sphenophyllum* foliage, a specimen from the Upper Carboniferous of Missouri, natural size.

The identification of species of *Sphenophyllum* foliage is actually rather precarious due to the great variation in size, and degree of dissection as well as over-all shape. This may occur within a single specimen; Abbott figures a shoot of *S. cuneifolium* in which the leaves are wedge-shaped with an entire margin on the upper parts of the branches, whereas somewhat lower the leaves are partially dissected and in the basal region they are divided to their base into slender linear segments. As many as eight different names have, in the past, been applied to the various leaf types which are now regarded as falling within the range of this species!

Cones

One of the more significant recent advances in Carboniferous paleobotany is the discovery of numerous cone types, some of which are remarkably well preserved, that have been attributed to the

Sphenophyllales. There is, in fact, such an interesting array of these fossils now known that it is difficult to decide what should be omitted in a summary of this sort.

In a monographic account of many presumed sphenophyllalean cones Hoskins and Cross have recognized 18 species in the genus *Bowmanites*. Some three or four have been found in organic connection with stems bearing typical *Sphenophyllum* foliage, whereas a few others have been found associated with *Sphenophyllum* twigs in a way that leaves little doubt as to where they belong in the plant kingdom. In most species of *Bowmanites* the spore-bearing organs are arranged in distinct cones consisting of whorls of bracts that are usually fused to form a basal disc; associated, immediately above the bract whorl, are sporangiophores with terminally borne sporangia. The degree to which the sporangiophores are actually fused with the bracts varies considerably and some authors have preferred to designate the latter, sporophylls. In view of the diversity in this genus it seems likely that several distinct genera will eventually be recognized.

Bowmanites dawsoni (Williamson) Weiss (Fig. 9-14) is one of the better known species. The cone is about 20 mm in diameter and

Fig. 9-14. *Bowmanites dawsoni,* slightly oblique section of a specimen from Belgium, 4X.

composed of numerous whorls of 14 to 20 bracts each, fused to form a shallow, funnel-shaped basal portion; the free distal limbs extend upwards for a considerable distance. About twice as many slender, terete stalks arise from the upper surface of the "funnel" near its junction with the axis. Each stalk (sporangiophore) bears a single sporangium in an anatropous position; that is, the stalk is reflexed so that the sporangium is directed toward the cone axis. Although there is apparently a single whorl of the sporangiophores associated with each bract whorl, their lengths are so variable that, in a median longitudinal section, the impression is gained of several rings of sporangia. The cone axis has a triangular stele similar to that of *Sphenophyllum*.

Bowmanites bifurcatus Andrews and Mamay (Fig. 9-15A) from the Upper Pennsylvanian of Illinois is noteworthy for its very small size. The cone is only 3 mm in diameter and the single specimen known is approximately 1.5 cm long. Through this length there were 17 whorls of appendages, one of which is shown in Fig. 9-15. Each whorl consists of six bracts which are fused to form a basal disc; the free segments ascend abruptly and each forks, resulting in 12 free tips. A sporangiophore arises from the upper surface of the proximal part of each bract and bifurcates at the distal end, bearing two sporangia.

In spite of its minute dimensions it was possible to determine the detailed vascular anatomy of the cone. The axis has three equidistant wood strands which may be compared with the apices of the primary stele of a *Sphenophyllum* stem. At the node a branch passes out horizontally from each of the three strands and immediately divides into an upper and lower segment. The upper one bifurcates horizontally and each of the two resultant strands supplies a sporangiophore; the lower segment does the same, and each strand forks again to supply a single one to each of the terminal divisions of the bract.

The occurrence of such tiny cones as this causes one to wonder how many others have been missed in routine examinations of coal ball petrifactions. The advantage of the peel technique (Chapter 17) is evident here where, in so small and complex a structure, numerous transverse and longitudinal sections are required to interpret its three-dimensional form.

Bowmanites fertilis (Scott) Hoskins and Cross is a remarkable cone (Fig. 9-15B) from the Coal Measures of England and Belgium; its organization is, in fact, so unique as to render usual morphological terms somewhat inadequate. It attained a length of 6 cm and a diameter of 2.5 cm. Six "appendage complexes" depart at a node, each one consisting of an upper unit that will be called a compound sporangiophore, and a lower pair of branches which may be regarded tentatively as homologous with a bract in the

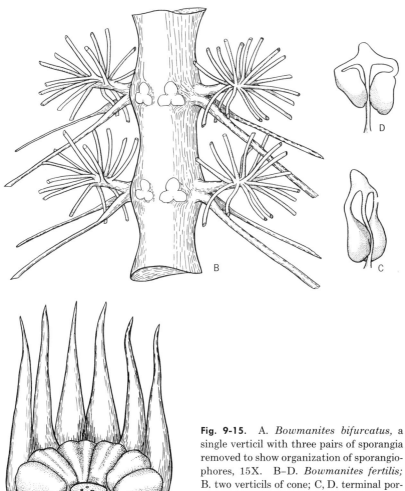

Fig. 9-15. A. *Bowmanites bifurcatus,* a single verticil with three pairs of sporangia removed to show organization of sporangiophores, 15X. B–D. *Bowmanites fertilis;* B. two verticils of cone; C, D. terminal portions of two sporangiophores, each with a pair of sporangia. (A from Andrews and Mamay, 1951; B–D from Leclercq, 1936.)

previously described species. The compound sporangiophore
erupts into a dense cluster of about 16 branches, each of which
terminates in a pair of sporangia (Fig. 9-15C, D). The lower pair
of branches ("bracts") extended downward and outward as long,
slender hornlike structures.

Bowmanites moorei Mamay is a small and much less complex
type. The total length is unknown, but it attained a diameter of
4 mm. There are three bracts or sporophylls at a node, but each
one divides to form a central and two lateral lobes. Two sporangiophores are borne above each median lobe and these terminate in
an inverted sporangium; there are thus six sporangia per whorl.

Litostrobus iowensis Mamay from the Des Moines series of Iowa
is another cone of rather simple construction and quite small (3.5
mm in diameter). At a node there are 12 basally fused bracts and
half as many sporangia, each being borne erect on a stout pedicel;
the latter apparently are oriented directly above every other bract.
The vascular system is poorly preserved, but in the whorled
arrangement of the appendages, basally fused bracts and multiple
of threes, we have characters that point to the sphenophylls.

Sphenostrobus thompsonii Levittan and Barghoorn diverges in
its structure from the other cones described in several ways. It is
about 9 mm in diameter and consists of whorled appendages (16 at
each node) which are termed sporophylls rather than bracts since
the sporangia, also 16 in number, are sessile and attached to the
disc a short distance from the cone axis. The stele differs from
that of the other sphenophyll cones in being tetrarch.

In *Sphenophyllum hauchecornei* (Weiss) Remy from the Westphalian of Germany the close association of the sterile and fertile
appendages of the cone is even more pronounced. Although
described from compression fossils the preservation is good enough
to reveal many critical features. The cone is 1 cm in diameter and
in excess of 12 cm long. The 12 sporophylls at each node are
united in their basal portion; they extend out laterally for 2 or 3
mm, then turn up abruptly and each of the 12 segments divides
into two slender, tapering, hairy lobes. The latter are nearly 3 cm
long and thus overlap five or six of the nodes above. Two or three
sporangia were probably attached by short stalks to the basal
(disc) part of each sporophyll; in a longitudinal view the sporangia
appear in two verticils, one above the other. These cones are
believed to have been borne on stems with the *Sphenophyllum
cuneifolium* type of foliage.

SOME PROBLEMATICAL ARTICULATES

There are a few fossils attributed to the articulate group that are of exceptional morphological interest, so much so as to deserve real consideration in a general account such as this, yet one can do little more than guess at their proper position in the Arthrophyta. *Eviostachya hφegi* Stockmans is based on cones from the Upper Devonian of Belgium which attain a length of 5.5 cm and a diameter of 0.8 cm although most of the specimens are somewhat smaller. Each cone consisted of a peduncle about 2.5 mm long with a single whorl of six bracts below the fertile region; above this are as many as a dozen verticils of sporangiophores. At each whorl there were three pairs of sporangiophores, a pair being derived from each angle of the triangular stele of the cone axis. Figure 9-15 illustrates one sporangiophore: the stalk is trisected at its apex into three branches and each of these divides into two laterals, and a shorter central one; each secondary branch terminates in a cluster of three sporangia, there being a total of 27 on each sporangiophore. There are no associated sterile appendages other than the single basal whorl. The spores range in size from 35 to 55 μ, the plant apparently being homosporous. Although

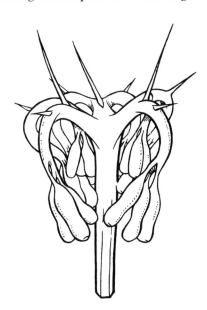

Fig. 9-16. *Eviostachya hφegi,* restoration of a single sporangiophore, about 15X. (From Leclercq, 1957.)

numerous specimens of *Eviostachya* have been found, most of the critical details were worked out by Professor Leclercq from a single specimen that was partially petrified with iron hydroxide. Since the entire sporangiophore is only 2 to 3 mm long it is remarkable that its organization could have been revealed in such precise detail.

In 1897 Scott described *Cheirostrobus pettycurensis* which I believe still holds the record as the most complex of all pterido-phytic sporangiate organs. It was derived from the well-known basal Carboniferous deposits at Pettycur, Scotland, and although only one specimen has ever been found it is well preserved and there is no reason to doubt the accuracy of Scott's restoration.

It is large as well as complicated (Fig. 9-17), measuring 3.5 cm in diameter and consists of whorls of appendages that have been called compound sporophylls. There are 12 such units in a whorl and each divides within a few millimeters of the cone axis, into a lower and upper segment. The lower segment divides horizontally into three slender stalks (bracts) which become laminate toward their distal end and form a conspicuous heel, whereas the ascend-ing limb bifurcates, resulting in a total of six free tips. The upper segment likewise divides horizontally into three slender stalks (sporangiophores), each of which is peltate at the distal end and bears four very elongate sporangia. Each sporangium is 1 cm long and 1 mm in diameter; the cone axis contains a 12-rayed stele.

A third plant may be included here, but it is perhaps well to caution that its position in the Arthrophyta is somewhat less cer-tain than the preceding two. *Prosseria grandis* Read is a compres-sion fossil found at an Upper Devonian horizon in Yates County, New York. The stem fragment that is preserved is 25 mm broad and the leaves are whorled as in the articulates but otherwise quite distinct; they are 6 mm wide, 33 cm long, and were probably attached in groups of three.

Summary Comments

In all probability it will always be difficult to recognize the earliest members of a great plant group; these are the plants in which characters, that will later be diagnostic, are in the formative stage and it is evident that the early Devonian and Silurian pteri-dophytes formed a vast melting pot from which certain clearly defined lines emerged in late Devonian times. Thus, one may be reluctant to accept a plant such as *Protohyenia* as an early arti-culate. Yet, in view of the evidence afforded by *Calamophyton*

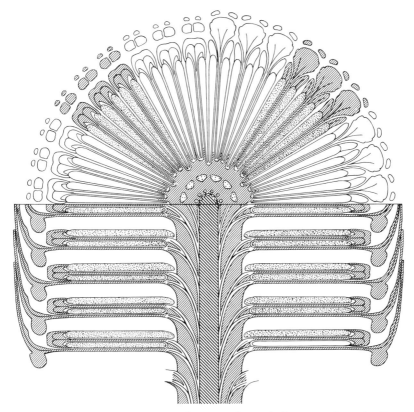

Fig. 9-17. *Cheirostrobus pettycurensis,* a complex articulate cone from the Lower Carboniferous of Scotland, 3X. (From Scott, 1897.)

from the Middle Devonian and Eviostachya from the Upper Devonian it is clear that the origins of the group must be found ultimately in the early Devonian or in the Silurian.

Cheirostrobus is especially perplexing; one almost wishes that it had been found in the Permian instead of at the base of the Carboniferous! We may fall back once again on the explanation that has been used several times, that this is a stray remnant from an upland flora in which articulate evolution had already attained a high degree of complexity.

The gap between any of the Middle Devonian articulates and the dominant lines of the Carboniferous, the calamites and sphenophylls, is also a broad one.

The discovery in the past decade of so many new cone types in the Carboniferous is quite encouraging; they indicate clearly that the Carboniferous forests were by no means monotonous even though the flowering plants were not present to add splashes of color. Yet in spite of these discoveries evolutionary trends are by no means crystal clear. Mamay has suggested a tendency toward increasing simplicity in the evolution of sphenophyll cones; *Cheirostrobus pettycurensis,* the most complex of all pteridophytic fructifications, comes from the early Carboniferous; *Bowmanites fertilis* is a rather complex cone from the English Lower Coal Measures; *Litostrobus,* the simplest sphenophyll cone, is from the Des Moines series; and *Bowmanites bifurcatus* occurs in the McLeansboro formation. Thus he notes:

> Although an absolute correlation between the geologic age and structural complexity of the various other sphenophyllalean fructifications intermediate between the extreme forms cannot be demonstrated, the wide stratigraphical separation seems to suggest a general evolutionary trend toward reduction rather than proliferation. Accordingly *Litostrobus* may best be interpreted as a highly advanced and reduced fructification, rather than a primitive one. (1954, p. 237)

This view seems to me acceptable at least as a working hypothesis, but of course we are left with the problem of how the older, complex sporangiate organs reached that state!

The modern *Equisetum* is certainly closely akin to the calamites yet we would like to know more about its exact relationships with that group. Did the equisetums split off in Carboniferous times as a semiherbaceous line and why are they the only articulates that have been able to survive to the present?

REFERENCES

Abbott, Maxine L. 1958. The American species of *Asterophyllites, Annularia* and *Sphenophyllum. Bull. Amer. Paleont.,* 38: 289–390.

Ananiev, A. R. 1957. New Lower Devonian fossil plants from the southeast of western Siberia. *Akad. Nauk USSR (Botany),* 42: 691–702.

Andrews, Henry N., Jr. 1947. *Ancient plants and the world they lived in.* Comstock Pub. Co., Ithaca. 279 pp.

————. 1952. Some American petrified calamitean stems. *Ann. Missouri Bot. Gard.,* 39: 189–218.

———— and Mamay, S. H. 1951. A new American species of *Bowmanites. Bot. Gaz.,* 113: 158–165.

Arnold, Chester A. 1958. Petrified cones of the genus *Calamostachys* from the Carboniferous of Illinois. *Univ. Michigan, Contrib. Mus. Paleont.,* 14: 149–165.

Baxter, Robert W. 1950. *Peltastrobus reedae:* a new sphenopsid cone from the Pennsylvanian of Indiana. *Bot. Gaz.,* 112: 174–182.

————. 1955. *Palaeostachya andrewsii,* a new species of calamitean cone from the American Carboniferous. *Amer. Journ. Bot.,* 42: 342–351.

Brush, Grace S., and Barghoorn, E. S. 1955. *Kallostachys scotti:* a new genus of sphenopsid cones from the Carboniferous. *Phytomorphology,* 5: 346–356.

Daugherty, Lyman H., and Stagner, Howard R. 1941. The Upper Triassic flora of Arizona with a discussion of its geologic occurrence. *Carnegie Inst. Washington Pub.,* 526: 1–108.

Delevoryas, Theodore. 1955. A *Palaeostachya* from the Pennsylvanian of Kansas. *Amer. Journ. Bot.,* 42: 481–488.

Hoskins, John H., and Cross, Aureal T. 1943. Monograph of the Paleozoic cone genus *Bowmanites* (Sphenophyllales). *Amer. Mid. Nat.,* 30: 113–163.

Jones, O. A., and de Jersey, N. J. 1947. Fertile Equisetales and other plants from the Brighton beds. Univ. Queensland Papers, Dept. Geology 3 (4): 1–16.

Kosanke, Robert M. 1955. *Mazostachys—*a new calamite fructification. *Illinois State Geol. Surv., Rept. Investigations,* No. 180: 1–22.

Lacey, William S. 1943. The sporangiophore of *Calamostachys. New Phyt.,* 42: 1–4.

Leclercq, Suzanne. 1936. A propos de *Sphenophyllum fertile* Scott. *Ann. Soc. Géol. Belgique,* 60: 170–172.

————. 1940. Contribution à l'étude de la flore du devonien de Belgique. *Acad. Roy. Belgique,* ser. 2, 12: 1–65.

————. 1957. Étude d'une fructification de Sphenopside à structure conservée du devonien supérieur. *Acad. Roy. Belgique,* ser. 2, 14: 1–39.

————, and Andrews, H. N., Jr. 1960. *Calamophyton bicephalum,* a new species from the Middle Devonian of Belgium. *Ann. Missouri Bot. Gard.* 47: 1–23.

Levittan, Edwin D., and Barghoorn, Elso S. 1948. *Sphenostrobus thompsonii:* a new genus of the Sphenophyllales? *Amer. Journ. Bot.,* 35: 350–358.

Mamay, Sergius H. 1954. A new sphenopsid cone from Iowa. *Ann. Bot.,* n.s., 18: 229–239.

————. 1959. A new bowmanitean fructification from the Pennsylvanian of Kansas. *Amer. Journ. Bot.* 46: 530–536.

Phillips, Tom L. 1959. A new sphenophyllalean shoot system from the Pennsylvanian. *Ann. Missouri Bot. Gard.* 46: 1–17.

Read, Charles B. 1953. *Prosseria grandis,* a new genus and new species from the Upper Devonian of New York. *Washington Acad. Sci.,* 43: 13–16.

Reed, Fredda D. 1952. *Arthroxylon,* a redefined genus of Calamites. *Ann. Missouri Bot., Gard.* 39: 173–187.

Remy, Winfried. 1955. Untersuchung von kohlig erhaltenen fertilen und sterilen Sphenophyllen und Formen unsucherer systematischer Stellung. *Abh. Deutschen Akad. Wissen.,* 1: 5–40.

Renault, Bernard. 1893, 1896. Bassin Houiller et Permien d'Autun et d'Epinac. Études des Gites Mineraux de la France. Paris. Atlas, 1893; Text, 1896.

Schultes, Richard E., and Dorf, Erling. 1938. A sphenopsid from the Lower Devonian of Wyoming. *Harvard Univ. Bot. Mus. Leaf.,* 7: 21–33.

Scott, Dunkinfield H. 1897. On *Cheirostrobus,* a new type of fossil cone from the Lower Carboniferous strata (Calciferous Sandstone series). *Phil. Trans. Roy. Soc. London,* 189: 1–34.

Seward, Albert C. 1898. *Fossil Plants.* Vol. I. Cambridge Univ. Press. pp. 1–452.

Thomas, H. Hamshaw. 1909. On a cone of *Calamostachys binneyana* (Carruthers) attached to a leafy shoot. *New Phyt.,* 8: 249–260.

————. 1911. On the leaves of Calamites (*Calamocladus* section). *Phil. Trans. Roy. Soc. London,* 202B: 51–92.

Walton, J. 1949. On some Lower Carboniferous Equisetineae from the Clyde area. *Trans. Roy. Soc. Edinburgh,* 61: 729–736.

Weiss, Christian E. 1876–1884. Beiträge zue fossilen Flora. Steinkohlen-Calamarien mit besonderer Berucksichtigung iher Fructificationen. Abh. geol. Specialkarte Preuss. Thüring. Staaten. Berlin. I. (1876), pp. 1–149; II. (1884), pp. 1–204.

Wilson, Leonard R. 1943. Elater-bearing spores from the Pennsylvanian strata of Iowa. *Amer. Mid. Nat.,* 30: 518–523.

10

the CYCADOPHYTA

The term *Cycadophyta* has been used to include three groups of gymnospermous plants, the Pteridospermophyta, the Cycadales, and the Bennettitales. The first was dealt with in Chapter 5 where it was pointed out that it is a large and varied assemblage of plants which in itself encompasses several distinct lines. It is not my intention to discredit the concept that the Cycadales and Bennettitales may have originated somewhere in the seed-fern complex, but there is actually a very broad gap between the pteridosperms and the other two orders, and we are far from understanding just how it may have been bridged. A consideration of the reproductive organs of the latter will make this apparent.

The problem of classification is further complicated by the fact that the seeds and microsporangiate organs of the Cycadales and Bennettitales are quite different and it is now known that, in spite of a gross similarity in the morphology of the foliage of the two, the structure and distribution of the stomates are not closely comparable. In his study of the Scoresby Sound fossils Harris, who has probably contributed more to our knowledge of the cycadophytes than any other recent worker, concludes that the Cycadales and Bennettitales arc probably quite remote from one another. I am inclined to feel that as evidence accumulates it will become apparent that this is the case and that we have been misled, largely by a superficial resemblance of the foliage. To obviate any misunderstanding the following terminology will be used here: the Cycadophyta are regarded tentatively as including two orders, the Cycadales and Bennettitales. The word "cycadophyte" will be applied to any plant in the entire assemblage. In the literature on these plants it has been common practice to call the Cycadales simply "cycads" and the Bennettitales "fossil cycads." This has resulted in a certain amount of confusion since the Cycadales are represented in the fossil record as far back as mid-Mesozoic times. Common names are useful if not unavoidable and the term "bennettite," rather than "fossil cycad," will be used here in reference to bennettitalean fossils.

Foliage that is probably referable to the Bennettitales makes its appearance in Permian rocks and the group apparently became extinct sometime during the Cretaceous. The Cycadales are known from the Triassic to the present; nine genera, with about 90 species, now live in the tropical and subtropical parts of the Americas, Asia, Australia, and South Africa. It is perhaps one of the vagaries of preservation that we know less about the general habit of such fossil Cycadales as have been found than we do about the Bennettitales.

Since study material is not always as readily available as for most of the extant groups described in previous chapters, it may be helpful to consider briefly a living cycad.

Zamia Floridana—a Living Cycad

Zamia floridana grows in central and southern Florida and several other species are found in the West Indies and South America. The stem of the mature plant (Fig. 10-1) is a rather massive and irregularly branching structure and only the apical portion appears above ground; the root system is sparsely branched but penetrates to a considerable depth. Specimens are not infrequently encountered in the Florida woods which branch much more profusely than the one illustrated; a cluster of a dozen or so leaves is found at the tip of each branch, thus several crowns of leaves in close proximity to each other usually indicate a single plant. The leaves are once pinnate.

Zamia is dioecious, that is, a plant may bear seed cones or microsporangiate ones but not both. The seed cones are stout barrel-shaped structures up to about 12 cm. long; the central axis bears a dozen vertical rows of appendages, each of which (Fig. 10-1B) consists of a stalk and a hexagonal head from which two seeds are suspended toward the axis. The microsporangiate cones are essentially the same in their general organization but are more slender and elongate; on the under side of each appendage (microsporophyll) there are 20 or 30 small spherical sporangia. (Fig. 10-1B).

In its cone and leaf structure *Zamia* is a representative cycad although in many of the other species these organs may be considerable larger. The stem structure is somewhat less typical, for in most of the living cycads it is an upright, sparsely branched, columnar trunk covered by a rough armor of the persistent basal portions of the petioles. A maximum size is attained by certain

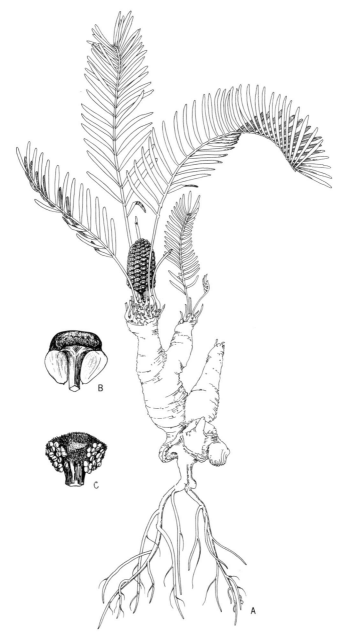

Fig. 10-1. A. A plant of *Zamia floridana;* one branch is shown bearing a seed cone, about ⅙X; B. a single seed-cone appendage (megasporophyll) with its two seeds; C. a single microsporophyll from the underside showing numerous sporangia. (From Andrews, 1947.)

species of *Cycas*, in the East Indies, in which the trunks reach a height of 23 meters and a diameter of nearly 1 meter. *Cycas* is represented by several species in the Asiatic-Australian region and is commonly cultivated in the extreme southern parts of the United States. It diverges notably from the other cycads in that the seed-appendages (megasporophylls) are not aggregated into a cone; some bear four or five seeds and may be pinnately divided, thus presenting a slight resemblance to the leaves.

CYCADOPHYTE FOLIAGE

Foliage that is usually attributed to the cycadophytes is found from the Permian to the present but is especially abundant in Triassic and Jurassic rocks. The leaves vary from a few centimeters long to well over 1 meter; for the most part they are unbranched and once pinnate, although some are entire or nearly so. Beyond these generalities they present one of the most difficult problems of identification in the whole field of paleobotany. Dozens of genera and scores of species of leaves have been described and the names of many have been changed several times in accordance with the views of different investigators. It was not until the discovery was made that cuticular structures were more dependable than the gross morphology that identifications began to take on some real biological significance. Comprehensive studies of the cuticle structure of these plants was initiated by Nathorst in the early years of the century, were greatly advanced by Thomas and Bancroft in 1913, and have been followed by several others in more recent years, one of the most comprehensive and useful being Harris' discussion in Part 2 of his Scoresby Sound flora—a *vade mecum* for so many paleobotanical topics.

Many of the cycadophytes seem to have had tough, leathery foliage with a heavy cuticular coating as do the living cycads; consequently, preservation of the epidermal cell-wall structure is occasionally quite good. With specimens in which these features are not preserved it may be quite impossible to determine whether the foliage belongs to a cycad or a bennettite, or whether it even lies within the limits of these two.

A few examples of cycadophyte foliage are briefly considered next with the intent of showing: (a) the difficulty of separating cycads from bennettites on the basis of gross morphology of the leaves; (b) the distinctive microscopic, epidermal details which seem to offer a more dependable way of delimiting the two.

General morphology of foliage

A few examples of cycad and bennettite leaves, or representative portions of them, are shown in the upper part of Fig. 10-2 (see also Figs. 10-3A, 10-11, and 14-2B, C). It will be evident from this small selection that it would be difficult to divide them into two groups on the basis of gross form. There is actually much more variety known than is indicated and the differences in size are tremendous.

Cycadean (Cycadales) Stomatal Structure

The extant *Zamia muricata* is taken as an example. The accompanying figures (10-2) show: a stomate in surface view (K); a section cut through the center and at right angles to the long axis (L); and a section cut lengthwise through one of the guard cells (J). In Fig. K the guard cells (*g*) are surrounded by a ring of somewhat irregularly shaped *subsidiary* cells (*s*) and it has been shown that these originate from independent mother cells and not from the guard mother cell. The guard cells are somewhat sunken and in the section cut parallel to the long axis (J) it is evident that the central part (which is associated with the actual opening) is more deeply depressed than the ends (poles) which turn up abruptly; these appear at *g'* in Fig. K. The outer surface of the guard cells is covered with a cuticle of quite even thickness.

Bennettitalean Stomatal Structure

In leaves of the Bennttitales there are two lateral subsidiary cells associated with each stomate (Fig. 10-2G, H). These are believed to have originated from the same mother cell that pro duced the guard cells; presumably the mother cell divided to form two, and each of these divided again in the same plane, the inner pair of the four becoming the guard cells and the outer two the subsidiary cells. The epidermal cells immediately adjacent to the poles are not modified in any special way. A second distinctive feature of the bennettitalean cuticle is the unique thickening of the outer and dorsal walls of the guard cells (following the terminology of Harris, the term ventral wall is applied to the one facing the aperture (stoma) and the dorsal wall is the one opposite to it).

The stomata in the Bennettitales are in general oriented transversely and the epidermal cell walls are conspicuously sinuous; in the Cycadales, and other pinnate-leaved fossil gymnosperms, the

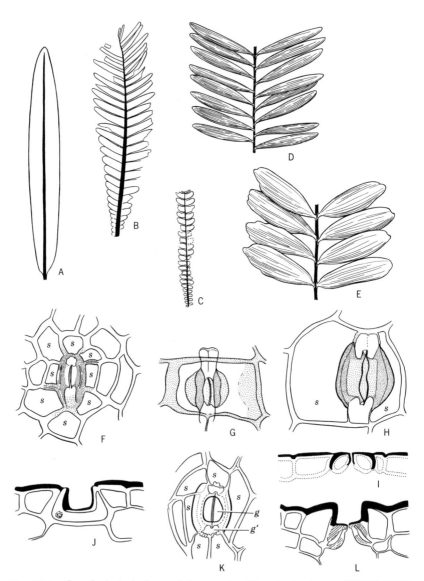

Fig. 10-2. Cycadophyte leaf morphology and epidermis structure. CYCADALES: D, E. portions of the leaves of *Zamia intergrifolia* and *Z. furfuracea,* respectively (both extant). J, K, L. *Zamia muricata* (extant), all 350X; J. median longitudinal section through guard cell; K. surface view; L. transverse section through middle of stomate. F. Surface view of stomate and surrounding cells of *Pseudoctenis spectabilis,* 350X. BENNETTITALES: A. *Nilssoniopteris vittata.* B, G. *Pterophyllum aster-tense;* B. leaf; G. stomate and subsidiary cells. C, H, I. *Pterophyllum rosenkrantzi;* C. leaf; H. stoma and subsidiary cells, 900X; I. median transverse section through same. *s,* subsidiary cells; *g,* guard cells; *g',* pole of guard cell. (All except A, D, E from Harris, 1932.)

294

stomata tend to be irregularly or longitudinally oriented and the epidermal cell walls are straight.

As might be expected there is a good deal of variation within the two groups. For example, in the cycads the guard cells may be depressed below the surface to varying degrees, whereas the subsidiary cells or others peripheral to them arch over the guard cells. The latter are quite deeply sunken in the living *Cycas revoluta*.

THE CYCADALES

Although the Cycadales have existed from Triassic times, or possibly earlier, well-preserved reproductive organs are scarce. One of the most informative and interesting suites of fossils referable to the order are leaves, seed organs, and microsporangiate cones that have been described under the generic names *Nilssonia, Beania,* and *Androstrobus* (Fig. 10-3). They are closely associated in the mid-Jurassic Cayton Bay beds on the Yorkshire coast. The foliage, *Nilssonia compta* Phillips, is one of the most abundant elements in the flora (Fig. 10-3A); specimens attain a length of 40 cm and consist of a strong rachis with a lamina that is divided into truncate segments which vary a good deal in size. *Beania gracilis* Carruthers is a loose "cone" with an axis up to 10 cm long and spirally borne appendages consisting of a stalk about 2 cm long which expands at the distal extremity to form a broadly ovate, recurved head with two seeds on its inner surface. The largest seeds measure about 16 × 13 mm and are well enough preserved so that the two layers of the integument can be recognized (Fig. 10-3C): an outer (possibly fibrous) zone and an inner stony one which enclosed a vascularized nucellus. The male cone, *Androstrobus manis* Harris, is about 5 cm long and 2 cm in diameter. On the under side of each of the numerous, spirally disposed sporophylls are several scores of cylindrical sporangia (Fig. 10-3E).

Androstrobus is closely comparable to the microsporangiate cones of the living cycads, and if one were to stretch out the axis of a *Zamia* seed cone the result would be something nearly identical to *Beania*. Moreover, the stomatal structure of *Androstrobus* and *Beania,* as well as the *Nilssonia* foliage, are sufficiently alike as to leave little doubt that they represent organs of a single species and that they are cycadean. Combined with their close association at certain spots along the outcrop on the Cayton Bay beach, the case seems to be a good one.

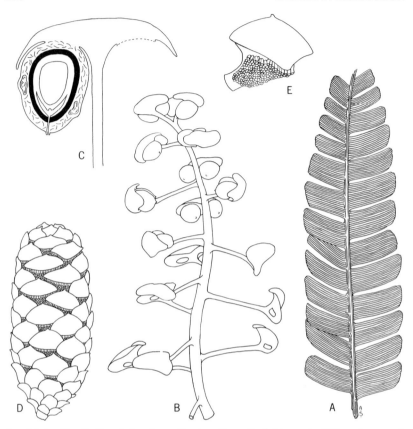

Fig. 10-3. Fossil Cycadales from Cayton Bay, Yorkshire. A. *Nilssonia compta.*
B, C. *Beania gracilis;* B. infructescence or seed "cone," reduced slightly; C. a single
megasporophyll with one seed, restored, in longitudinal section. D, E. *Androstrobus
manis;* D. the microsporangiate cone, about natural size; E. a single sporophyll.
(B–E from Harris, 1941.)

The mere association of fertile and vegetative parts of plants
that are suspected of belonging together has long been known to
constitute dubious evidence and has been the subject of much
criticism. In beds where large numbers of plants are found, many
of which are obviously unrelated, the evidence from association
may be worthless. The distribution of fossil plants at Cayon Bay
is, however, not an ordinary one, for distinct assemblages may be
encountered every few yards. On two visits to this memorable
locality I gained the impression that the association patterns of
the plants cannot be the result of indiscriminate deposition; others
have arrived at the same conclusion, including Dr. H. H. Thomas

whose several seasons of collecting led to the recognition of the
Caytoniales.

THE BENNETTITALES

Petrified cycadophyte trunks, most of which are probably referable to this order, have been found in many Mesozoic horizons and
are widely scattered geographically.   They are unique as paleobotanical specimens and have long been prized by collectors.   In
fact, a trunk of *Cycadeoidea etrusca* has the distinction of being
the oldest known fossil plant specimen collected by man.   According to Wieland it was "Placed with vases and other objects of
superstitious reverence on one of the supulchral chambers of the
ancient Necropolis at Marzabotto by the Etruscans more than
four thousand years ago" (1906, p. 12).   It was rediscovered about
a century ago and given the binomial indicated above by Capellini
and Solms.

Many trunks have been found in Jurassic rocks of the Freezeout
Hills of Wyoming, smaller numbers have come from the Potomac
formation of Maryland, from other Mesozoic horizons in Kansas,
Colorado, Prince Edward Island, and several other North American areas.   Numerous specimens have also been found on the Isles
of Wight and Portland off the south coast of England and a few
have turned up in France, Belgium, the USSR, Poland, Germany,
and India.

The most prolific collections by far, however, have been made in
the Black Hills of South Dakota and are of early Cretaceous age.
The silicified trunks from this area were first brought to the attention of botanists in 1893 and although many specimens are said to
have been damaged or destroyed through the depredations of
curiosity hunters, considerable numbers were acquired and preserved through the efforts of T. H. MacBride of the State University of Iowa, by Lester F. Ward and O. C. Marsh, and particularly
G. R. Wieland whose vast collections are deposited at Yale
University.

For the most part these petrified trunks are heavy, bulky specimens, hard and quite difficult to prepare for study.   Many of them
make attractive and unique museum specimens and there seems to
have been a general reluctance to cut them up sufficiently to allow
comprehensive study.   The result is that few have actually been
investigated in their entirety.

Fig. 10-4. *Cycadeoidea marshiana,* shown from above. (From Wieland, 1906.)

The Black Hills area was rich in species as is evinced by the variety in trunk shape and reproductive organs. The trunks may be of upright columnar form, of conical shape, and under a meter in height, or of a low growing and profusely branching habit (Fig. 10-4). These last are particularly striking and give the effect of several pineapples aggregated into a dense bunch. It is estimated that the columnar *Cycadeoidea jenneyana* may have attained a height of several meters, but most of them were appreciably shorter. Numerous illustrations of these unique plants are given in Wieland's *American Fossil Cycads* and in Ward's *Status of the Mesozoic Floras.*

Some of the silicified trunks have been found with partially developed leaves still intact, so that one may readily visualize them as they lived on the Cretaceous landscapes with a crown, or several crowns in the case of the branching species, of pinnate leaves presenting an appearance much like that of some modern cycads. In the organization of their reproductive organs they were, however, very different. As an introduction to these bizarre and often well-preserved fossils, *Cycadeoidea dacotensis* will be taken as an example (Fig. 10-5).

Note. A brief comment may be in order concerning the terminology that is used here and elsewhere. The term *cone* or *strobilus* is certainly applied to far too many spore and seed-bearing organs in which the comparative morphology is often obscure. The term *flower* has been used rather generally for the reproductive organs of gymnospermous plants as well as the angiosperms (Anthophyta);

Fig. 10-5. *Cycadeoidea dacotensis.* Semidiagrammatic sketch of a flower in longitudinal section; *h,* hairy bract; *m,* expanded microsporophyll (one at the left is shown in immature condition); *c,* central axis with numerous seeds and interseminal scales. (From Wieland, 1906.)

although it seems to me preferable to reserve this for the latter group its usage is rather well established in bennettitalean literature. One is torn between a choice of setting up an almost endless series of terms to apply to spore, pollen, and seed organs in the various groups or to hold to a bare minimum. The latter course is perhaps preferable so long as the plant structures involved are understood.

Unlike the living cycads in which the cones are borne at the apex of the stem, they appear as rosettelike bodies in the cycadeoids, scattered among the leaf bases; they are generally much smaller than those of the cycads and in some species prodigious numbers were formed—up to several hundred on a plant.

The flower itself presents some similarity to a magnolia in its general organization, and, although some botanists have chosen to see an evolutionary sequence here, a close comparison between the two leaves a rather great gap. However, in the bennettites as in the magnolias there is a basal aggregation of microsporophylls surmounted by a receptacle that may be flattened, dome-shaped or conical, bearing on its outer surface seeds and interseminal scales.

The flower of *C. dacotensis* is, therefore, described as being bisporangiate in that it includes both microsporangiate and seed parts. It is 5 to 10 cm in diameter and 6 cm long, being borne on a peduncle of comparable length. The upper two-thirds of the peduncle bore a close spiral succession of 100 to 150 elongate hairy bracts which enclosed the young reproductive organ.

There are hundreds of interseminal scales on the receptable of *C. dacotensis,* with seeds distributed among them except in the basal and apical regions. The seeds of the cycadeoids are generally quite small, being little more than 1 mm long in this species. The structure and relationships of the seeds and associated scales are more clearly shown in Fig. 10-6 which is a median section through the flower of *C. wielandi;* it differs from *C. dacotensis* in the shortened receptacle and greatly elongate stalks of the seeds and interseminal scales.

The microsporangiate portion of *C. dacotensis* is composed of 19 or 20 once-pinnate appendages with their bases fused to form a disc. In the restoration (Fig. 10-5) the one on the left is shown in an immature condition and the one on the right fully expanded. Two rows of complex synangia are arranged along each primary branch. Each synangium is a flattened, tear-shaped organ attached by a short, stout stalk; a single one is shown at a much higher enlargement in Fig. 10-7. The outer wall is a single layer of large,

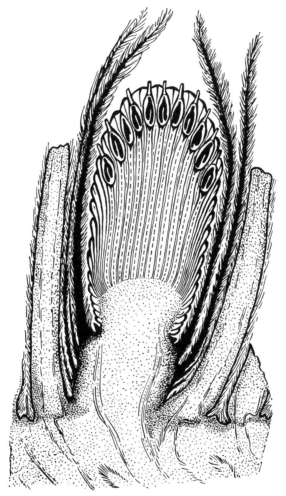

Fig. 10-6. *Cycadeoidea wielandi.* Longitudinal section through a flower. (From Wieland, 1906.)

thick-walled cells; within this is a thin zone of parenchymatous cells followed by a ring of sporangia extending around the periphery. These dehisced longitudinally emptying the spores into the center of the synangium which in turn opened at maturity.

At first glance the cycadeoids appear to have been exceptionally prolific seed producers. The occurrence of several scores of strobili on a trunk is not uncommon and, as an extreme case, some 500 to 600 were found on a single trunk of *Cycadeoidea dartoni* Wieland. Since the strobili are numerous when present at all and remnants

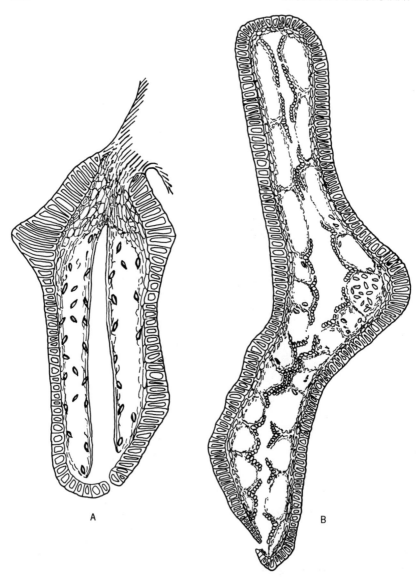

Fig. 10-7. *Cycadeoidea dacotensis.* A. Longitudinal section through a synangium; B. transverse section. (From Wieland, 1906.)

of older ones have not been observed among the leaf bases, Wieland concluded that the plants probably were monocarpic, that is, they fruited only once and died shortly thereafter. The evidence does suggest that an individual plant fruited heavily and at long intervals, but whether only once in its life is uncertain.

It is possible to compute the age of the living cycads by dividing the number of leaf bases on the trunk by the number of leaves formed each year; allowing for some variation in the number put out each year, and the fact that fewer were formed in the crown in the earlier life of the plant, at least an approximate age may be computed. In his book on the living cycads Chamberlain figures a Mexican specimen of *Dioon edule* 5 feet in height that is approximately 1000 years old. Assuming that the growth of the cycadeoids went on at about the same rate as the modern cycads one may estimate an age of several centuries for the larger trunks. It is a little difficult to believe that they would have fruited but once over so long a life span.

The morphology of the unique interseminal scales has perplexed most serious students of the Bennettitales and in his study of the Scoresby Sound fossils Harris encountered several fragmentary strobili (Fig. 10-8) which offer a possible clue. *Vardekloeftia conica* Harris differs from other bennettitalean seeds in the presence of an additional integument or cupule, and in *Bennetticarpus crossospermus* Harris the micropyle of the seed is surrounded by a "micropylar plate" which may represent the distal portion of such an outer integument. Harris points out that the micropylar plate is closely comparable with the adjoining interseminal scales and notes that this "supports the view . . . that the interseminal scales are homologous with seeds and are in fact formed by the diverted development of seed initials" (1932, p. 116).

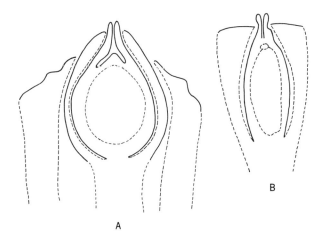

B

A

Fig. 10-8. Reconstructions in longitudinal section of bennettitalean reproductive organs from East Greenland; A. *Vardekloeftia conica;* B. *Bennetticarpus crossospermus.* (From Harris, 1932.)

We will return once again to the beach at Cayton Bay to consider a bennettite with reproductive organs that are basically comparable with those of some of the Dakota fossils although the general habit of the plant was quite different. The flowers, known as *Williamsoniella coronata* Thomas (Fig. 10-9A), were probably borne in the axils of leaves that have long been known under the binomial *Taeniopteris vittata,* but more recently have been given the revised name *Nilssoniopteris vittata* (Brongniart) Florin (Fig. 10-2C).

The flower, borne on a peduncle about 3.5 cm long, is bisporangiate; a ring of 12 or 14 separate microsporophylls is attached at the base of a central column on which the seeds and interseminal scales are arranged. The entire organism was enclosed by a perianth of hairy bracts like those of the cycadeoids considered previously.

Each microsporophyll is shaped like an orange segment with fingerlike branches in the central portion (facing the axis) which partially enclose two pairs of *capsules.* Each capsule is two-valved and was probably constructed of several sporangia; this term is suggested by Harris (who has revised the original description of Thomas) as preferable to *synangium* since it avoids the implication that they evolved by a fusion of separate sporangia. The microsporophylls were separate from one another and probably were shed following spore dispersal; they are frequently found isolated in the rock.

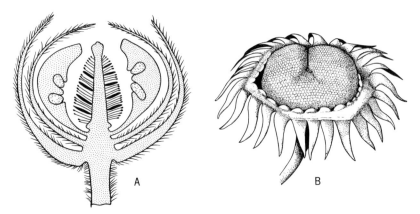

Fig. 10-9. A. A flower of *Williamsoniella coronata* showing: the central, elongate column bearing seeds (in solid black) and interseminal scales (with bulbous head); two microsporophylls; the enclosing hairy bracts, about 3X. B. A flower of *Sturiella langeri* enlarged several times. (A from Harris, 1944; B from Kräusel, 1948.)

The gynoecial (seed) axis is only about 10 mm long, but bore some 300 minute ovules interspersed among 1200 interseminal scales, the latter being characterized by a very slender stalk and a terminal bulbous head. The abundance of the *Nilssoniopteris* leaves in the Cayton Bay beds implies that the plant was well represented in this mid-Jurassic flora. They are entire, lanceolate to narrowly ovate, and virtually all sizes may be found up to 20 cm long and 3 cm wide. Fragments of slender stems have been found associated with the leaves. Thus in the organization of the vegetative organs the plant was quite different from the Dakota cycadeoids with their massive stems and pinnate foliage.

Some years ago Nathorst described a delicately branching bennettite plant, *Wielandiella angustifolia,* from the Rhaetic of Scania, in southern Sweden. In its general habit it was probably quite like *Williamsoniella* and certainly was in contrast to the massive-stemmed cycadeoids. The stems (Fig. 10-13) for the most part do not exceed 1.5 cm in diameter and divided by what is referred to as a false dichotomy; apical growth was apparently terminated by flower formation and two nearly equal branches developed as shown in the restoration. Since it is not possible to determine the exact mechanics of branching in a plant that has been dead for 150 million years, it is hardly advisable to distinguish dogmatically between a dichotomy and a false dichotomy.

Leaves, about 8 cm long, classified as *Anomozamites minor* are believed to have been borne on the stems. Such leaves are abundant in the same beds in which the stems are found, and there is a close comparison between their winged petioles and the bracts which enclosed the flower. The latter are not especially well preserved; the structure of the microsporangiate organs is obscure while the central ovulate axis bore numerous small seeds and ovuliferous scales as in other bennettitalean flowers. The plant is of particular interest for the slender, forking stems and the tendency of the leaves to be concentrated in the area of branching.

Another bennettitalean plant of apparently small and delicate dimensions, *Sturiella langeri* Kräusel, comes from the Triassic of Lunz, Austria. It is known from an axis 4 cm long; slender side branches terminate in a "cone" (Fig. 10-9B) that bears a superficial resemblance to a miniature sunflower head. The distal end of the branch flares out into a shallow cup, and arranged around the outer edge are 25 to 30 raylike lobes; a sporangium (or possibly two that are coalesced) is attached to the inner edge opposite the base of each ray. The central part of the organ appears to have been ovulate, possibly similar to that of *Williamsoniella* but much shortened; there is, however, very little structure preserved.

The diversity of the Bennettitales is emphasized by the varied types of microsporangiate organs that have been referred to the group. Some of these almost certainly represent distinct male organs, the corresponding female or seed axis having been borne on another part of the plant or possibly a separate plant. Although some of them are exceptionally interesting themselves, much remains to be learned about the general habit of the plants and the manner in which the reproductive organs were borne.

The genus *Williamsonia* includes several Jurassic species based on male flowers. They are rather large structures with a cup-shaped base opening into numerous flaring lobes, with the sporangiate organs proper arranged along the upper (inner) surface. *Williamsonia spectabilis* Nathorst (Figs. 10-10, 10-11) is a decidedly striking fossil that measured about 9 cm across the greatest diameter; the basal cup is 3 cm in diameter and expanded into about a dozen segments which tended to bend out horizontally and then turn abruptly upward, probably arching over the center of the cup in life. Numerous slender branches are arranged on the inner side of each sporophyll which in turn bear two rows of synangia. The latter probably were similar in organization to the synangia of *C. dacotensis;* they are divided into several chambers, but spores have not actually been found in them.

Williamsonia whitbiensis Nathorst (Fig. 10-10, B, C) is of similar size and shape although the stalk, a conspicuous feature of *W. spectabilis,* is not present, and the synangia are borne in two rows directly on the inner surface of the sporophylls and not on branches arising from them. Another variant in bennettite microsporophyll morphology has been described by Sitholey and Bose as *W. santalensis,* which is known from incomplete specimens found in the Jurassic Rajmahal series of Bihar, India. The authors caution that their restoration (Fig. 10-10D) is tentative, but the several specimens that they figure seem to substantiate it. The basal cup or disc is perhaps somewhat more shallow than the two Yorkshire species described previously, but the most novel feature of this Indian species lies in the apparent forking of each microsporophyll into a blunt terminal portion and a narrower side branch which bears two rows of tapering fingerlike appendages. Each one includes two rows of chambers which in all probability contained spores although they were not preserved. It is thus not certain whether the unique appendages should be regarded as synangia or whether they enclosed the synangia.

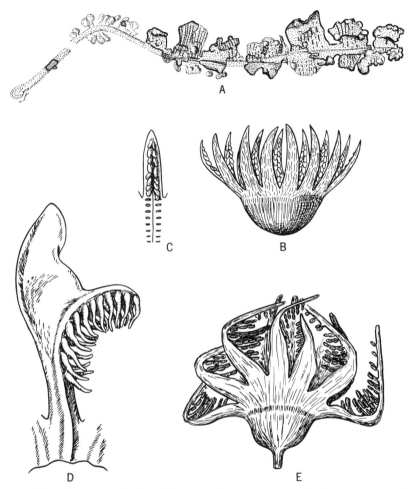

Fig. 10-10. Some bennettitalean microsporangiate organs. A. A single microsporophyll of *Wonnacottia crispa.* B, C. *Williamsonia whitbiensis;* B. restoration of the male flower; C. a single sporophyll showing the two rows of synangia. D. *Williamsonia santalensis,* reconstruction of a microsporophyll. E. A flower of *Williamsonia spectabilis.* (A from Harris, 1942; B, C from Nathorst, 1911; D from Sitholey and Bose, 1953; E from Thomas, 1913.)

In contrast to the more or less funnel-shaped organs considered above, in which the microsporophylls are united basally, *Wonnacottia crispa* Harris (Fig. 10-10A) is a very different type of apparent spore-bearing organ. It was discovered several years ago in lower Middle Jurassic deposits of Yorkshire and is essentially leaflike, being about 12 cm long and 1.5 cm broad with nearly opposite segments (pinnules) which tend to taper outward slightly.

Fig. 10-11. A fine specimen of *Williamsonia spectabilis* from Whitby, England, about ⅔X. (Swedish Natural History Museum.)

Numerous globose bodies containing pollen are partially sunken in the under surface and the proximal segments are considerably reduced, consisting of little more than the pollen-containing bodies. Typical bennettitalean stomata clearly point to the general affinities of the fossil. Although originally described as a sporophyll, and presumably a very primitive one, Professor Harris has informed me that it is possible that this may actually be a leaf of *Anomozamites nilssoni* that has been heavily attacked by a gall mite and that the pollen has simply drifted into the pouches. It seems unlikely to this writer that so many pollen would drift into a gall cavity (as many as 100 have been found), but if it is in fact a leaf and was borne immediately below a flower, this may be the case. *Sagenopteris* leaves, according to Harris, are sometimes found literally covered with *Caytonanthus* pollen grains.

A *Williamsonia* (*W. sewardiana*) from Jurassic rocks of the Rajmahal Hills, India, affords the best evidence we have as to the general habit of these plants. The restoration prepared by Professor Sahni (Fig. 10-12) is based on fertile branches that he described in 1932 and trunk specimens that had been recorded previously under the name *Bucklandia indica.*

The branch shown on the left side of the trunk terminated in a female flower. The axis of the cone is dome-shaped and bore many

interseminal scales with somewhat fewer seeds, the general organization being similar to that of the cycadeoids.   The difference in
habit is most pronounced, for the cone of *W. sewardiana* is borne
at the apex of a conspicuous branch instead of being sunken among
the leaf bases of the trunk.   The seeds are also distinctive in that
they seem to have been composed of a short stalk, a greatly along-
ate nucellar region, and a distally extended integument called an

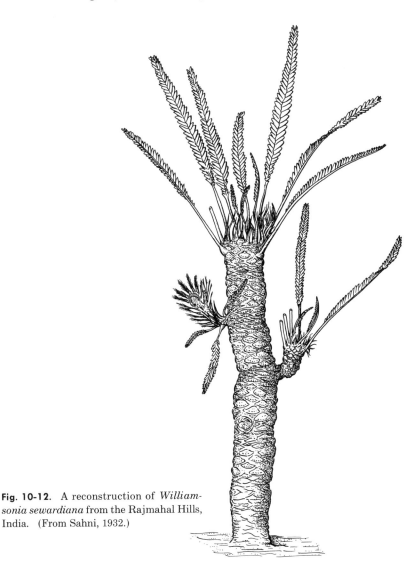

Fig. 10-12. A reconstruction of *Williamsonia sewardiana* from the Rajmahal Hills,
India. (From Sahni, 1932.)

"apical funnel." The cone as a whole was enclosed by long bracts covered with a dense hairy ramentum above and scales below. The microsporangiate organ is not known and there is no evidence that one was ever attached at the base of the female flower. It may be noticed that the fertile branch bears a somewhat smaller sterile one; it seems probable that growth was terminated by the cone and when it had shed its seeds the secondary branch assumed dominance.

The restoration is based on the known presence of columnar trunk fragments from the same locality which bear round scars that apparently represent the point of attachment of the branches; some of these trunk specimens have been found with leaves of the *Ptilophyllum cutchense* type.

Summary Comments

One might conclude from the distribution of the living cycads and certain of their morphological features that they are of very ancient lineage. Two genera enjoy an extensive geographical range: *Zamia*, with 28 species, is found from Florida through the West Indies and Mexico to northern South America and down the west coast to Chile; *Cycas* stands as a sort of eastern counterpart, its 16 species being distributed from Australia through the islands to southern Japan, and the range swings west to include China, India, and Madagascar. The two are as far apart in the organization of their seed-bearing organs as they are geographically, with closely compacted cones in *Zamia* and leafy megasporophylls in *Cycas*. The latter is often cited as the most primitive on the basis of this character, but the wood anatomy does not correlate; the tracheids are pitted in *Cycas* and scalariform in *Zamia*. The explanation may lie in the underground habit of the *Zamia* stems and they are perhaps best regarded as persistent juveniles, in this character, as Arnold suggests. The staminate cones of these two genera are, as in other living cycads, quite uniform, offering a sort of "neutral" position as far as phylogeny is concerned. Of the other genera, *Macrozamia* and *Bowenia* are confined to Australia; *Encephalartos* and *Stangeria* are found only in South Africa; *Dioon* and *Ceratozamia* occur in Mexico; *Microcycas* is limited to western Cuba. This scattered distribution combined with the fact that the nine genera are sufficiently distinct as to render interrelationships obscure is indicative of considerable age.

A survey of the Mesozoic fossils leaves no doubts concerning the diversity of the Cycadophyta. Two points in particular stand out from studies of the foliage: there is a great deal of gross variation in form suggesting considerable diversity; this is confirmed by studies of the cuticles, which reveals two distinct types of stomatal structure, and some leaves of cycadean aspect have been studied which have a stomatal morphology that indicates they may not be closely allied to either. As to specific and generic identifications the situation is close to that of the ginkgophyte foliage; without adequate cuticle material a name in many cases means little.

When one contrasts the trunk structure of *Cycadeoidea marshiana* (Fig. 10-4) or *C. gigantea* (a stout columnar type from the Isle of Portland, England) with the delicate shoot system of *Wielandiella angustifolia* (Fig. 10-13) it is evident that a broad gap separates them. Closely related plants may, however, be quite different in certain of their vegetative organs and it is perhaps more convincing to compare microsporangiate parts such as *Wonnacottia crispa, Cycadeoidea dacotensis,* and *Williamsonia santalensis;* quite clearly these represent diverse branches of a great race of plants concerning which we still know but little.

The similarity of *Beania* and *Androstrobus* to the seed and microsporangiate organs of a modern cycad such as *Zamia* suggests that

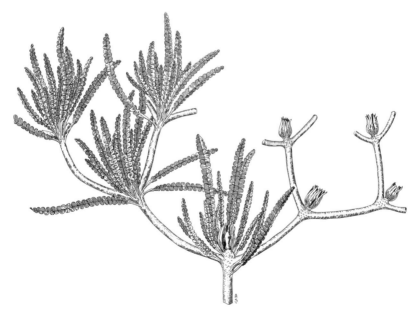

Fig. 10-13. Restoration of *Wielandiella angustifolia* from the Rhaetic of Scania, Sweden. (From Nathorst, 1909.)

the Cycadales was a well-established group in the Jurassic. From a somewhat earlier horizon (Rhaetic) in southern Sweden Florin has described a cycad, *Bjuvia simplex,* with a stout columnar trunk bearing a crown of pinnately veined leaves and a terminal cluster of megasporophylls, each with two pairs of ovules. His restoration suggests a plant very much like that of a modern *Cycas. Bjuvia* and *Beania* would thus seem to indicate that the *Cycas* and *Zamia* lines have been distinct for a very long time.

It seems more than likely that the cycadophytes existed as a distinct group or cluster of groups by late Paleozoic times, but precise details become obscure below the Triassic. Several cycado-phyte-like leaves have been reported from the Upper Carboniferous. In his Permian Shansi (China) flora Halle describes *Dioonites densinervis* from pinnate frond fragments 4 cm wide and of unknown length. Of interest also in this flora is an abundance of *Taeniopteris* species; one of them, *T. nystroemii,* was a large leaf attaining a breadth of 20 cm; these are possibly of cycadean affinities.

The assumption made by many authors that the cycadophytes had their origin in the pteridosperm complex seems to me to be unwarranted. It is true that the stems have certain features in common with the medullosan seed-ferns, but so far as I am aware the leaves and more especially the reproductive organs of the two are so different as to render such an origin for the cycadophytes highly improbable. The view is therefore taken here that they evolved independently as seed plants and very possibly along at least two different lines from a very early stage, resulting in the groups we refer to as cycads and bennettites.

REFERENCES

Arber, E. A. N. 1919. Remarks on the organization of the cones of *Williamsonia gigas* (L. and H.). *Ann. Bot., 33:* 173–179.

Andrews, Henry N., Jr. 1947. *Ancient plants and the world they lived in.* Comstock Pub. Co., Ithaca. 279 pp.

Arnold, Chester A. 1953. Origin and relationships of the cycads. *Phytomorphology,* 3: 51–65.

Chamberlain, Charles J. 1919. *The living cycads.* pp. 172. Univ. Chicago press.

Harris, T. M. 1932. The fossil flora of Scoresby Sound East Greenland. Part 3. Caytoniales and Bennettitales. *Meddelel. om Grönland,* 85 (5): 1–130.

————. 1941. Cones of extinct Cycadales from the Jurassic rocks of Yorkshire. *Phil. Trans. Roy. Soc. London,* 231B: 75–98.

————. 1942. *Wonnacottia*, a new Bennettitalean microsporophyll. *Ann. Bot.,* 6: 577–592.

————. 1944. A revision of *Williamsoniella. Phil. Trans. Roy. Soc. London,* 231B: 313–328.

Kräusel, R. 1948. *Sturiella langeri*, nov. gen., nov. sp., eine Bennettitee aus der Trias von Lunz (Nieder-Osterreich). *Senckenbergiana,* 29: 141–149.

Nathorst, A. G. 1909. Paläobotanische Mitteilungen. No. 8. *Kungl. Svenska Vetenskapsakad. Handl.,* 45 (4): 1–33.

————. 1911. Paläobotanische Mitteilungen. No. 9. *Kungl. Svenska Veteskapsakad. Handl.,* 46 (4): 1–33.

Sahni, B. 1932. A petrified *Williamsonia (W. sewardiana,* sp. nov.) from the Rajmahal Hills, India. *Mem. Geol. Surv. India, Paleont. Indica,* n.s., 20 (3): 1–19.

Seward, A. C. 1900. The Jurassic flora. I. The Yorkshire coast. Catalogue Mesozoic plants, Brit. Mus. Nat. Hist., pp. 1–341.

————. 1917. *Fossil Plants.* III. Cambridge Univ. press. Pp. 1–656.

Sitholey, R. V. and M. N. Bose. 1953. *Williamsonia santalensis* sp. nov.—a male fructification from the Rajmahal series, with remarks on the structure of *Ontheanthus polyandra* Ganju. *The Palaeobotanist,* 2: 29–39.

Stopes, Marie C. 1918. New Bennettitean cones from the British Cretaceous. *Phil. Trans. Roy. Soc. London,* 208B: 389–440.

Thomas, H. H. 1913. The fossil flora of the Cleveland District. *Quart. Journ. Geol. Soc. London,* 59: 223–251.

————. 1915. On *Williamsoniella,* a new type of Bennettitalean flower. *Phil. Trans. Roy. Soc. London,* 207B: 113–148.

———— and Nellie Bancroft. 1913. On the cuticle of some recent and fossil cycadean fronds. *Trans. Linnean Soc. London,* ser. B, Botany, 8: 155–204.

Ward, Lester F. 1905. Status of the Mesozoic floras of the United States. *U. S. Geol. Survey Mon.,* 48, pts. I, II, pp. 1–616.

Wieland, G. R. 1906. *American fossil cycads.* Vol. I. *Carnegie Institution Washington Pub.,* 34: 1–295.

————. 1916. American fossil cycads, Vol. II. *Carnegie Institution Washington Pub.,* 34: 1–277.

————. 1920. Distribution and relationships of the Cycadeoids. *Amer. Journ. Bot.,* 7: 125–145.

————. 1934. Fossil cycads. with special reference to *Raumeria reichenbachiana* Goeppert sp. of the Zwinger of Dresden. *Palaeontographica,* 79B: 85–130.

11

the CONIFEROPHYTA
and GINKGOPHYTA

Introduction

As used here the Coniferophyta includes two subdivisions, the Cordaitales and Coniferales. The former was a dominant arborescent group of Carboniferous seed plants and the Coniferales evolved from them in the latter part of that period and became quite diversified in early Permian times. Our knowledge of this sequence of development has been revealed in recent years largely through studies of the seed-bearing organs by the Swedish paleobotanist Rudolf Florin and it stands as one of the great triumphs of paleontology. That the two orders are thus closely related seems to me to be certain. There are, however, many gaps to be filled in and it is expedient to retain the two ordinal names for the present.

In view of the very general application of the term *cone* in referring to spore and seed-bearing organs of many groups of plants, it may be noted that the use of the name *conifer* here and elsewhere in this book will imply plants of the Coniferales. The group includes many familiar evergreen trees such as pine, fir, spruce, redwood, juniper, as well as a few that are deciduous such as bald cypress and larch. There are in total some 550 extant species, some of which belong to genera that are of very ancient vintage. It seems to me safe to assert that the Coniferales, as a major group of plants, exceeds all others in the combination of the length of time that it has existed and its dominance over vast areas of the earth's surface. The oldest members of the order (from the Lower Permian) flourished approximately 200 million years ago; their evolution appears to have been rapid during the late Paleozoic and early Mesozoic with relatively minor deviations since that time. Until man began cutting away the coniferous forests for their choice lumber they continued to compete favorably with other groups.

The cordaite-conifer line appears to have originated independently as seed plants; there is no reason to believe that they have any common ancestors, at the seed plant level, with the pteridosperms, cycads, or bennettites.

The Ginkgophyta confronts us with many vexatious but intriguing problems. Only one species survives today, the maidenhair tree (*Ginkgo biloba*), and it is truly the dwindling remnant of a race of plants that was quite diverse and, in places, a dominant feature of the Mesozoic landscapes. The wood anatomy of *Ginkgo* compares closely with that of the cordaites and conifers. Unfortunately, we know almost nothing about the stem structure of the dozen and a half extinct genera of Mesozoic ginkgophytes, and it is significant that the earliest presumed ginkgophyte in which reproductive organs are well preserved, the Lower Permian *Trichopitys heteromorpha,* does not bear a close comparison with the contemporaneous cordaites or conifers. In summarizing the relationships of this fossil Florin has noted that

The Ginkgoinae, Cordaitinae, Coniferae, and Taxinae undoubtedly belong to the same natural group of higher order . . . but they constitute parallel evolutionary lines which probably were already separated from each other in Upper Devonian or Lower Carboniferous times. At all events, a clear differentiation can be seen as far back as the available fossil records go. (1949, pp. 101–102)

I am inclined to go a little farther than this and proffer the opinion that the ginkgophytes will ultimately be shown to have evolved as a distinct and independent line of seed plants.

THE CONIFEROPHYTA

In order to demonstrate the relationships of the plants that are usually classified in the Cordaitales and Coniferales it is not feasible to consider them under two separate headings. Thus, in the pages that follow, certain aspects of the two, notably the seed organs and wood structure, are described together although I have tried to make clear the taxonomic status of the genera that are dealt with.

The cordaitales, habit and vegetative structure

The cordaites formed one of the most imposing elements of the Carboniferous and Permian forests. They were trees of monopodial habit and attained a height of at least 100 feet; Seward has recorded a trunk found near Newcastle, England, that was 72 feet long and it was not complete. The branches bore spirally arranged leaves (Fig. 11-1) that must have presented a close superficial resemblance to those of an iris. They were elongate, strap-shaped foliage organs which ranged, in different species, from 15 or 20 cm to as much as a meter long and attained a maximum width of about 15 cm. On a recent trip to the Minto coal field in south central New Brunswick, Canada, I encountered fine specimens up to 10 cm broad. These huge cordaite leaves along with fern, pteridosperm, and articulate fossils bore vivid evidence of the contrast with modern coniferous forests of that region.

The genus *Cordaites* was established for these foliar organs and has also been applied to the stems; the latter, however, offer special problems of nomenclature that will be considered separately. The leaves were probably of a tough, leathery texture in life; in some species the parallel veins alternate with I-shaped girders of fibrous cells and in addition there may be a considerable development of fibrous tissue immediately within the epidermis. The vascular bundle organization is essentially identical with that of a cycad; the primary wood is predominantly centripetal (develops toward the center) but is accompanied by a few centrifugal tracheids.

Of any extant trees the closest comparison in general habit seems to be with certain species of *Podocarpus* and *Araucaria;* according to Chamberlain the leaves of *Podocarpus wallichianus* may be 12.5 cm long and 3.5 cm broad, whereas the leaves of some species of *Araucaria* are nearly as long and somewhat broader. However, they fall short by a considerable margin of the larger leaved species of *Cordaites.*

The stems that bore the *Cordaites* foliage have a transversely chambered pith up to about 1.5 cm in diameter; it consisted of discs of parenchyma alternating with lens-shaped gaps; this feature is not confined to the cordaites, being present in some living plants such as the walnuts. Beginning at the periphery of the pith there is a transition in the primary wood from spiral, through scalariform, to pitted tracheids; this may encompass a dozen cells. The primary wood merges into the secondary wood uniformly so that the distinction between primary and secondary wood is evident only in a radial section. The pitting on the radial walls of the secondary wood consists of quite regular vertical rows of closely compacted hexagonal pits. The secondary wood is simple as far as the number of cell types are concerned, there being only the vertically aligned tracheids and parenchyma cells of the rays; the latter are usually only one cell wide but may vary considerably in height. The wood will be discussed further in a later section of this chapter.

Reproductive organs of the cordaites and certain conifers

The genus *Cordaianthus* has been used for both pollen and seed-bearing organs of the cordaites; in view of the close similarity in the morphology of the two there seems to be justification for this, but to the best of my knowledge well-preserved male and female organs have not yet been correlated in a single species. There is actually much less known about the pollen organs of both cordaites and conifers than is known about the seed organs.

Some years ago Grand'Eury described cordaitean shoots with leaves and male inflorescences attached. Figure 11-1 is taken from his work. In addition to the leaves and a large vegetative bud there is a male inflorescence apparently associated with each leaf. Cordaitean inflorescences are common in coal-ball petrifactions of Iowa and Kansas, but unfortunately well-preserved specimens are of infrequent occurrence. Delevoryas has described specimens from the Des Moines series of Kansas under the name *Cordaianthus concinnus* consisting of a slender axis (Fig. 11-2A) 1 to 2 mm in diameter which bore two rows of dwarf shoots that have been referred to as cones or budlike bodies. It seems to me that the phrase *dwarf shoot* is descriptive and morphologically accurate and it will thus be used here for structures that are apparently homologous. It may be noted that each shoot is borne in the axil of a prominent bract that actually exceeds it in length.

Each dwarf shoot is about 6 mm long and consists of 25 to 40 closely imbricated scales. The latter are uniformly arranged on the shoot axis in a spiral, although the distal fertile ones are somewhat more slender than the others. There are usually six sporangia, each 1 mm long, borne terminally on a fertile appendage; the six sporangia are fused at the base.

Cordaianthus penjoni Renault which comes from the middle Stephanian of France was of quite similar organization; the appendages or scales of each dwarf shoot were more slender than those of the American species and a higher number seem to have been fertile. Florin, who has restudied them and added to Renault's original description, notes that all the scales are homologous as in *C. concinnus.*

Before describing the seed-bearing organs a few words are appropriate concerning the history and significance of the investiga-

Fig. 11-1. A cordaitean branch with inflorescences in the axils of the leaves. (From Grand'Eury, 1877.)

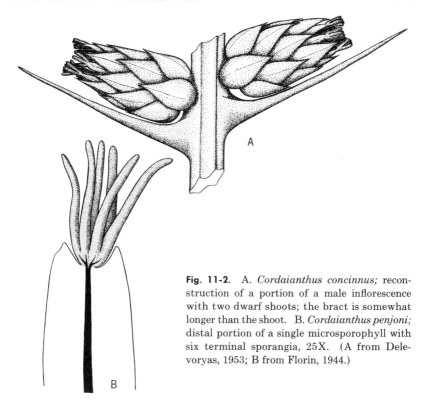

Fig. 11-2. A. *Cordaianthus concinnus;* reconstruction of a portion of a male inflorescence with two dwarf shoots; the bract is somewhat longer than the shoot. B. *Cordaianthus penjoni;* distal portion of a single microsporophyll with six terminal sporangia, 25X. (A from Delevoryas, 1953; B from Florin, 1944.)

tions that have led to our present knowledge. The conifers have long attracted the interest of layman and botanist; their economic worth as a source of wood and pulp as well as their great esthetic value have been widely appreciated. Among the most distinctive features of many of them are their conspicuous seed-bearing cones, and a great deal of effort has been expended in an attempt to understand their morphology. Although there is considerable diversity in the group, the seed cone of the Abietaceae (according to the classification given by Chamberlain in his *Gymnosperms, Structure and Evolution*) consists of a central axis with spirally arranged "double appendages" which consist of a stiff, woody *ovuliferous scale* bearing several seeds on its upper surface, and a generally smaller and more delicate *bract.* The bract is quite conspicuous in the cones of the fir (*Abies*) and larch (*Larix*) and is readily detected with the naked eye when a cone is broken open (Fig. 11-3); in the Douglas tree (*Pseudotsuga taxifolia*) the bract is a conspicuous three-pronged appendage and is appreciably longer

than the ovuliferous scale which it subtends. The same organization is found in a pine cone although the bract is small and not as
readily discernible.

The question of the origin of this distinctive orientation of seeds,
ovuliferous scale, and bract has led to many studies on the comparative morphology of the cone, and numerous theories have been
proposed which would require almost a page of print just to list.
Although there are still gaps in our knowledge, the elucidation of
the problem from a series of upper Paleozoic and lower Mesozoic
cordaites and conifers constitutes one of the great achievements in
evolutionary studies.

Seed organs of the cordaites

The female inflorescence of the cordaites has essentially the same
organization as the male one; it may reach a length of 30 cm and
it bears two rows of dwarf shoots, each in the axil of a bract. Two
main types are recognized:

Cordaianthus pseudofluitans Kidston is an older one (Westphalian) in which the dwarf shoots (Fig. 11-4) consist of numerous
scalelike appendages spirally arranged on the shoot axis. They are
all apparently homologous, but several of the distal ones are
greatly elongated, dichotomize several times, and bear two or more
terminal seeds that tend to be recurved toward the main axis of
the inflorescence.

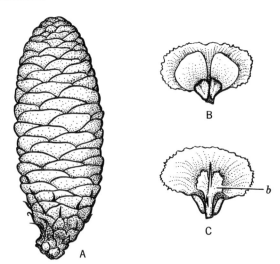

Fig. 11-3. A. A modern fir cone, natural size; B. a single ovuliferous scale with the
two-winged seeds on the upper (adaxial) surface; C. the under side of the scale showing
the bract (*b*).

Fig. 11-4. *Cordaianthus pseudo-fluitans.* Part of an inflorescence showing the axis and two rows of dwarf shoots in several of which the seeds are still intact, natural size. (From Florin, 1944.)

In a geologically younger species, *Cordaianthus zeilleri* Renault from the Stephanian, the general organization of the inflorescence is similar, but each dwarf shoot bears not more than four fertile appendages, and occasionally only one. They are unbranched and produce but one terminal, erect ovule; the fertile appendages are short and the ovule or ovules tend to be concealed among the sterile appendages.

Seed organs of the conifers

Although the cordaite-conifer sequence in general seems to me to be an established fact, it may be admitted that there is a conspicuous break between the two; this is evident in both the foliage and the seed organs.

Until rather recently much of the leafy twig material attributed to the conifers, found in Upper Carboniferous and Lower Permian horizons, was assigned to the genus *Walchia*. Although this is still in use Florin has recognized, on the basis of cuticular structure and seed cones, 14 species which are now placed in *Lebachia* (Fig. 11-5A) and one in *Ernestiodendron* (Fig. 11-5B); he notes that there still remain 11 species of *Walchia,* most of which will ultimately prove referable to *Lebachia*. The leaves of *Lebachia* tend to be more closely appressed to the stem axis than are those of *Ernestiodendron* where they extend out at about 90° to the axis and give the impression of having been quite stout and stiff. In their general habit these plants were probably similar to the extant *Araucaria excelsa,* the Norfolk Island pine. Several other genera, known only from their foliage, are recognized from the Lower Permian; in *Carpentieria frondosa* (Goeppert) Florin, for example, the leaves are slender and deeply forked, but in their general organization the leafy twigs appear "coniferous."

The seed organs of the conifers are aggregated into distinct cones in contrast to the open inflorescences of the cordaites. The cone of *Lebachia piniformis* (Schlotheim) Florin is 6 or 7 cm long and is composed of spirally arranged, closely imbricated bracts with a forked tip. In the axil of each bract there is a dwarf shoot

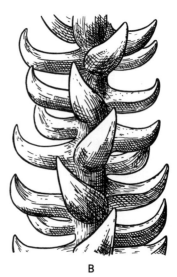

A   B

Fig. 11-5. Foliage of two Lower Permian conifers: A. *Lebachia piniformis;* B. *Ernestiodendron filiciforme.* Both about 5X. (From Florin, 1944.)

(Fig. 11-6) apparently homologous with the dwarf shoots in *Cordaianthus;* it consists in turn of several closely imbricated scales or sterile appendages, with one bearing a terminal erect ovule. This fertile appendage is usually so oriented that it is on the inner side of the dwarf branch and not visible from the outside of the cone as a whole.

It is apparent that a good deal of paleobotanical research is needed to bridge the gap between the kind of foliage borne by the cordaites and that of the *Walchia* type. The same holds true for the female inflorescences; there are missing links between *Cordaianthus* with its dwarf shoots arranged on the central axis in two rows and *Lebachia* in which the shoots are spirally arranged and aggregated into a compact unit (cone); yet the homology is apparent. From *Lebachia* on we may observe a reduction (with apparently numerous evolutionary side lines) in the dwarf shoot to the relatively simple ovuliferous scale of the modern *Pinus* and its immediate relatives. For the most part the following descriptions consider only the dwarf shoot.

Lebachia goeppertiana Florin (Lower Permian) has dwarf shoots (Fig. 11-7B) with numerous sterile appendages and a single fertile one, with an erect ovule, on the inner side. Florin points out that there is a suggestion here of the shift to the flattening and bilateral symmetry characteristic of the later conifers. Another Lower Permian fossil, *Ernestiodendron filiciforme* (Schlotheim) Florin, had cones about 2 cm long which were borne terminally on a next-to-ultimate leafy branch. The bracts are forked as in *Lebachia* and enclose dwarf shoots that are composed of several fertile appendages only (Fig. 11-7D), each with an erect ovule. *Walchia* (*Ernestiodendron?*) *germanica* Florin is very similar except that the ovules are inverted (Fig. 11-7C). Under the name *Walchiostrobus* sp. Florin has figured dwarf shoots that are somewhat flattened. They include 20 to 30 sterile scales and 4 to 6 with terminal ovules; in one specimen the ovules were found erect and in another they were inverted.

In the Upper Permian *Pseudovoltzia liebeana* (Geinitz) Florin each dwarf shoot (Fig. 11-7G) is found as usual in the axil of a bract. The axis of the shoot is quite rudimentary and there are only five scales; three of which are larger than the other two, the unit being distinctly flattened.

The ultimate in reduction of the dwarf shoot is found in *Ullmannia bronnii* Goeppert (Upper Permian); here the "shoot" is a disc-shaped structure (Fig. 11-7H) consisting of five wedge-shaped and

Fig. 11-6. *Lebachia piniformis.* A. Portion of a leafy shoot with several ultimate branchlets terminated by seed cones; B. a single dwarf shoot considerably enlarged, showing the forked bract, the tips of which are missing; C. the dwarf shoot from the inner (adaxial) side with one appendage bearing a terminal, erect seed, 10X. (From Florin, 1944.)

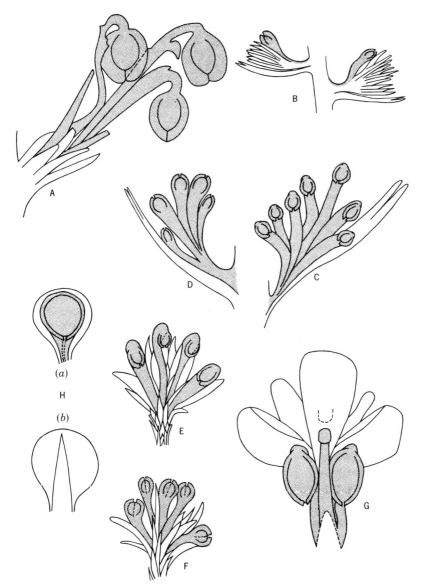

Fig. 11-7. Individual dwarf seed shoots of a cordaite and several conifers, all enlarged. A. *Cordaianthus pseudofluitans,* 2X; B. *Lebachia goeppertiana,* 3.5X; C. *Walchia (Ernestiodendron?) germanica,* 3X; D. *Ernestiodendron filiciforme,* 2X; E. *Walchiostrobus* sp., 3X; F. *Walchiostrobus* sp., 3X; G. *Pseudovoltzia liebeana,* 1.6X; H. *Ullmannia bronnii* showing upper side (*a*) and under side (*b*), 1.4X. (From Florin, 1944.)

fused appendages. The fusion is so complete, however, that it appears to be a single, nearly circular scale. One fertile appendage is borne immediately above it with an inverted ovule.

Glyptolepis, an Upper Permian and Triassic genus, has shoots consisting of five or more (depending on the species) sterile scales which are even more strongly flattened than those of *Pseudovoltzia* and there are only two fertile appendages. In the Triassic *Voltzia* the sterile appendages are also strongly flattened and it differs from previously noted genera in that the fertile appendages are fused to the upper surface of the sterile ones. The dwarf shoot of the Lower Jurassic *Schizolepis* consists of only three sterile appendages; these are partially fused and three seed stalks or modified fertile appendages are fused to their upper surface. Finally, the Lower Jurassic *Hirmeriella rhätoliassica* Hörhammer appears to consist of a single sterile scale with two inverted ovules borne on its upper surface. There is, however, some question as to whether this is actually a simple (single) scale or composed of two or three completely fused ones.

In summary, the fossil evidence indicates quite clearly that the seed cones of the modern conifers, such as fir, larch, and pine, are a much reduced inflorescence in which the ovuliferous scale and seeds borne on its upper surface are the remnants of a radially symmetrical dwarf shoot. As this evolutionary trend progressed, numerous side lines branched out as is evident from the fact that all the examples cited do not fit a perfect sequence.

The dwarf shoots, each in the axil of a bract, first appear aggregated into a distinct cone in the Lower Permian; these in turn appear to have evolved from the Upper Carboniferous *Cordaianthus* in which the dwarf shoots are arranged in two rows along a central axis. Racial development appears to have progressed rapidly, for by Upper Permian times the "dwarf shoot" of *Ullmannia* is a simple disc-shaped structure not unlike the ovuliferous scales of the modern abietinean conifers. Evolution in the Coniferales since the Permian appears to have been a matter of minor proliferations of certain lines and extinction of others. That some of them have long been and still are a hardy lot is evident, but their "morphological progress" essentially faded out some 150 million years ago.

Isolated seeds attributed to the Cordaitales

Numerous authors have figured and described *Cordaianthus*-like inflorescences from compression specimens which show winged, bilaterally symmetrical seeds attached.   This evidence combined with the frequent association of such seed compressions with cordaite foliage, and the occurrence of similarly shaped petrified seeds with cordaitean stem and leaf remains in coal balls, is the basis for the supposed relationship.   Thus although one may safely assign most petrified Carboniferous seeds that are radially symmetrical in cross section to the pteridosperms and bilaterally symmetrical ones to the cordaites, it must be regarded as a tentative classification.   The seeds described below are probably of cordaitean affinities.

The genus *Cardiocarpus,* as defined by Seward in 1917, is based on bilaterally symmetrical, petrified seeds from the Upper Carboniferous which are quite common in Europe and in this country. *Cardiocarpus spinatus* Graham is found in some abundance in coal balls of Kansas and Iowa; the seed is about 15 mm long, and in a median transverse section the diameters are approximately 15 × 10

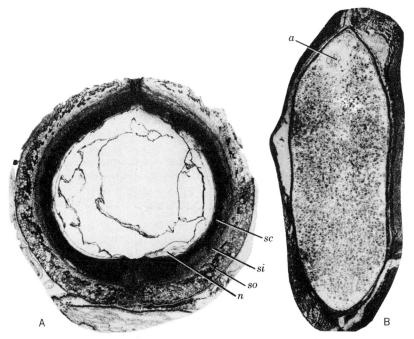

Fig. 11-8. *Cardiocarpus spinatus.* A. Longitudinal section through the broad diameter of a seed, 4X; B. longitudinal section through narrow diameter of a specimen containing a gametophyte, 5X; *a,* archegonium; *n,* nucellus; *sc,* sclerotesta; *si,* inner sarcotesta; *so,* outer sarcotesta. (A photograph courtesy R. W. Baxter; B from Andrews and Felix, 1952.)

mm. Thus, facing the broad side, the seeds are almost perfectly circular, whereas from any other angle they appear broadly oval. The integument is remarkable in its complexity and in Kansas specimens recently described by Roth it is beautifully preserved (Fig. 11-8). He recognizes five distinct tissue layers: A sarcotesta composed of inner and outer zones; the cells of the outer layer are large (about 165 μ in diameter) and thin-walled and are usually filled with a brownish substance which gives them a rich amber color; the inner sarcotesta differs from the outer only in the much smaller size of the cells. The sclerotesta also consists of two zones, the outermost one being conspicuous because of the dark-colored, elongate, thick-walled cells and its tendency to proliferate outward in the form of spines which occasionally extend to the epidermis. In the seed that is illustrated they are not strongly developed, this specimen having been selected for the well-preserved sarcotesta. Another layer of thick-walled cells follows, the inner sclerotesta, and within this is a tissue of thin-walled cells, the endotesta.

In mature seeds the nucellus appears as a thin, inconspicuous band and in a few specimens rather well-preserved gametophytes have been found (Fig. 11-8); the cells composing it are quite uniform and thin-walled and at least two archegonia were formed at the micropylar end. In his study of fossil seeds, published in 1881, Brongniart has described several with gametophytes preserved; it is certainly one of the outstanding works of the nineteenth century and should be consulted by any student of Paleozoic plants.

The nucellus was free from the integument except at the base. In European specimens of *Cardiocarpus* Seward describes the vascular system as consisting of a central strand in the base of the seed which gives off two branches that pass up through the inner layer of the sarcotesta, and a second pair that penetrate the peripheral tissue of the nucellus. So far as I know the nucellar strands have not been observed in the American specimens.

The sclerotesta of these seeds was undoubtedly hard and highly resistant to decay; in coal balls specimens are often found with only this part remaining. There can be little doubt that most or all of the Carboniferous seeds, known from impressions or compressions that are described as winged, are of this type; that is, the tough sclerotesta was responsible for the impression of the central body of the seed, whereas the fleshy sarcotesta formed the peripheral outline or "wing"—since the sarcotesta was a uniform tissue around the entire seed the term wing is misleading. The number of seeds of this type that have been described from impressions is very considerable and suggests a greater diversity for the Cordaitales than is indicated by the leaves and wood.

Kamaraspermum leeanum Kern (Fig. 11-9) offers several unique features and I would emphasize that it is only tentatively classified as a cordaite seed. Several specimens were found in coal balls from the Upper Carboniferous (Des Moines series) of Iowa; they are approximately 11 mm long and in a median transverse section (Fig. 11-9B) measure 11 by 3 mm. In its gross aspects two features are especially striking: a basal cavity located below the nucellar chamber; and a micropyle that consists of two distinct zones, a massive one directly above the nucellus and a much narrower elongate terminal part which had the form of a gently tapering, flattened funnel.

The sequence of tissues composing the integument is unlike that of other seeds in consisting of outer and inner sclerotic zones separated by one of apparently more delicate cells. The inner sclerotic layer may be seen to continue upward to form the inner part of the proximal portion of the micropyle, and downward to form a conspicuous lining for the basal chamber. It is thought that this was an air chamber adding buoyancy which would have aided dissemination by water. A seed with a similar basal chamber has been reported by Renault (*Codonospermum olivaeforme*) from France, but it is a radially symmetrical seed and quite different in the structure of its integument from *Kamaraspermum*.

Some cordaitean and coniferous woods

Petrified logs and various sized fragments thereof that are attributable to the cordaites, conifers, and ginkgo are frequently encountered from the Devonian to the present. For the most part the anatomy is rather stereotyped; many genera of extant conifers are extremely difficult to identify on the basis of their secondary wood alone, and although numerous studies have been devoted to this type of fossil material it is my impression that they have not entirely rewarded the effort involved. Since such wood specimens are common and the paleobotanist is occasionally pressed for an identification, a few comments seem appropriate. It is my intention to mention a few plants in which distinctive characters are present and to cite some of the problems presented by the remainder.

Most authorities have recognized three families under the Cordaitales: Poroxyleae, Pityeae, Cordaiteae.

The genus *Poroxylon* (Poroxyleae) is known from slender stems which are rarely more than 2 cm in diameter, with large exarch primary bundles around the periphery of the pith; the rays of the

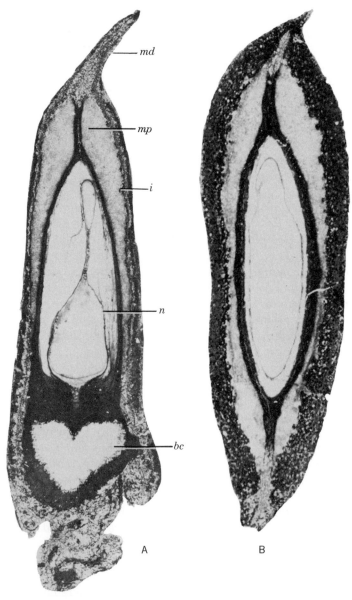

A   B

Fig. 11-9. *Kamaraspermum leeanum.* A. Longitudinal section through narrow diameter; B. median transverse section; *bc,* basal chamber; *i,* integument; *n,* nucellus;
md, distal part of micropyle; *mp,* proximal part of micropyle. Both 13X. (From
Kern and Andrews, 1946.)

secondary wood are quite high and several cells wide; hypodermal sclerenchyma bands are present in the outer cortex. There is, therefore, a resemblance to the stem of *Lyginopteris,* but the leaves that are believed to have been borne on the *Poroxylon* stems are of the *Cordaites* type and quite unlike those of the Carboniferous pteridosperms. The viewpoint is expressed quite strongly elsewhere in this book that the coniferophytes and pteridosperms are not at all closely related, but it seems only fair to mention this plant which may possibly represent a link between the two great groups.

The Pityeae includes several genera, some of them of great size, such as *Callixylon,** in which the primary bundles are mesarch and the secondary wood is for the most part of the coniferous type; that is, the rays are usually not more than two cells broad, and the tracheids have multiseriate-bordered pits on their radial walls. The genus *Callixylon* is of unusual importance on several counts; it includes several species distributed through the Upper Devonian and they are found in numerous localities in the United States and Europe. Most of our information on the American species has come from the researches of C. A. Arnold who has described a specimen from Indiana that is 3 feet broad at the base and tapers, through a length of 9 feet, to about 18 inches at the upper end. Trunks 5 feet in diameter have been reported from Oklahoma, a size greatly exceeding that of any other Devonian plant. The secondary wood is similar in its general organization to that of the cordaites (compare Figs. 11-11A, B), consisting of tracheids and rays only. In the specimen illustrated the rays are considerably larger than those of the cordaitean wood. Particularly distinctive is the grouped arrangement of the pits in the radial walls of the secondary tracheids (Fig. 11-10). The foliage and reproductive organs are not known and although it is a reasonable guess that *Callixylon* should be regarded as of cordaitean affinities, and a seed plant, this has not been proven.

In the Cordaiteae several genera have been recognized on the basis of stem anatomy: *Cordaites,* in which the primary wood is entirely centripetal, that is, the protoxylem is endarch, and *Mesoxylon* in which some centripetal wood is present. These two can be distinguished only with especially well-preserved specimens.

In view of the lack of information concerning the reproductive organs of the Poroxyleae and Pityeae the above classification is of course a tentative one.

* See footnote on page 412.

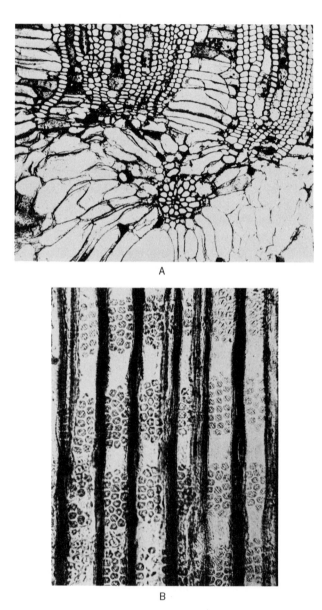

Fig. 11-10. *Callixylon* sp. A. Peripheral region of pith showing mesarch primary bundle and some secondary wood, 60X; B. secondary tracheids in radial view showing grouped arrangement of pits, about 150X.

Of special concern in connection with the problem of wood identification is the homogeneity of the secondary wood of plants in the Pityeae and Cordaiteae. Fragmentary pieces of secondary wood cannot be assigned with any confidence to a genus of these two groups nor can it be separated from that of some coniferous genera, notably *Araucaria*. (*Callixylon* with its grouped pits may be regarded as an exception.) It is evident from the photographs that there is little to differentiate between the cordaitean wood (Upper Carboniferous) shown in Fig. 11-11A and the Cretaceous araucarian in Fig. 11-11C. As a supplementary note it is of interest to record the presence of a persistent leaf trace in *Araucaria,* that is, the trace tissue continues to be formed by the cambium long after the leaf has fallen off. They tend to become quite widely separated in older branches.

Thus, in dealing with fragments of secondary wood of the "cordaitean" type it has been general practice to use the generic name *Dadoxylon* for Paleozoic specimens and *Araucarioxylon* for woods of similar organization from the Mesozoic and Tertiary. Many paleobotanists, including this writer, have been tempted to describe new species of these nebulous genera. The result is that there are a great many, but I am afraid that they contribute but little to the sum total of our knowledge of the plant life of past ages.

A few of the extant genera of the Coniferales possess features of their secondary wood that are quite distinctive. For example, *Pinus* (pine), *Picea* (spruce), *Pseudotsuga* (Douglas tree) and *Larix* (larch) have conspicuous resin canals; the peripheral secretory cells of those in pine are thin-walled whereas they are thick-walled in the other three. In a few genera, including redwood, resin canals are not normally present but may be produced as a result of wounding. In *Pseudotsuga* and *Taxus* (yew) delicate tertiary spiral thickenings characterize the tracheids. In *Araucaria* and *Agathis* the pits in the radial walls of the tracheids tend to be crowded and angular. However, the vast majority of living conifers present secondary wood in which generic distinctions are by no means clear cut. Thus with much of the fossil coniferous wood from Mesozoic and Tertiary horizons the names that have been applied are little more than numbers which offer no clues to the real identity of the plants concerned. For a detailed account of coniferous woods the student is referred to Greguss' recent study.

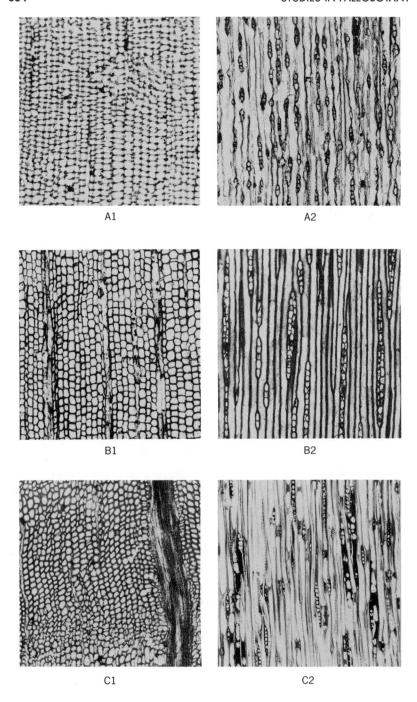

A1

A2

B1

B2

C1

C2

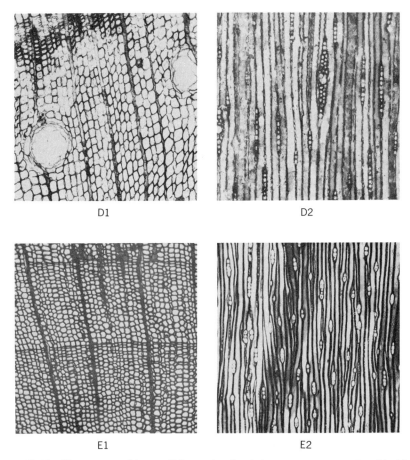

D1   D2

E1   E2

Fig. 11-11. Transverse and tangential aspects of certain gymnosperm woods. A1, A2. *Dadoxylon* sp., Upper Carboniferous; B1, B2. *Callixylon newberryi*, Upper Devonian; C1, C2. *Araucarioxylon* sp., Upper Cretaceous; D1, D2. *Pityoxylon* sp. (probably a pine), Eocene; E1, E2. *Ginkgo biloba* (extant). (All 37X.)

THE GINKGOPHYTA

Probably no other plant has so effectively captured the imagination of botanists and laymen, from the standpoint of its past history, as has *Ginkgo biloba*. There are other extant plants with as old or older fossil records but several factors seem to conspire to give priority to ginkgo as the foremost "living fossil." The tree is unique in its appearance, possessing an abundance of characters that set it off sharply from any other; for the first 20 or 30 years it tends to do little other than grow straight up and the lower branches gradually begin to spread out, a few of them becoming quite massive in the course of a century. Slow in growth, it is a shade tree to be planted for future generations, but even to a non-botanist its novel mode of growth in the early decades offers some compensation. The fan-shaped leaves, unlike those of any other plant, are borne singly along the terminal branches which may shoot out a half meter in a single season; in their axils branch buds are formed which for the most part grow very slowly, forming the characteristic short shoots. Each one bears a dense cluster of leaves at its apex (Fig. 11-12); the short shoots occasionally sprout out to form new long shoots.

The plant is dioecious, that is, the short shoots of a tree may produce a cluster of pollen-bearing catkins, whereas the seeds are

Fig. 11-12. A branch of the extant *Ginkgo biloba* with three short shoots each bearing leaves and seeds, about ½X. (From Andrews, 1947.)

found on a different tree (Fig. 11-12). The mature seeds are about the size of a small apricot and the integument is composed of an outer orange-colored fleshy portion and an inner, hard stony layer. The fleshy coat is rich in butyric acid and when crushed emits an odor that is unpleasant to some people. However, the seeds are often produced in great abundance and tend to hang on for a month or two after the leaves are shed; a female tree in this late fall attire is a thing of great beauty and certainly outweighs the disadvantage of the somewhat unsavory seeds when they have fallen to the ground.

The ginkgo is known to have been cultivated for many centuries in Chinese gardens and it is very probable that man's interest in the plant saved it from extinction—somewhat the reverse of the usual fate of living things as human occupation has spread over the earth. It has been reported as a native plant in a small area in the Province of Chekiang in eastern China, but there is some uncertainty as to whether or not such specimens are escapes from gardens.

Western travelers first encountered the ginkgo in eastern China and introduced it into Europe in the early eighteenth century; it was brought to the United States a few decades later. It has since been widely planted throughout the country and is slowly but surely becoming one of our most valuable shade trees. One of ginkgo's great assets is the lack of any serious diseases, having apparently outlived any that may have bothered it in the past.

The fossil record of the ginkgophytes is an especially disconcerting one for two reasons:

Foliage has been found at many horizons and geographical points in the northern hemisphere (less is known of southern records) from the late Paleozoic to the present; some genera are quite distinct but many of the fossils consist of leaves that fit into an essentially continuous sequence beginning with deeply dissected ones in the early Mesozoic to the nearly entire ones of a living ginkgo. The difficulties of delimiting genera and species have perplexed many a botanist.

In contrast to the abundance of foliage, reproductive organs are almost nonexistent in the fossil record.

In view of the paucity of reproductive structures it is clearly not possible to set up a satisfactory classification or to attempt to establish evolutionary sequences other than on a tentative basis.

In order to deal with the evidence at hand it is proposed to consider fossil remains that are ginkgophytes or probable ginkgophytes under the following headings:

Mesozoic and Tertiary fossils that seem closely related to the extant *Ginkgo biloba.*

Mesozoic and Paleozoic fossils that are ginkgophytes but clearly distinct at the generic level from *Ginkgo.*

Some Paleozoic fossils that are suggestive of being ancestral ginkgophytes.

Mesozoic and Tertiary fossils probably closely related to G. biloba

It is well known that leaves on a living ginkgo may offer considerable variation in form, especially so with trees in the first decade or two of growth. A mature plant displays much less variation and although it is, therefore, reasonable to suppose that the fossil record includes many species that are close to *G. biloba* a clear-cut delimitation is not always possible. It seemed to me it might prove most informative to consider approaches that certain authorities have taken in their studies of these vexing fossils.

In dealing with fossils chiefly from western United States that he regarded as "unmistakably assignable to *Ginkgo*" R. W. Brown has suggested a classification of three species, as follows (see Fig. 11-13):

Ginkgo adiantoides (Unger) Heer. These are of reniform outline, usually with a conspicuous apical notch; found most abundantly from the Paleocene onward.

Ginkgo lamariensis Ward. The outline is wedge-shaped and if notched, only slightly so; predominantly of late Mesozoic age.

Ginkgo digitata (Brongniart) Heer. The general outline is wedge-shaped, but the blade is deeply dissected; found in early Mesozoic horizons.

Such a classification as applied to western United States is of stratigraphical value provided one has reasonably abundant material from any one horizon. Brown emphasizes that this grouping is one of expediency and that the three do not represent "biologic species in the strict sense." Perhaps the most obvious difficulty in applying this grouping is that it is not easy, if indeed it is possible, to distinguish in all cases between what is "unmistakably" a *Ginkgo* and what is not. This is brought out in the next example.

Many students of the ginkgophytes have tended to classify fossil leaves that appear to be closely allied to *Ginkgo biloba* in the

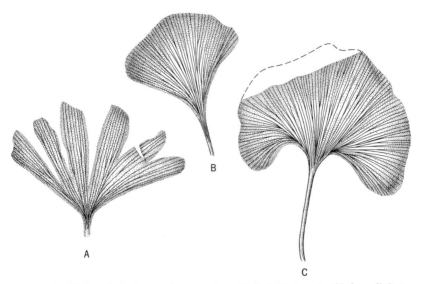

Fig. 11-13. Ginkgophyte leaves from western United States; A. *Ginkgo digitata;*
B. *Ginkgo lamariensis;* C. *Ginkgo adiantoides.* (From Brown, 1943.)

genera *Ginkgoites* and *Baiera*. A remarkable assemblage of these
plants has been found in the Rhaetic-Jurassic rocks of East Greenland and I have chosen to deal with them in some detail, as constituting a representative and carefully studied collection, rather
than attempt a general survey of these two important genera.
Although it seems to be true in a general way that species of the
early to mid-Mesozoic, with deeply dissected leaves, gradually gave
way to ones with entire or nearly entire leaves, considerable
variety may be encountered within a narrow time range; this is
strikingly borne out by the Scoresby Sound collections.

Harris separates the two as follows:

Ginkgoites. A distinct petiole is present and the outline of the
lamina is semicircular.

Baiera. A distinct petiole is absent and the leaf is wedge-shaped.

From a detailed study of some 10 species of these two genera,
found in the Jurassic of Japan, Oishi cites the following characters
as diagnostic: the stomates occur on both upper and lower surfaces
but are much more numerous on the latter; guard cells are sunken
below the surface and the median slit (stoma) has no definite
orientation; the five to seven subsidiary cells form a circular group
which arch over the guard cells. Oishi's investigations agree with

those of Harris in that the epidermal structures are indispensable in delimiting species within the two genera, but the genera as such cannot be separated in that way.

A few examples will indicate something of the diversity of these plants in the Greenland flora.

Ginkgoites obovata (Nathorst) Seward. The lamina is entire or nearly so and judging from the larger fragmentary specimens the leaves attained a breadth of at least 16 cm.

Ginkgoites fimbriata Harris (Fig. 11-14A). The lamina is small, about 1.5 cm broad and 2 cm wide, and is borne on a petiole about 4.5 cm long; quite distinctive are the delicate pointed teeth of the terminal margin of the lobes. As to epidermal structure, there are four to seven subsidiary cells surrounding the guard cells; the surface of the subsidiary cells is excessively thickened and forms an irregular ring around the stomatal aperture.

Ginkgoites acosmia Harris (Fig. 14-2E). This is a dominant species at certain horizons; the lamina generally tends to be deeply divided into two halves and each one is subdivided into three lobes which are in turn partially dissected.

Ginkgoites minuta (Nathorst) Harris (Fig. 11-14E). This is a finely divided leaf of a type that has frequently been assigned to *Baiera;* it would be impossible to distinguish it from that genus were it not for the conspicuous petiole.

Ginkgoites hermelini (Hartz) Harris (Fig. 11-14F). Most of the leaves tend to be divided into six lobes which are elongate-oval; the stoma is surrounded by four to six subsidiary cells, each with a papilla which points upward and does not project over the aperture in which the guard cells are sunken.

Baiera spectabilis Nathorst (Fig. 14-2D). This species seems to have been very variable in size, degree of dissection, and width of the segments; Harris figures one specimen that is 22 cm long.

Baiera boeggildiana Harris (Fig. 11-14C). The leaves are typically 3 to 4 cm long with an apex that is divided into two rounded lobes; the guard cells are located at the bottom of an oval pit and the subsidiary cells bear papillae that project over the aperture.

Some other ginkgophytes of the Mesozoic and late Paleozoic

The most ancient fossil that may with some confidence be attributed to the ginkgophyte line, and which offers some significant clues concerning the evolution of the living species, is *Trichopitys heteromorpha* from the Lower Permian of Lodeve in southern

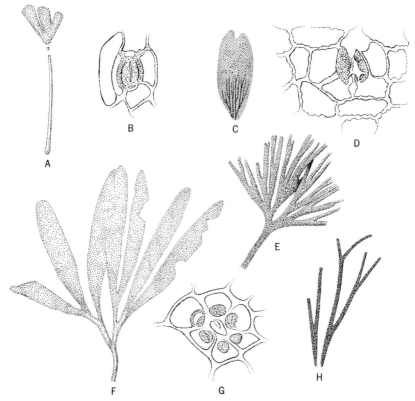

Fig. 11-14. Ginkgophyte leaves from Scoresby Sound, East Greenland. A, B. *Ginkgoites fimbriata;* A. leaf, 0.5X; B. stomate, 300X. C, D. *Baiera boeggildiana;* C. leaf, 1X; D. stomate, 400X. E. *Ginkgoites minuta,* ⅔X. F, G. *Ginkgoites hermelini;* F. leaf, 0.5X; G. stomate, 250X. H. *Baiera leptophylla,* 0.6X. (From Harris, 1935.)

France. Originally described by Saporta, our present understanding of the plant is based on more recent investigations by Florin; judging from photographs, his restoration (Fig. 11-15) offers an accurate view of the fertile branches. They are known from fragments about 8 mm in diameter which bore leaves that probably were spirally arranged. The leaf is thought to have been more or less terete and dichotomized several times. It is perhaps not amiss to regard these foliar organs as representing a stage in ginkgophyte evolution comparable to that presented by *Calamophyton* in the articulate series.

Seed-bearing branches are found in the axils of some of the leaves; these have been described by Florin as "sporangial trusses"

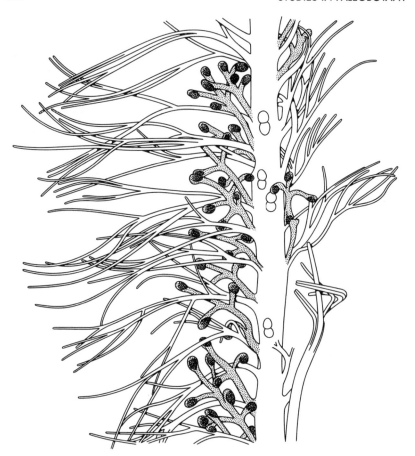

Fig. 11-15. *Trichopitys heteromorpha,* a shoot fragment with leaves bearing seed-branches in their axils, 0.7X. (From Florin, 1949.)

or "sporangiophoric complexes"; they probably represent a some-what "overtopped" dichotomous branch system with a single inverted ovule borne at the end of each branch. Four to six ovules per truss seems to have been most common, but as many as 20 have been recorded.

The short shoot habit apparently had not evolved at this time although it was established in some of the Mesozoic ginkgophytes. Florin suggests that the reproductive complexes, which are present in *Trichopitys* on long shoots (and this is occasionally so in *Ginkgo*), were gradually transferred to the perennial short shoots so characteristic of the extant species. In the latter only one or two seeds usually mature on a single "complex," but numerous instances are on record in which several were initiated, forming a branch system or truss very much like that of *Trichopitys*. In brief, the case for regarding this fossil as an early member of the ginkgophyte line seems to be a rather good one.

Sixteen or more genera of ginkgophytes have been recognized from Mesozoic horizons ranging from the Triassic to the Upper Cretaceous. One of the most interesting is *Sphenobaiera*, a genus that extends from the Lower Permian to the Lower Cretaceous. *S. furcata* (Heer) Florin (Fig. 11-16D) from the late Triassic near Basle, Switzerland, bore foliage similar to that of *Trichopitys*, but the leaves were arranged on short shoots as well as long shoots. Microsporangiate organs have been found in organic connection and appear to have been borne on short shoots; the central axis divided to form branches which in turn bifurcated once or twice and terminated in clusters of three to five sporangia.

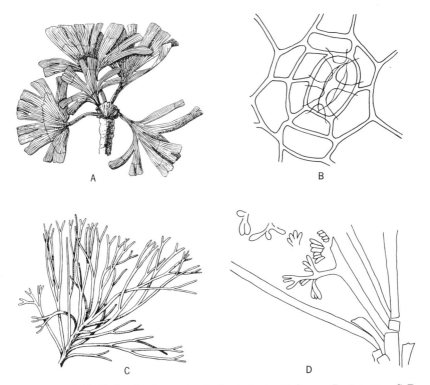

A   B

C   D

Fig. 11-16. A, B. *Ginkgoites lunzensis;* A. short shoot with leaves; B. stomate. C, D. *Sphenobaiera furcata;* C. long shoot with leaves; D. portion of short shoot with fragmentary microsporophyll. (From Kräusel, 1943.)

It is also of interest to find fossils in the Triassic that are of quite advanced aspect; *Ginkgoites lunzensis* (Stur) Florin bore leaves on short shoots that resemble those of some Jurassic species and the epidermal structure tends to confirm this. Figure 11-16B shows the stomates characteristically sunken and the papillae of the subsidiary cells overhanging the aperture.

A considerable assemblage of ginkgophytes has been described from high Arctic Jurassic rocks of Franz Joseph Land. The leaves of *Windwardia crookallii* Florin are long, almost linear and entire; they measure about 12 cm by 5 mm broad and have a nearly truncate apex. The leaves of *Stephanophyllum solmsi* (Seward) Florin are similar in form but have a rounded apex and were larger, probably attaining a length of 30 cm and a breadth of about 1 cm. The leaves of *Arctobaiera flettii* Florin are also of the same general type, but are quite small, being about 6 cm long, 4 mm broad, and with a bluntly rounded apex; some are entire and others are bifurcated for about one-third of their length. In all of the above three the leaves are borne in clusters on short shoots.

Some Paleozoic fossils, possibly representing early ginkgophytes

Halle has described a species of *Saportaea* (*S. nervosa*) from the Permian of Shansi, China (Fig. 11-17) in which the larger leaves have petioles about 5 cm long and 4 mm broad; the blade attained a breadth of at least 14 cm and was probably entire or nearly so in life. The most distinctive feature of the leaf lies in the venation; the vascular bundles of the petiole (probably two), instead of dichotomizing uniformly as in a *Ginkgo* leaf, follow along the lateral margins and produce secondary veins that take a nearly parallel course toward the distal margin. This type of venation was

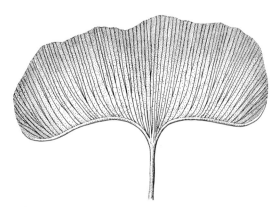

Fig. 11-17. *Saportaea nervosa*, a leaf restored. (From specimens in the Swedish Natural History Museum.)

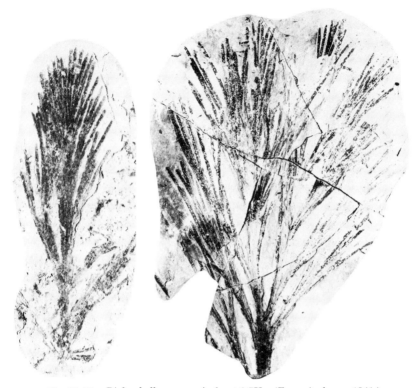

Fig. 11-18. *Dichophyllum moorei,* about 0.7X. (From Andrews, 1941.)

recorded previously for leaves from the Jurassic of Japan known as *Ginkgodium nathorsti* Yokoyama; they are somewhat spoon-shaped, about 6.6 cm long and 2 cm broad, entire or lobed by a median sinus and with a short petiole.

Ginkgodium was regarded by Seward in volume 4 of his *Fossil Plants* as of uncertain affinities, but it is accepted by Harris as ginkgoalean. Basically, both *Ginkgodium* and *Saportaea* seem to me to be very close to the *Ginkgo-Baiera* type; the difference in venation, although superficially rather striking in the former two, is certainly not fundamental.

It is with some hesitation that the fossil considered next is included in this section and it should be noted that it is by no means certain that it is a ginkgophyte ancestor. Some years ago I spent several days digging in an Upper Carboniferous shale bed in eastern Kansas for specimens of *Dichophyllum moorei* Elias, a fossil that displays a particularly interesting branching pattern (Fig. 11-18).

The available specimens show no distinct segregation of stems and leaves; the main axis divides by more or less equal dichotomies which ultimately terminate in slender, linear subdivisions. In some cases they are rather densely clustered, which obscures the precise details of the branching pattern. The plant was probably not a fern, as a distinct cuticle is present although poorly preserved; the only other evidence bearing on the affinities of the plant is the association of seeds that are characterized by two prominent horn-like projections at the micropylar end.

It is only in the ferns that we have begun to obtain a reasonably clear understanding of the evolution of the distinction between stems and leaves; possibly *Dichophyllum* is an early ginkgophyte, or more appropriately a "preginkgophyte" in which the characters of the group are becoming established.

It certainly would be interesting to know what brought about the decline of this unique assemblage of plants. Having flourished quite profusely over a period of some 150 million years it would seem they had been able to compete favorably with many other groups and had become acclimatized to much climatic fluctuation. But by Oligocene time only 2 out of a total of some 19 genera remain. Although absent from western United States during the early Tertiary, apparently due to the warm climate that prevailed, *Ginkgo* reappears in the Oligocene, but the sole surviving species disappeared at the close of the Miocene. It apparently held on a little longer in Europe, having been recorded as occurring in late Pliocene deposits.

REFERENCES

Literature—Coniferophytes

Andrews, Henry N. Jr. and Charles J. Felix. 1952. The gametophyte of *Cardiocarpus spinatus* Graham. *Ann. Missouri Bot. Gard.*, 39: 127–135.

Arnold, Chester A. 1930. The genus *Callixylon* from the Upper Devonian of central and western New York. *Papers Michigan Acad. Sci. Arts Letters*, 11: 1–50.

————. 1931. On *Callixylon newberryi* (Dawson) Elkins et Wieland. *Univ. Michigan, Contrib.* Mus. Paleont., 3: 207–232.

————. 1934. *Callixylon whiteanum* sp. nov., from the Woodford chert of Oklahoma. *Bot. Gaz.*, 96: 180–185.

Beck, George F. 1945. Tertiary coniferous woods of western North America. *Northwest Science*, 19: 67–69; 89–102.

Brongniart, Adolphe. 1881. *Recherches sur les graines fossiles silifiées*. Paris. Pp 34.

Chamberlain, Charles J. 1935. *Gymnosperm Structure and Evolution.* Univ. Chicago press. 484 pp.

Delevoryas, Theodore. 1953. A new male cordaitean fructification from the Kansas Carboniferous. *Amer. Journ. Bot.,* 40: 144–150.

Florin, Rudolf. 1944. Die Koniferen des Oberkarbons und des Unteren Perms. *Paleontographica,* 85B, sechstes heft: 365–456; siebentes heft: 457–654.

————. 1950. On female reproductive organs in the Cordaitinae. *Acta Horti Bergiani.* 15: 111–134.

————. 1951. Evolution in Cordaites and Conifers. *Acta Horti Bergiani.* 15: 285–388.

Grand'Eury, Cyrille. 1877. Flore Carbonifere du Department de la Loire et du centre de la France. *Mem. Acad. Sci. Institut France,* 24: 624 pp. Paris.

Greguss, Pál. 1955. Identification of living gymnosperms on the basis of xylotomy. Budapest. 263 pp.

Jeffrey, Edward C. 1917. *The Anatomy of Woody Plants.* Univ. Chicago press. 478 pp.

Kern, Ellen M. and H. N. Andrews, Jr. 1946. Some petrified seeds from Iowa. *Ann. Missouri Bot. Gard.,* 33: 291–306.

Roth, Elmer A. 1955. The anatomy and modes of preservation of the genus *Cardiocarpus spinatus* Graham. *Univ. Kansas Sci. Bull.,* 37: 151–174.

Seward, Albert C. 1917, 1919. *Fossil Plants,* vol. III, 656 pp; vol. IV, 543 pp. Cambridge Univ. press.

Torrey, Ray E. 1923. The comparative anatomy and phylogeny of the Coniferales. Part 3. Mesozoic and Tertiary coniferous woods. *Mem. Boston Soc. Nat. Hist.,* 6: 41–106.

References—Ginkgophytes

Andrews, Henry N., Jr. 1941. *Dichophyllum moorei* and certain associated seeds. *Ann. Missouri Bot. Gard.,* 28: 375–384.

Brown, Roland W. 1943. Some prehistoric trees of the United States. *Journ. Forestry,* 41: 861–868.

Dorf, Erling. 1958. The geological distribution of the Ginkgo family. *Bull. Wagner Free Inst. Sci.,* Philadelphia. 33(1): 1–10.

Florin, Rudolf. 1936. Die Fossilen Ginkgophyten von Franz-Joseph-Land. I. Spezieller Teil. *Palaeontographica,* 81B: 71–173.

————. 1949. The morphology of *Trichopitys heteromorpha* Saporta, a seed plant of Paleozoic age, and the evolution of the female flowers in the Ginkgoinae. *Acta Horti Bergiani,* 15(5): 79–109.

Harris, Thomas M. 1935. The fossil flora of Scoresby Sound, East Greenland. Pt. 4. Ginkgoales, etc. *Meddelelser om Grönland,* 112 (1): 1–176.

Kräusel, Richard. 1943. Die Ginkgophyten der Trias von Lunz in Nieder-Österreich und von neue Welt bei Basel. *Paleontographica,* 87B: 59–93.

Oishi, Saburo. 1933. A study on the cuticles of some Mesozoic gymnospermous plants from China and Manchuria. *Sci. Rept. Tohoku Imperial Univ., Sendai (Geol. ser.),* 12: 239–252.

Seward, A. C. 1938. The story of the Maindenhair tree. *Sci. Progress,* 32: 420–440.

12

some gymnospermous plants
of uncertain affinities

It has been pointed out in previous chapters that the cordaite-conifer and the pteridosperm-cycadophyte lines in
all probability originated independently as seed plants; and within
the latter assemblage the morphological diversity is so great that
some paleobotanists regard it as composed of more than one division. I am strongly inclined toward the view that the seed has
evolved several times, for which further supporting evidence comes
from certain highly problematical fossils of gymnospermous affinities
that have been described in recent years. A few of these are considered in the following pages, and although they cannot be assigned
with any degree of confidence to a recognized group of seed plants,
they are still important for several reasons. Some were numerically
significant elements in the landscapes of their day, all present intriguing morphological features, and they indicate quite dramatically
the diversity of gymnospermous groups that is not revealed by the
extant flora.

It is safe to predict that many more such problematical fossils will
be excavated in the years to come; some may help to fill in noticeable gaps in the record as we know it and others will bring to light
previously unknown groups that flourished in the past and left no
close living relatives as a clue to their existence.

The plants included here are, therefore, a miscellaneous assemblage and certainly not closely related to each other. Where a
reasonable guess can be made concerning their affinities it will be
offered, but it seems to me more scientific and more honest to admit
our ignorance rather than force fossils into some established taxonomic category that is highly questionable.

The order in which the several plants or plant groups are considered is of no significance.

The Pentoxyleae

The name *Pentoxyleae* has been given to associated stems, leaves, seed-bearing, and microsporangiate organs from a Jurassic horizon in the Rajmahal Hills of India. They include a combination of coniferous and cycadean characters, as well as others that are wholly unique.

The stems, *Pentoxylon sahnii* Srivastava (Fig. 12-1A), reach several centimeters in diameter and contain a ring of five, or occasionally six, closely aggregated steles; each of these has a tangentially elongated strand of primary wood enclosed in secondary wood that is strongly endocentric in its development; that is, most of the

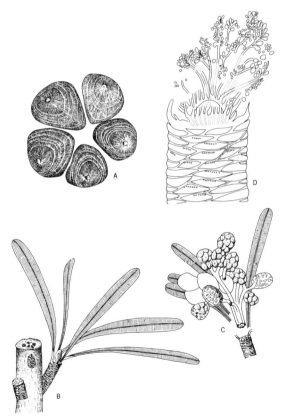

Fig. 12-1. The Pentoxyleae. A. Stelar system of stem showing endocentric development of secondary wood. B. Foliage-bearing branch. C. *Pentoxylon sahni,* female cone. *Sahnia nipaniensis,* male flower. (A–C from Sahni, 1948; D from Vishnu-Mittre, 1953.)

cambial activity took place on the inner face of the steles. The wood is typically coniferous, consisting of tracheids and small, uniseriate rays; the tracheids have one or more rows of somewhat crowded, circular-bordered pits on their radial walls. Well-defined growth rings are present. Alternating with these steles are five much smaller ones which consist chiefly of secondary wood.

Another genus of stems has been recognized (*Nipanioxylon*) in which the number of steles is greater and the endocentric growth is less strongly pronounced.

The stems bore branches 5 to 7.5 mm thick which are covered with an armor of closely aggregated leaf cushions, and it is thought that leaves, described under the name *Nipaniophyllum raoi* Sahni, (Fig. 12-1B) were attached to these shoots; they attained a length of 7 cm and were a little less than 1 cm broad. The vascular bundles compare quite closely in structure with those of a modern cycad, whereas the stomata are of the bennettitalean type.

Two species of seed-bearing organs (Fig. 12-1C) have been described, *Carnoconites compactum* Srivastava and *C. laxum* Srivastava; the peduncle of the infructescence divides into several branches or pedicels, each of which terminates in a cone. The cone axis contains a ring of five vascular strands and bore several closely compacted seeds with a thick, fleshy integument and with the micropyle directed out. No appendages are associated with the seeds.

The microsporangiate organs or "male flowers" (*Sahnia nipaniensis* Vishnu-Mittre) were terminally borne on shoots (Fig. 12-1D) resembling those of *Pentoxylon sahnii* and consist of a ring of filiform, spirally branched appendages which are fused at the base to form a disc. Unilocular sporangia terminate short subdivisions of the appendages and contain monocolpate or boat-shaped pollen grains.

Our knowledge of the Pentoxyleae is the result of the work of several Indian paleobotanists and in a summary account Sahni made the following comment:

> Some discoveries in science help, or appear to help, in the solution of the old outstanding problems; others—and these are perhaps the most interesting—seem to create new difficulties in our path. My object here is to draw attention to a recently recognized group of plants which defies classification and presents a new problem in our understanding of the evolution of Gymnosperms. (1948, p. 47)

The supposition that these plant organs represent a single genus or closely related genera of plants seems well founded. Assuming this to be the case the Pentoxyleae are of particular interest for their unique combination of characters; the wood is coniferous in its detailed structure but is unlike any conifer in the gross organization of the steles; the leaves show both cycadean and bennettitalean features, whereas the seed-bearing and microsporangiate organs are peculiar unto themselves. Since the evidence as cited in previous chapters indicates that there is no close relationship between cycadophytes and coniferophytes, it seems most appropriate to regard the Pentoxyleae as a wholly distinct group of gymnosperms.

The Vojnovskyales

The greater part of the readily accessible knowledge of the later Paleozoic floras comes from work that has been done in northwestern Europe during the past 150 years; to a lesser degree it is based on American fossils and still less is known of the vast areas of the USSR, China, and other parts of eastern Europe and Asia. Unfortunately, much of what has been published on the latter areas has not been read extensively by western paleobotanists; but even a hasty perusal of the Soviet literature dealing with Carboniferous and Permian plants indicates the presence of many interesting fossils and it is not surprising that wholly new types should be encountered.

One of the most fascinating of these discoveries is *Vojnovskya paradoxa* Neuburg, shown about natural size in Fig. 12-2. It was probably shrubby or arborescent in habit, but the actual size of the plant is not known. The leaves were fairly large fan-shaped organs of the *Nephropsis rhomboidea* Neuburg type; scars where several of these were attached are shown scattered over the surface of the shoot. Several fertile branches or cones are preserved; these are about 2.5 cm in diameter and include closely compacted, spirally arranged microsporophylls, each of which bears two pairs of sporangia. Scattered among them are megasporophylls described as being similar to seeds known as *Samaropsis;* these are about 1 cm long, bilaterally symmetrical, and with a notched tip.

Through the courtesy of Professor Neuburg I had the good fortune to see this specimen on a recent visit to the Geological Institute in Moscow. Additional information is particularly needed to elaborate the details of the reproductive organs, but it immediately gives one the impression of being totally unlike any other seed plant. The close association of microsporangiate organs and seeds,

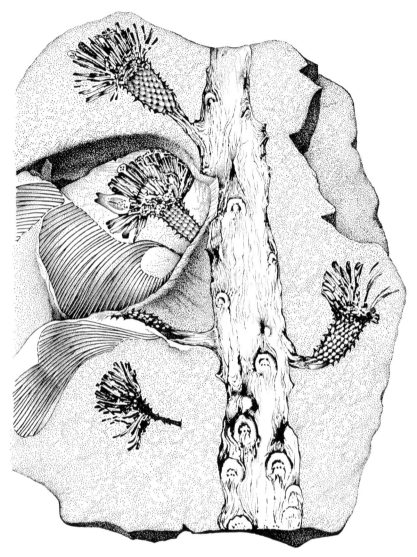

Fig. 12-2. *Vojnovskya paradoxa;* three branches with microsporangiate organs and seeds. (From Neuburg, 1955.)

as well as the form of the leaves, are particularly distinctive and, along with other unique details, segregate it from other Paleozoic seed plants.   One may see some resemblance to the bisexual cones of the Bennettitales, but it is hardly a close one.   The new order Vojnovskyales has been proposed by Neuburg for the plant.

Czekanowskia and Associated Reproductive Organs

Czekanowskia is a rather well-known genus based on leaves that are widely distributed in Jurassic rocks. Taking as an example *C. nathorsti* Harris, from Greenland, we find that the leaves are arranged in fascicles of about 15 on a short shoot. They reach a length of 15 cm, dichotomize three or four times, the segments being narrowly linear and only about 1 mm broad. They have long been regarded by most paleobotanists as the foliage of a ginkgophyte. Recently, however, Harris has described a unique seed-bearing organ, *Leptostrobus,* associated with the leaves. Although not found in organic connection, the two occur together in Jurassic rocks of East Greenland, in Yorkshire, and in eastern and southwestern Siberia. Thus the evidence from association supports the supposition that they may represent one and the same plant, and there are also similarities in the cuticle structure.

Leptostrobus (Fig. 12-3) is an infructescence or loose cone with a slender, unbranched axis; the appendages or "fruiting capsules" are arranged at intervals of about 5 mm and consist of a pair of slightly lobed valves up to 5 mm long and 5 mm wide with three to five seeds borne on each inner surface. Harris notes that a single one of these valves might be compared with the seed-bearing structures of a number of other gymnospermous groups, such as a

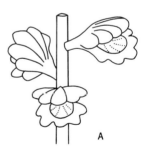

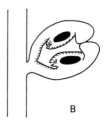

Fig. 12-3. *Leptostrobus longus.* A. Restoration of a part of the cone; B. a "capsule" in longitudinal section, megaspores in solid black. (From Harris, 1951.)

single *Caytonia* fruit, a *Cycas* megasporophyll, a *Cupressus* cone scale, or a several-seeded cupule of a Lower Carboniferous pteridosperm, but the two-valved nature of the appendages of *Leptostrobus* places it in a morphological category of its own and seemingly unrelated to any other group.

Glossopteris and Certain Allied Fossils

To do full justice to the literature that has developed around *Glossopteris* and certain fossils that are considered as related to it would require almost a separate volume. They have been the subject of much discussion in very recent years because of problematical reproductive organs that are found attached to the leaves (Fig. 12-4, 16-5B, C).

The name *Glossopteris* was first proposed by Brongniart in 1828 and it was considered for some time to be a fern. It is the key plant of the so-called *Glossopteris* flora being distributed through South Africa, India, Australia, and South America in rocks of late Paleozoic and early Triassic age. The leaves are spathulate, ovate, or linear-lanceolate and vary in length from 3 to 40 cm; in some species a distinct petiole is present whereas in others it is lacking. The side veins that pass out toward the margin from the conspicuous midrib form an anastomosing network, and in some species the leaves have been found arranged in whorls on slender stems. Smaller specimens are quite similar in appearance to the leaflets of *Sagenopteris* and on several occasions have been mistaken for them.

Recent studies of the epidermal structure of several species leave no doubt as to its being a seed plant and several types of unique reproductive organs have been discovered attached to the leaves. None of these is as well preserved as might be desired and they have given rise to much speculation.

Mrs. Plumstead has described a considerable collection (Fig. 12-4) from the Lower Permian of Vereeniging in the Transvaal in which the leaves bear reproductive organs thought to be seed-bearing cupules. Five such species are described under the name *Scutum*. The structures in question are described as a round, lanceolate or ovate, bilaterally symmetrical, cupules attached by a pedicel to the midrib of the leaf; evidence has been offered to support the view that the cupule was closed after fertilization, thus allowing the possibility of angiosperm relationships.

Fig. 12-4. Reproductive organs attached to or associated with *Glossopteris* leaves: A. *Scutum stowanum;* B. *Lanceolatus lerouxides;* C. *Scutum leslium;* D. *Scutum dutoitides.* (From Plumstead, 1952, 1956.)

What is referred to as the fertile half of an open cupule bears a central head in which are embedded several small oval sacs which range from 1 to 2 mm in size, with the entire head surrounded by a wing. In a later study of these plants Mrs. Plumstead described specimens which suggest that the fertile appendages may have been bisexual, that is, with pollen organs as well as seeds. No reasonable doubt exists that these are reproductive bodies, but beyond that there is considerable divergence of opinion. It has been suggested, for example, that the fructification is not a cupulate head but rather a strobilus, the supposed cupule actually consisting of laminate extensions of the closely aggregated "seeds."

More recently H. H. Thomas has described sterile and fertile leaves of the *Glossopteris* type from Natal; these are from a somewhat younger Permian horizon than *Scutum* and because of the distinctive nature of the reproductive bodies they have been given the name *Lidgettonia africana* Thomas. The leaves, although complete specimens are not known, range up to 15 cm long and 3 cm broad; the secondary veins are numerous, crowded, and anastomosing as is usual in *Glossopteris* foliage. Associated with these are much smaller, so-called scale leaves, which bear several umbrella-shaped structures. The latter are borne in two rows, there being a total of six or seven of the appendages, each consisting of a stalk about 5 mm long and a peltate disc 5 mm in diameter. Both isolated seeds and sporangia occur associated with the "fertile" leaves and it seems very likely that they were borne by the peltate appendages but had been shed prior to fossilization.

Gangamopteris is another characteristic leaf genus of the *Glossopteris* flora; it is similar in shape to *Glossopteris* and, indeed, some authorities have questioned the validity of a generic distinction. *Gangamopteris* leaves, however, lack a midrib and according to Seward the venation, although reticulate, shows greater uniformity in the size and shape of the meshes than does *Glossopteris;* they are also somewhat larger, reaching a length of nearly 40 cm.

Specialized, apparently reproductive, appendages have recently been found attached to the *Gangamopteris* leaf at its petiolar region. These were originally described by Zeiller a half century ago under the name *Ottokaria,* and the more recently discovered specimens consist of a stalk several centimeters long which bore a pair of disc-shaped appendages at its distal end; the margins of these had a row of flat appendages, lending a superficial appearance to a sunflower head with its conspicuous ray florets. Reproductive bodies, possibly seeds, were borne on the inner face of one disc and the other may have served as a protective device.

Although none of the presumed reproductive bodies found attached to *Glossopteris* and *Gangamopteris* leaves can be said to be well preserved, they are sufficiently so as to indicate an assemblage of plants that are not as closely related as the leaf structure might imply. *Glossopteris* has been classified in some accounts as a pteridosperm; Plumstead has suggested the new class Glossopteridae to include plants that bore *Ottokaria, Scutum,* and related fructifications. In the light of the cited facts, interesting but tantalizing, one can only say that we are dealing with a diverse group and one that is not closely related to any other.

Rhexoxylon

The genus *Rhexoxylon* is based on stems of rather complex anatomy which attain a diameter of 25 cm. Three species are known; *R. africanum* Bancroft, and *R. tetrapteroides* Walton from the Triassic (Molteno beds) of South Africa, and *R. priestleyi* (Seward) Walton from South Victoria Land, Antarctica. The following notations are based on Walton's description of *R. tetrapteroides* (Fig. 12-5).

The largest specimen is a trunk fragment 10 feet long which varies from 2 to 7 inches in diameter. A rather large pith occupies the central part of the stem and consists of a parenchymatous ground tissue which includes secretory ducts, sclerotic nests, and a

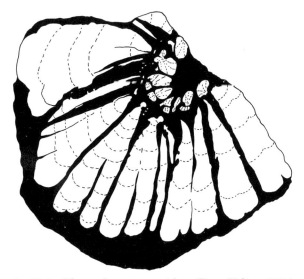

Fig. 12-5. *Rhexoxylon tetrapteroides.* (From Walton, 1923.)

few small irregularly scattered vascular strands. Around the outside of the pith is a ring of strands or small steles which make up what is known as the perimedullary system. Each strand usually consists of two segments, an outer one that developed centrifugally and an inner centripetal one; these are composed of only secondary wood and are separated by a narrow gap which probably was occupied by parenchyma in life.

Outside the perimedullary steles is another ring of strands of much greater radial extent; there is a small amount of primary wood at their inner apex and the rest is of secondary origin.

In its detailed structure the wood, which is mostly secondary throughout, is typically coniferous. The tracheids have one or two rows of circular-bordered pits on their radial walls; the rays are uniseriate and up to 15 cells high. Vascular strands have also been observed running horizontally between the wood sectors; these apparently originate from the perimedullary bundles and may represent leaf traces.

The general organization of the wood in *R. tetrapteroides* is reminiscent of certain modern angiospermous lianas (vines) such as *Tetrapteris, Banisteria,* and *Wilbrandia.* There is, however, no reason to believe that the fossil was angiospermous, being appreciably larger than any of the modern lianas.

There is considerable variation in the three species. In *R. priestleyi* the outer ring of wood was quite compact, indeed almost as much so as a coniferous stem, and the perimedullary system is not as strongly developed as in *R. tetrapteroides.* The opposite tendency is with *R. africanum;* in a specimen from Portuguese East Africa the "pith" region (internal to the outer ring of steles) is 12 cm in diameter and contains numerous bundles of the perimedullary type, whereas the steles of the outer ring are widely separated. There is thus some reason to suspect that the three may actually represent growth stages of but one species.

When first discovered it was suggested that *Rhexoxylon* might be related to the medullosas, but the detailed organization of the stems is wholly different. It seems reasonable to suppose that these are gymnospermous stems but no more can be said with certainty.

Hydropteridangium

Few, if any, paleobotanical investigations have ever brought to light so many previously unknown plants as Harris' account of the

Rhaetic-Liassic deposits of East Greenland. Of the many problematical fossils that he has described *Hydropteridangium marsilioides* Halle seems especially appropriate to this chapter. It is a microsporangiate organ (Fig. 12-6A) with a main axis about 4 cm long and 2 to 3 mm thick. The axis gives off lateral branches in all planes and these in turn divide irregularly in a three-dimensional pattern, ultimately terminating in capsules 3 × 2 mm. The capsule is a two-valved structure with about seven elongate sporangia (probably embedded) on the inner face of each valve; the sporangia contain numerous winged microspores.

On the basis of specimens found at Bjuf, Sweden, the plant was originally likened to the water-fern *Marsilia*, but Harris points out

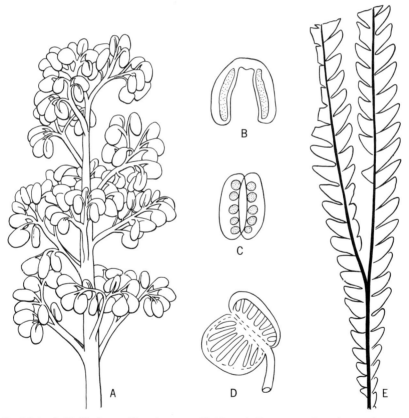

Fig. 12-6. A–D. *Hydropteridangium marsilioides.* A. Reconstruction of part of sporophyll, about 2X; B. longitudinal section of capsule; C. transverse section of unopened capsule; D. reconstruction of entire capsule. E. *Ptilozamites nilssoni,* about ½X. (From Harris, 1935.)

that the branching pattern, a gymnospermous type of cuticle, presence of only one kind of spore (*Marsilia* and the other water-ferns are heterosporous), and the association of these organs with the cycadlike foliage known as *Ptilozamites nilssoni* Nathorst (Fig. 12-6E), all render such a classification untenable.

The *Ptilozamites* fronds are unique in that the rachis dichoto-mizes twice, but otherwise they have the gross aspect of cycadophyte foliage; the stomates are of the gymnospermous type but with peculiarities unto themselves, and in general the cuticle structure compares closely with that of *Hydropteridangium.*

Perhaps the most likely possibility, though admittedly little more than a guess, is that *Hydropteridangium-Ptilozamites* repre-sent a unique group of gymnospermous plants derived from a com-mon ancestor with the cycadophytes.

Nucellangium

It is perhaps verging on the indiscrete to include a discussion of *Nucellangium* in a textbook. I have done so because it presents morphological features that may be of some importance and be-cause they epitomize the more vexing types of problems encountered in paleobotanical investigations.

Coal ball collections from an Upper Carboniferous horizon near Des Moines, Iowa, contain abundant "seedlike" bodies of two dis-tinctly different kinds, yet there is reason to suppose that they were borne by a single species of plant. They are abundantly represented in the flora of that horizon; in a petrifaction 6 inches in diameter a half dozen or more specimens are often revealed in a saw cut and, because of the distinctive preservation of the epider-mal cells, they can sometimes be broken out of the coal balls intact. For reasons that will be pointed out they will be called "normal" and "proliferated" reproductive bodies in the following description.

The "normal seeds" (Fig. 12-7A) are about 12 mm long and broadly ovate, the diameters in mid-section being approximately 6 × 9 mm. They were attached at one end as indicated by a small circular scar 0.5 mm in diameter, whereas the distal end tapers to a blunt point. In cross section (Fig. 12-7B) the following sequence of tis-sues is encountered. Outermost is an epidermis of radially elongate, thick-walled cells; it is probable that the original chemistry of the walls is but little changed and the sharp physical and chemical difference between them and the surrounding matrix is responsible for the ease with which they can be broken out intact. Next there is a broad zone of nearly isodiametric, thin-walled cells which composes a major portion of the wall; then follows a prominent layer of thicker walled cells that are at least twenty times as long as they are broad, bearing numerous oval-shaped pits. The end walls are transverse and do not appear to be truly tracheidal but seem

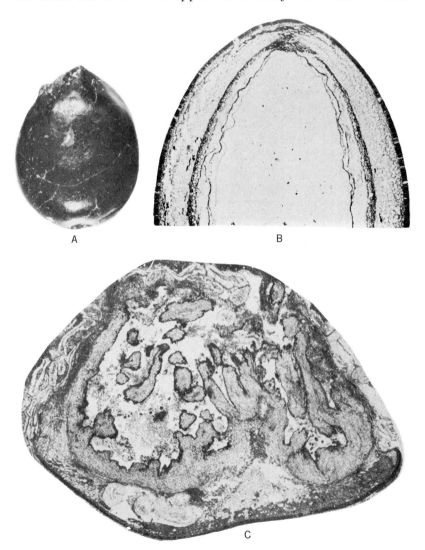

A   B

C

Fig. 12-7. *Nucellangium glabrum;* A. a "normal seed" showing point of attachment at the base, 4X; B. median transverse section showing half of the organ; C. section through a "proliferated" organ, 8X. (From Andrews, 1949.)

to be designed to serve as conducting elements. Finally there is an innermost layer of thin-walled cells that is usually poorly preserved. Within this tissue system is a distinct yellow band which is identified as a megaspore membrane.

A small, vascular strand enters the fossil at the point of attachment and divides to produce two strands which extend up to the distal end.

The so-called proliferated bodies (Fig. 12-7C) are slightly larger than the normal ones, measuring about 10×13 mm, and they possess an epidermis and outer parenchymatous layer that are identical with the normal ones. The resemblance ends at this point, for the parenchymatous tissue proliferates into numerous fingerlike lobes which are directed in a general way toward the center. It may be noted that the lobes are not preservation artifacts as they have a clearly defined epidermis; moreover, the peripheral portion of this parenchymatous tissue is vascularized and a very slender strand extends out into each lobe.

The normal bodies may be interpreted as a complex sporangium, with a single megaspore, or as a true seed. These possibilities will be considered further in Chapter 13, but the former interpretation may be tentatively accepted for two reasons: There is no evidence of a micropyle at the distal end and it has not been possible to distinguish between integument and nucellus.

The morphology of the proliferated bodies is even more vexing. It seems virtually certain that these were borne on the same plant, this conclusion being based on the close association of the two, their similarity in size, and particularly the identity of the peripheral tissues as described. It is therefore suggested that the proliferations are an aposporous growth of the parenchymatous tissue and that the body functioned as a kind of "gemma," or in somewhat simpler terms, that they represent a specialized asexual reproductive structure.

There is little doubt that future investigations will turn up a growing number of these "taxonomically difficult" plants adding interest and complexity to our knowledge of the plant world. Some of them will eventually be fitted into the evolutionary scheme of things, whereas others may always remain enigmatic.

REFERENCES

Andrews, Henry N., Jr. 1949. *Nucellangium,* a new genus of fossil seeds previously assigned to *Lepidocarpon. Ann. Missouri Bot. Gard.,* 36: 479–504.

———— and Mamay, S. H. 1955. Some recent advances in morphological palaeobotany. *Phytomorphology,* 5: 372–393.

Harris, Thomas M. 1951. The fructification of *Czekanowskia* and its allies. *Phil.
Trans. Roy. Soc. London,* 235B: 483–508.

————. 1935. The fossil flora of Scoresby Sound, East Greenland. Part 4. Ginkgoales, coniferales, lycopodiales and isolated fructifications. *Meddelelser om Grönland,* 112 (1): 1–176.

Neuburg, Maria F. 1955. New representatives of the Lower Permian Angara flora.
Doklady Acad. Sci. USSR, 102 (2): 613–616.

Pant, D. D. 1958. The structure of some leaves and fructifications of the *Glossopteris* flora of Tanganyika. *Bull. Brit. Mus. Nat. Hist. (Geology),* 3: 127–175.

Plumstead, Edna P. 1952. Description of two new genera and six species of fructifications borne on *Glossopteris* leaves. *Trans. Geol. Soc. South Africa,* 55: 281–328.

————. 1956. Bisexual fructifications on *Glossopteris* leaves from South Africa.
Palaeontographica, 100B: 1–25.

————. 1956. On *Ottokaria,* the fructification of *Gangamopteris. Trans. Geol.
Soc. South Africa,* 59: 211–236.

Sahni, Birbal. 1948. The Pentoxyleae: a new group of Jurassic gymnosperms from
the Rajmahal Hills of India. *Bot. Gaz.,* 110: 47–80.

Sen, J. 1955. On some fructifications borne on *Glossopteris* leaves. *Bot. Not.
(Lund),* 108: 244–252.

————. 1955. A *Glossopteris* bearing sori-like structures. *Nature,* 176: 742–743.

Thomas, H. Hamshaw. 1958. *Lidgettonia,* a new type of fertile *Glossopteris. Bull.
British Mus. Nat. Hist. (Geology),* 3: 179–189.

Vishnu-Mittre. 1953. A male flower of the Pentoxyleae with remarks on the structure of the female cones of the group. The *Palaeobotanist,* 2: 75–84.

Walton, John. 1923. On *Rhexoxylon,* Bancroft—a Triassic genus of plants exhibiting
a liana-type of vascular organization. *Phil. Trans. Roy. Soc. London,* 212B: 79–109.

————. 1956. *Rhexoxylon* and *Dadoxylon* from the Lower Shire region of Nyasa-
land and Portuguese East Africa. *Colonial Geol. Min. Resources, London,* 6:
159–168.

13

heterospory and the evolution of the seed

There are several attractive problems centered around the origin of the seed which seem worth considering in a separate chapter. For the purpose of discussion three questions may be cited as being particularly pertinent, although it is neither possible nor desirable to segregate them as wholly distinct from one another.

Are seeds of the gymnospermous groups strictly homologous; that is, was the seed evolved once, with later modifications producing the divergence in morphology that is now known, or has it evolved several times?

What is the significance of the many heterosporous pteridophytic groups in relation to seed evolution?

How early do seed groups appear in the fossil record?

Some exceptionally interesting discoveries have been made in recent years which bear on these questions; they are not wholly answered, but the accumulated evidence is impressive.

One point in particular should be emphasized—in all of the discussion that follows, relative to seed evolution, it is the pteridosperm seed (or seeds that are presumed to be referable to that group) that is chiefly alluded to. What bearing this may have on the corresponding organ of the angiosperms is beyond the scope of this account. Thus, in an attempt to answer the questions, at least in part, the order of discussion will be as follows:

Heterospory in the pteridophytic groups
Earliest evidence of heterospory
Earliest seeds or supposed seeds
Seed evolution in the pteridosperms
Telomic concept
Nucellar modification concept

Heterospory in the Pteridophytic Groups

There seems to be general agreement that there has been an evolutionary sequence from homospory to heterospory to the seed habit. "General agreement" does not necessarily mean that we are dealing with a truth, founded on clear-cut evidence, but in this case there is a certain amount of support for such a viewpoint and it is accepted as a reasonable working hypothesis. Homospory is the rule with all of the earlier land plants; so far as I am aware heterosporous plants do not appear in the fossil record until Upper Devonian (or possibly upper Middle Devonian) times, and unquestioned seeds have not been found below the Lower Carboniferous. It is very possible that heterosporous plants may ultimately be found in the mid-Devonian and that seeds will be identified in the late Devonian, but even if such discoveries are made in due course, the chronological order will remain unchanged.

It is perhaps not generally appreciated that heterospory is now known in quite a number of highly divergent pteridophytic groups. Most of them have been discussed in earlier chapters and they are only listed below; a few cases that have not found an appropriate place previously are described in more detail.

Arthrophyta. Heterospory is known in at least two Upper Carboniferous species of *Calamostachys: C. casheana* (Fig. 9-10) and *C. americana.* The large numbers of megaspores per megasporangium and relatively slight size difference (as compared, for example, with certain lycopods) indicate that heterospory exists here in a rudimentary stage.

Lycopodophyta. It does not seem amiss to say that heterospory has run rampant in this group. Many petrified and compression cone specimens have been recorded which reveal a general trend, starting in the Lower Carboniferous, from large numbers of megaspores to only one per sporangium, and heterospory is known in both herbaceous and arborescent forms. Enclosure of the megasporangium by the sporophyll resulted in a *seed* in both herbaceous lycopods (*Miadesmia*) and arborescent ones (*Lepidocarpon*).

In a study of 33 extant species of *Selaginella,* Duerden has reported much variation in megaspore number. In five species the numbers (of megaspores per megasporangium) were found to be in excess of four; seven species showed numbers less than four, whereas in a few both less and more than four were encountered. A few examples will illustrate the variability: in *S. willdenowii* four cases were encountered in which a megasporangium produced only one megaspore, one case of 16 megaspores, two cases of 36 megaspores, and four cases of 42 megaspores; in *S. lobbii* the following numbers were encountered: 8, 12, 14, 16, 20 and 24; in *S. inaequalifolia* v. *perelegans* it was found that a megasporangium usually contained two large spores and two small ones, and more rarely one large and three small. In a summary comment Duerden notes:

> The occurrence in *Selaginella* of megasporangia containing many comparatively small spores suggests a condition possibly not far advanced beyond the homosporous state, and on the other hand, the sporangia with fewer, comparatively large spores, indicate an advance in the direction of the seed habit." (1929, p. 456)

Archaeopteris. Heterospory is reported in one species, *A. latifolia,* from the Upper Devonian.

Stauropteris. The Lower Carboniferous *S. burntislandica* (Fig. 13-5) presents a unique type of heterospory in which only two megaspores matured. This is now known to have been a widely distributed plant.

Enigmophyton. The cones attributed to this very curious fossil (Fig. 2-11) from Spitsbergen were heterosporous; since they were not found in organic connection with the leafy shoots some doubt exists. The age is possibly as low as Middle Devonian.

Noeggerathia. Heterospory is clearly established here (Fig. 4-17) and the noeggerathias are certainly not closely related to any other pteridophytic plants.

Protopitys. This case has not been described in previous chapters and is therefore given in some detail at this point. What may represent an early stage in the evolution of heterospory has been reported in *Protopitys scotica* Walton from the Lower Carboniferous of Dunbartonshire, Scotland. *Protopitys* was originally established on petrified stems from Silesia (*P. buchiana* Goeppert) in which there is a characteristic elongate-oval pith with leaf traces departing from the opposite ends, the leaf arrangement being distichous. A good deal of secondary wood was formed resulting in stems up to a foot in diameter. The tracheids have elongate, almost scalariform, pits, and the rays are uniseriate and but a few cells high. It has enjoyed a varied taxonomic career, having been assigned tentatively to the cordaites, pteridosperms, and Filicales by different authors.

Protopitys scotica is based on a small branch fragment 7 cm long and 6 mm in diameter which presents the same general type of anatomy as *P. buchiana,* although only a small amount of secondary wood is formed. The specimen bears two lateral branches which subdivide by several unequal dichotomies and these are terminated by sporangia. The sporangia are all about 3 mm long and show no distinct annulus. The contained spores are 82 μ in diameter, but in some they are 147 μ, and in a few others the spores are more or less intermediate in size, measuring about 98 μ in diameter.

Evidence from isolated spores. Chaloner has recently described seven new species of megaspores from the Upper Devonian of Ellesmere Island, Arctic Canada. Two of these are similar to certain Carboniferous megaspores attributed to the lycopods, but the assemblage displays great range in morphology. For example, *Ocksisporites* is a triangular spore of 560 μ broad with a wide equatorial flange which is scalloped into sharp teeth; *Nikitinsporites,* which attains a diameter of 610 μ, is covered with exceptionally stout spines 200 μ long and 30 μ in diameter. It is improbable that such spores are lycopods, and being unlike those of any previously known plant, they suggest a greater variety of heterosporous plants than was formerly known to exist in the Upper Devonian.

Nikitin has reported several distinct spore types from the Voronezh region, some 500 kilometers north of the Sea of Azov; of particular interest are associated megaspores and microspores called *Kryshtofovichia africani.* The megasporangia are broadly ellipsoidal, measuring 2 to 3 mm long and 1.7 to 2.5 mm thick. The mature free megaspores are quite striking; the main body is spherical with three long and broad valves extending outward, forming a chamber called the androcamera. In addition the spores bore about a dozen stout spines up to 0.7 mm long with a terminal anchorlike hook. Microsporangia were found closely appressed to the megasporangia. They are somewhat smaller and are described as being occupied by a gelatinous, granulous material in which clusters of microspores were embedded. The microspores were found in some abundance attached to the megaspores and in the androcamera. The nature of the spore-bearing fructification as a whole is not known, but the unique form of the megaspores suggests a plant group that is distinct from any cited previously.

The diversity of heterosporous plants in the Lower Carboniferous has been demonstrated by Dijkstra and others by discoveries of an amazing variety of megaspores (Fig. 13-1). Although only a very few localities have been investigated, 60 species have been described chiefly from Egypt and the Moscow basin. The spores vary in size

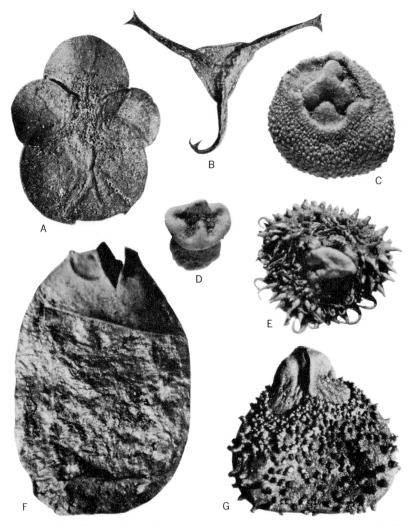

Fig. 13-1. Some Lower Carboniferous megaspores: A. *Cystosporites barbatus;* B. *Triletes hamatus;* C. *Triletes agninus;* D. *Cystosporites strictus,* abortive form; E. *Triletes furius;* F. *Cystosporites strictus,* fertile form; G. *Triletes acuminatus.* (From Dijkstra and Pierart, 1957.)

from about ¼ mm in diameter to almost 3 mm; some are smooth-walled, others have minute warty protuberances, massive bulbous, or long spiny ones, whereas still others display a distinct equatorial fringe of branching appendages. Many of these spores are probably lycopods, but the affinities of some are uncertain.

In view of the great diversity of the plants enumerated above, it seems certain that heterospory has arisen independently many times and in many different groups of plants. It also seems axiomatic that it has evolved from generally simpler homosporous plants in the articulates and lycopods; of the other heterosporous examples we have little or no information concerning their ancestry.

The Earliest Seeds or Seedlike Fossils

Although the presence of seeds in the Devonian has not been positively demonstrated, it is possible that they did exist in the latter part of that period or that they were actively evolving at that time. The reasons for this supposition lie in the well developed and, in some cases, rather elaborate seeds in the Lower Carboniferous, and the presence in the Devonian of fossils that are suggestive of being such and yet are not sufficiently well preserved to settle the matter. Several of these intriguing structures have been found in the Upper Devonian of Belgium:

Moresnetia zalesskyi Stockmans (Fig. 13-2A) is a rather slender dichotomizing branch system, with the ultimate divisions terminated by an elongate, deeply lobed structure. The latter tends to be rather strongly divided into two main lobes which are in turn lobed less deeply.

Condrusia rumex Stockmans (Fig. 13-2C) is an ovoid body which seems to have been partially enclosed by two narrow, spoon-shaped structures attached at the base. Another species, *C. minor* Stockmans, is equally interesting and problematical and I am inclined to believe should be placed in a distinct genus. It is an ovoid body which was attached by a fairly stout stalk, very little of which remains. At the other end there are two conspicuous hornlike projections and a greatly elongate central spine.

Xenotheca bertrandi Stockmans (Fig. 13-2D, E) is a slender, cupulelike organ with several lobes that tend to flare rather slightly. This genus was originally described from the Devonian of Devon, England.

It seems probable that all of these fossils were reproductive structures, but beyond this one can only speculate. *Condrusia rumex* is suggestive of a sporangium (nucellus?) partially enclosed by two sterile lobes. *C. minor* is a most perplexing structure. It may be recalled, however, that a number of Carboniferous seeds are bicornute in this fashion or have a lobed integument that might produce such an effect. It may be suggested that the long terminal "spine" is

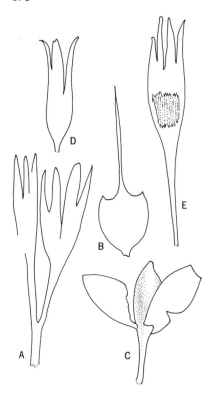

Fig. 13-2. Some problematical Upper Devonian fossils from Belgium. A. *Moresnetia zalesskyi;* B. *Condrusia minor;* C. *Condrusia rumex;* D, E. *Xenotheca bertrandi.* All about 3X. (Natural History Museum, Brussels.)

the attenuate apex of the micropyle, although Dr. Stockmans has interpreted this as the peduncle. As to *Xenotheca,* it seems most likely that this was either a cupule which enclosed one or more seeds or a microsporangiate organ. One of the specimens in the collections of the Brussels Museum reveals a rather heavy carbonaceous film in the basal portion that is suggestive of a seed. *Moresnetia* may have been simply a foliar organ, but it would have been a strange kind of leaf with such small "blades" borne out at the extremeties of a slender dichotomous branch system.

Seed Evolution in the Pteridosperms—with Particular Reference to the Nature of the Integument

There is admittedly a great gap in our knowledge between any of the heterosporous organs considered above and the seeds of the pteridosperms. Basically the pteridosperm seeds are regarded as being composed of a nucellus, within which a megaspore develops, which is enclosed by an integument, and in some this is in turn partially surrounded by the cupule. Petrified seeds that have been attributed to the pteridosperms may be separated into two groups on the basis of the relationship between nucellus and integument; in one group the two are said to be fused together and in the other they are separate or nearly so. An understanding of these structures, and their relationship to the cupule (or "outer integument"), seems fundamental to a consideration of the evolution of the seed and will be explored in the following pages. First, certain representative examples will be described briefly as a basis for discussion.

Seeds with nucellus and integument united

Lagenostoma lomaxi has been described in Chapter 5 (Figs. 5-2, 6-2A); its major features may be summarized briefly: the nucellus and integument are free only at the micropylar end; the integument is penetrated by a ring of nine vascular strands and is not lobed; the cupule is divided for about two-thirds of its length into several lobes which are also vascularized.

Sphaerostoma ovale (Williamson) Benson (Fig. 13-3B) from the Lower Carboniferous of Pettycur, Scotland, has a very closely

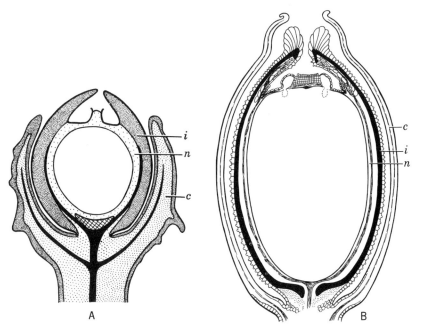

Fig. 13-3. A. *Tyliosperma orbiculatum*, 12X; B. *Sphaerostoma ovale*, 2.5X; both in median longitudinal section; *c*, cupule; *i*, integument; *n*, nucellus.

investing cupule. In the drawing I have taken the liberty to show it spread out slightly in order to render it clearly recognizable from the integument, the distal portion of which is lobed. The cupule is vascularized and an inner ring of strands are described as passing up through the integument; however, judging from the original illustrations of Miss Benson it is not easy to be sure whether these strands actually are running through the innermost part of the integument or within the nucellus.

Tyliosperma orbiculatum (Fig. 13-3A) presents us with features that are particularly cogent to the problem at hand. The distal part of the integument is lobed and the cupule, which is more deeply divided, is not as completely investing as is the case in *Sphaerostoma* or *Lagenostoma*. A set of vascular strands is present in the outer part of the nucellus rather than the integument. The epidermis extending over the apical portion of the nucellus is continuous with the epidermis of the integument—to allow complete freedom of thought in making morphological interpretations it may be more appropriate to note that there is a continuous epidermis over the tissues that would ordinarily be interpreted as integument and nucellus!

Seeds with integument and nucellus separate

The seeds included under this heading are ones that are classified, in many previous accounts, as the Trigonocarpales. Since, as noted in Chapter 5, there is some reason to believe that they were borne on medullosan plants the seeds are accordingly dealt with here as plants belonging to the Medullosaceae. The cupule (or a structure recognized as such) is not present; the integument is often massive, complex, and is free from the nucellus except in the basalmost part of the seed. Numerous vascular strands run through the length of the integument and the nucellus is heavily vascularized. *Pachytesta illinoense* (Fig. 5-13) and *Stephanospermum elongatum* may be regarded as representative; descriptions of both are given in Chapter 5.

There are two obvious problems of comparative morphology involved in an interpretation of the seeds considered above: What are the nature and origin of the integument and cupule in the first group? What are the nature and origin of the vascularized nucellus of the seeds of the second group (the medullosan seeds)?

The telomic concept

As long ago as 1904 Miss Benson suggested that the pteridosperm seed may be interpreted as a synangium in which all but one sporangium has become sterilized. This implies that the "seed state" was preceded by one in which a sporangium became closely invested by an encircling ring of sporangia which lost their spore-producing capacity and fused together to form the integument. More recently this concept has been somewhat simplified by Walton who proposes that the encircling elements (telomes) were sterile, that is, ordinary branch tips. There is now a great deal of evidence available in support of this theory.

Reference may be made first to the Silurian *Hedeia corymbosa;* the restoration given in Fig. 2-5 is based on specimens, discovered by Cookson, which demonstrated the three-dimensional nature of the terminal branchlets of the plant. It is not implied that this is a direct ancestor of the pteridospermous plants of the Carboniferous but merely as evidence of such a terminal clustering of sterile and fertile telomes. It seems pertinent to divert our attention for a moment to the pteridospermous microsporangia; Halle has suggested that the *Whittleseya* type synangium (see Chapter 5) is in fact a ring of fertile telomes which fused to form a "cyclic fertile syntelome." *Codonotheca,* in which the sporangia are only partially fused, may be regarded as an earlier stage, whereas *Aulacotheca,* in which there is only a small spical opening, may be a more advanced stage than *Whittleseya.* Thus in this sequence of fructifications the pattern of evolution is as clear-cut as may be expected from paleontological evidence.

Returning to the seeds, we find it significant to observe in *Lagenostoma* that the integument, although not lobed, is distinctly chambered and a vascular strand penetrates each unit; in *Tyliosperma* the integument is partially lobed. An important link in this line of evidence has been supplied by Long who has found Lower Carboniferous seeds in which the integument is free from the nucellus and divided into eight lobes for the greater part of its length. I am indebted to Mr. Long for allowing me to include this especially pertinent piece of information since the results of his investigations were in press at the time my manuscript was completed. Several stages in this proposed evolutionary sequence are shown in Fig. 13-4.

Evidence of a similar nature indicates a nearly identical evolutionary pattern for the cupule. In *Tyliosperma* it is deeply lobed; in *Lagenostoma* and *Calathospermum* fusion has taken place through the basal third of its length; and in *Sphaerostoma* there is only slight lobing, if any, at the distal extremity.

It seems appropriate to add a few correlating notes concerning the origin of the integument in the cordaite seeds. The latter are

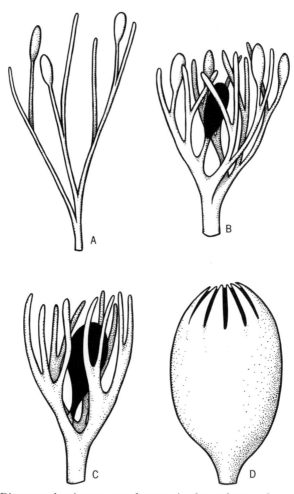

Fig. 13-4. Diagrams showing suggested stages in the enclosure of a sporangium to form nucellus and integument of a pteridosperm seed. A. Terminal branchlets (telomes), based on a plant of the *Rhynia* type. B. Corymbose clustering of fertile and sterile telomes with one sporangium assuming a central location; suggested by *Hedeia corymbosa*. C. Hypothetical stage; one central enlarged sporangium is surrounded by telomes that are tending to web together. D. A seed with sporangium (nucellus) enclosed by lobed integument. (Adapted in part from Walton, 1940.)

generally distinguished from pteridosperm seeds by their bilateral symmetry, that is, they appear strongly flattened when observed in cross section. It is not certain, however, that this character holds true in all cases.

Florin has observed that the fertile dwarf shoots of *Cordaian-thus williamsoni* bear some appendages that are forked at the tip;

a few of these have the primordium of a megasporangium, or what may be interpreted as an aborted nucellus, between the two lobes. It is therefore concluded that the integument of the flattened cordaitean seed originated by the forking of the megasporophyll and the two resultant branches ultimately completely enclosed the megasporangium.

It is evident here that the two lobes cannot be homologized with sporangia. Furthermore, in view of the fact that the cordaitean plant as a whole is so completely different from any pteridosperm, however primitive, there can be no doubt that this apparent enclosure of a megasporangium by a ring of telomes (in the pteridosperms) or a pair of telomes (in the cordaites) took place independently in the two groups.

The nucellar modification concept

There are several features of the pteridosperm seeds that are not readily explained by the telomic concept. One is the problem of the vascularized nucellus of the medullosan seeds; another is the fusion of integument and nucellus in the seeds of the Lyginopteridaceae.

In a recent discussion on the evolution of the seed Walton has pointed out that the so-called integument of the medullosan seeds may really be equivalent to the cupule of *Lagenostoma, Sphaerostoma*, and *Calathospermum* (and *Tyliosperma*, which has played an important role in our discussion, may be included); and in turn the outer vascularized part of the nucellus may represent an (inner) integument that is fused with a much reduced nucellus.

While admitting this as a valid explanation another possible approach is offered which may be termed the "nucellar modification concept." As a starting point reference may be made to the remarkable heterosporous coenopterid fern, *Stauropteris burntislandica*, that was introduced in Chapter 3. The megasporangium of this plant (Fig. 13-5) presents several exceptional features that are pertinent to the present discussion:

A slender vascular strand runs through the sterile portion; so far as I am aware this is the most extensive vascularization of any fern or fernlike sporangium.

In a few specimens the tapering apex has been observed to terminate in a minute opening.

Much of the sporangium (the basal half or more) is sterile, possibly serving as a food reservoir.

Quite clearly this is no ordinary megasporangium, yet it is not quite a seed. However, a seed might conceivably have evolved

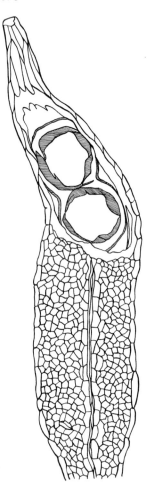

Fig. 13-5. *Stauropteris burntislandica,* a megasporangium in median longitudinal view showing apical pore, two megaspores, and slender vascular strand in center of lower half, about 80X.

from it along the lines indicated in Fig. 13-6. The first figure (A) represents a diagrammatic section through the *Stauropteris* megasporangium showing the apical opening, the two megaspores (enclosed in a cuticular envelope, the nature of which is problematical), and the central vascular strand. Figures B and C are hypothetical steps involving:

The reduction of two megaspores to one, and increase in size of the latter.

The "sinking" of the megaspore toward the basal part of the sporangium.

The displacment of the vascular strand and its division into two or more branches which are forced to grow up around the megaspore.

From this point (Fig. 13-6C) two possible lines of development are postulated:

With only a slight modification of the sporangium wall and apex, the vascularized nucellus of the medullosan type seed (Fig. D) is reached. It might also be noted that this appears to be the kind of organ that is represented by the "normal" *Nucellangium* seeds described in the previous chapter.

With somewhat more extensive modification or specialization of tissues in the peripheral wall of the megasporangium (C) a seed of

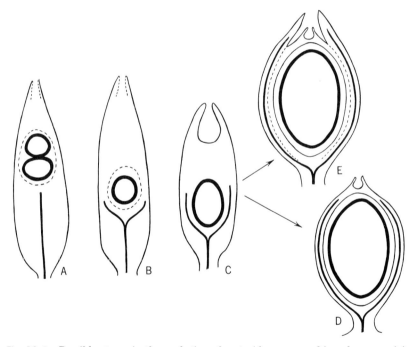

Fig. 13-6. Possible stages in the evolution of a pteridosperm seed based on an origin from a vascularized megasporangium. A. Diagrammatic longitudinal view of *Stauropteris burntislandica*. B, C. Reduction to single megaspore, "sinking" of megaspore, and branching of vascular strand. D. Seed of *Pachytesta* type with vascularized nucellus; integument not shown. E. Seed in which megasporangium wall is differentiated into "integument" (vascularized) and "nucellus." (Based in part on suggestions by John Walton and Sergius H. Mamay.)

the lyginopterid type (with "fused" integument and nucellus) is the end result (Fig. 13-6E).

In brief, this concept seems to offer a reasonable explanation for the vascularization of the nucellus of the medullosan seeds, the apparent "fusion" of integument and nucellus in the lyginopterid seeds, and the particularly vexing matter of the continuous apical epidermis in *Tyliosperma,* a feature that is difficult to account for if the structures interpreted as nucellus and integument actually developed as separate entities.

In presenting this concept I have drawn upon suggestions that have been offered by Dr. S. H. Mamay and Professor John Walton. It is not intended as a cure-all to explain the evolution of the pteridosperm seed; in fact, it is obvious that at best it only answers some of the problems that confront us.

REFERENCES

Arnold, Chester A. 1935. On seedlike structures associated with *Archaeopteris* from the Upper Devonian of northern Pennsylvania. *Univ. Michigan, Contrib. Mus. Paleont.,* 4: 283–286.

————. 1948. Paleozoic seeds. *Bot. Rev.,* 14: 450–472.

Benson, Margaret J. 1904. *Telangium scotti,* a new species of *Telangium (Calymmatotheca)* showing structure. *Ann. Bot.,* 13: 161–177.

————. 1914. *Sphaerostoma ovale (Conostoma ovale* et *intermedium,* Williamson), a Lower Carboniferous ovule from Pettycur, Fifeshire, Scotland. *Trans. Roy. Soc. Edinburgh,* 50 (1): 1–15.

Chaloner, William G. 1958. Isolated megaspore tetrads of *Stauropteris burntislandica. Ann. Bot.,* 22: 197–204.

————. 1959. Devonian megaspores from Arctic Canada. *Palaeontology,* 1: 321–332.

Cookson, Isabel. 1949. Yeringian (Lower Devonian) plant remains from Lilydale, Victoria, with notes on a collection from a new locality in the Siluro-Devonian sequence. *Mem. Nat. Mus., Melbourne,* 16: 117–130.

Dijkstra, S. J. 1956. Lower Carboniferous megaspores. *Mededelingen Geol. Stichting (Heerlen),* n.s., No. 10: 1–18.

———— and Piérart, P. 1957. Lower Carboniferous megaspores from the Moscow basin. *Mededelingen Geol. Stichting (Heerlen),* n.s., No. 11: 5–19.

Duerden, H. 1929. Variation in megaspore number in *Selaginella. Ann. Bot.,* 43: 451–457.

Hoskins, John H. and Aureal T. Cross. 1946. Studies in the Trigonocarpales. Part II. Taxonomic problems and a revision of the genus *Pachytesta. Amer. Mid. Nat., 36:* 331–361.

Mamay, Sergius H. 1954. Two new plant genera of Pennsylvanian age from Kansas coal balls. U. S. Geol. Survey Prof. Paper, 254-D: 81–95.

Martens, P. 1956. Un facteur évolutif négligé: le bec nucellaire de l'ovule. *Rev. Gen. Bot.,* 63: 529.

Nikitin, P. A. 1934. Fossil plants of the Petino horizon of the Devonian of the Voronezh Region. I. *Kryshtofovichia africani* nov. gen. et sp. *Bull. Acad. Sci.,* USSR No. 7: 1079–1092.

Stewart, Wilson N. 1954. The structure and affinities of *Pachytesta illinoense* comb. nov. *Amer. Journ. Bot.,* 41: 500–508.

Stockmans, François. 1948. Végétaux du Dévonien Supérieur de la Belgique. *Mem. Mus. Roy. Hist. Nat. Belgique, Mem.* 110: 1–85.

Thomas, H. Hamshaw. 1958. Fossil plants and evolution. *Journ. Linn. Soc. London, Zoology and Botany,* 56: 123–135.

Walton, John. 1953. The evolution of the ovule in the pteridosperms. *Adv. Sci.,* 10: 223–230. (British Assn. Adv. Sci., No. 38)

————. 1957. On *Protopitys* (Göppert): with a description of a fertile specimen *"Protopitys scotica"* sp. nov. from the calciferous sandstone series of Dunbartonshire. *Trans. Roy. Soc. Edinburgh,* 63: 333–339.

14

fossil plants of the
Arctic and Antarctic

Among the peculiar traits of mankind are his curiosity in a state of affairs other than what he finds here and now, his tendency to discredit phenomena that fall beyond the limitations of his own experiences, and his struggle to interpret the events of long ago and far away that he cannot hope to experience himself. A classic example of this is our interest in climates of the past and particularly of the polar regions. Some of the most bitter tales of man's effort to extend his influence over the surface of the globe come from his explorations beyond the Arctic circle, and what we know of that territory has been won very largely in the past century and a half. It is perhaps not surprising that, since only men of high intelligence as well as tremendous physical endurance could penetrate the most northerly latitudes up until a few years ago, they should have been intent on acquiring knowledge from all possible sources. Thus many expeditions within the past century have collected animal and plant remains where fossiliferous horizons were exposed. I think it is safe to say that one of the greatest surprises that the Arctic has yielded is the clear-cut evidence of lands richly forested in past ages which now are almost completely barren of plant life, or lie covered with snow and ice most of the year.

It has seemed to me that the paleobotany of the Arctic is of rather special interest and worth considering in a separate chapter, but in so doing it is not possible to avoid a little repetition here and there. It is clear, from the paleobotanical record as well as other evidence, that the present rigorous climate of the northern latitudes is very much the exception for that area during the past three or four hundred million years. We have a good deal of evidence from many geological horizons, but the difficulties and expense involved in Arctic travel have resulted in small collections, and these have been acquired, for the most part, quite incidentally to the particular expedition's main objectives. Only in certain parts of Greenland, Spitsbergen, and Alaska have paleobotanists actually done the collecting and carried out studies that are comprehensive. It is my impression, gained from the literature and from studies of the magnificent collections in the Swedish Natural History Museum, that we are only on the threshold of understanding the fossil floras of the Arctic; here is a vast expanse of the earth's surface awaiting paleobotanical investigation.

It has not been possible for the most part to pinpoint the age of fossil plant collections from the far north as accurately as in the more readily accessible parts of the earth, some having been identified, for example, as simply Tertiary or Carboniferous.

Tertiary

Fossil plants of many ages have been found in Spitsbergen. An older Tertiary flora has been described recently by Schloemer-Jäger from the Brögger Peninsula in which *Sequoia langsdorfii, Metasequoia occidentalis,* and *Cercidophyllum arcticum* (Katsura tree) are the dominant elements. Two other conifers, *Taxodium* and *Taiwania schaeferi,* were present in the flora and *Ginkgo* was well represented. An aquatic monocot, *Acorus brachystachys,* is reported, as well as the following dicots: *Cercidophyllum crenatum, Hamamelis clarus* (witch-hazel), *Acer spitzbergense* (maple), *Aesculus longipedunculus, Planera ulmifolia* (water-elm), doubtful species of *Alnus* (alder), and *Vitis* (grape) plus a few indeterminate dicot remains. Although some of the fossils in this flora are fragmentary, it is noted that the branches of *Metasequoia* and *Taxodium* are quite well preserved and could not have been transported any great distance. It is suggested that the plants indicate a temperate climate with fairly high rainfall.

Little has been done as yet with the microfossil floras of the Arctic, but initial investigations indicate that they will ultimately contribute a great deal. Spore studies of Paleocene-Eocene coals of West Spitsbergen have revealed 52 different types, some of which correlate with plants previously known from macrofossils and others are new. Where Nathorst (1910) reported 6 pteridophytes, Manum (1954) found 12 spore types, most of which are referred to the Osmundaceae and Polypodiaceae, with one possible lycopod. Numerous gymnosperms are represented by pollen, including ones that are tentatively identified as *Abies* and *Sciadopitys,* genera which are not known from macrofossils. The maple, water-lily, heath, sweet gale, and palm families are evident from the fossil pollen; the presence of palms in the Tertiary of Spitsbergen had not been known from the macrofossil record.

In 1911 a very readable account of Arctic floras appeared from the pen of A. G. Nathorst. In describing Tertiary plants from Harön, near Waigattet, Greenland, he notes that:

> In one of the deposits the fossil flora consisted almost exclusively of leaves of the Maple (*Acer*), crowded like those which cover the ground in autumn, and among these leaves large samaras, like those of *A. otopteryx* Gp., occur. In another bed the tuff was formed of cinders and small lapilli, and the way in which the vegetable fragments were embedded leads one to suppose that the branches, leaves, and fruits of the trees were broken off by a shower of cinders and lapilli. A medley of silicified branches of different sizes occurs, and among them are the cones of the Spruce, the nuts of the Walnut (*Juglans*), and the Hickory (*Carya*), with the leaves of *Ginkgo,* etc. (1911, p. 225)

Nearly a century ago Oswald Heer, in the course of his studies of Arctic fossils, described a fairly extensive flora from Iceland; it included *Equisetum,* several species of pine, alder, birch, hazelnut, oak, elm, sycamore, grape, tulip tree (*Liriodendron*), maple, walnut, and several others. Many of Heer's fossils are described from fragmentary specimens and some of the identifications may be in error. However, there are some quite well-preserved Tertiary leaves from Iceland in the Swedish Natural History Museum, apparently collected at a later date, and it is evident that the Tertiary flora of that island was more luxuriant than the present one.

Tertiary plants have been found in numerous localities in Alaska, a summary account having been given by Hollick in 1936. He recognized 327 species plus 42 that were identified only to the genus. Many elements in the floras are suggestive of a temperate climate. One hundred and twenty-six species (33%) are distributed in 12 amentiferous genera: *Populus* (poplar), *Salix* (willow), *Myrica, Juglans* (walnut), *Betula* (birch), *Carpinus* (hornbeam), *Alnus* (alder), *Fagus* (beech), *Castanea* (chestnut), *Quercus* (oak), and *Ulmus* (elm). A few of the dicots, however, indicate warmer conditions, as:

Piper (Piperaceae)
Engelhardtia (Juglandaceae)
Artocarpus, Ficus (Moraceae)
Grevillea (Proteaceae)
Magnolia (Magnoliaceae)
Cinnamomum, Laurus, Persea (Lauraceae)

Sophora, Canavalia (Papilionaceae)
Semecarpus (Anacardiaceae)
Cissus (Vitaceae)
Elaeocarpus (Elaeocarpaceae)

Convincing evidence of a mild climate is found in the palm, *Flabellaria florissanti*, from the Matanuska coal field in south central Alaska, at about 62° N. latitude. Other plants that seem to point strongly to a milder climate than the present one are the cycads *Dioon* and *Ceratozamia*, the conifers *Taxodium*, *Sequoia*, and *Metasequoia*, and several ferns.

These Alaskan floras come from many localities scattered through latitudes 57° to 65° and further study is needed to establish their exact position in the Tertiary.

Cretaceous

An extensive fossil flora has been found in western Greenland from Disco and Upernivik Islands and the Nugssuak Peninsula, an area within 70° to 71° north. This was studied in detail by the Swiss paleobotanist Heer many years ago and has been treated more recently by Seward who made collections in the summer of 1921. One of the results of his trip was a delightful little book entitled *A Summer in Greenland*, a short but informative account of the history, geology, living flora, and something of the fossil plants of this great ice-capped continent.

Two ferns are included in the Greenland Cretaceous flora, *Gleichenites* and *Laccopteris*. Among the gymnospermous elements are several cycadophytes, *Pseudocycas* and *Ptilophyllum*. Ginkgophyte leaves are abundantly represented and of diverse form, including specimens assigned to *Ginkgoites, Baiera,* and *Phoenicopsis*. Among the conifers are: *Sequoia concinna,* known from foliage shoots and cones, which is rather similar to the living California bigtree; *Pagiophyllum,* a supposed araucarian relative; abundant leaves of *Sciadopitytes* which compares most closely with the modern umbrella pine (*Sciadopitys*) of Japan. Among the dicots are:

Quercus (Fagaceae)
Artocarpus (Moraceae)
Menispermites (Menispermaceae)
Magnoliaephyllum (Magnoliaceae)
Laurophyllum, Cinnamomoides (Platanaceae)
Dalbergites (Leguminosae)

The identification of many of the conifers and dicots in this flora is admittedly questionable, but in this respect it is in no way unusual as a Cretaceous flora. Regardless of the accuracy of the names the assemblage presents striking contrasts with the extant flora of Greenland. The Cretaceous flora was more diverse, including trees of respectable size, and it is of a distinctly warmer climatic relation. The presence of an *Artocarpus* leaf and fruit (Fig. 14-1), which seem very similar to modern breadfruit (this is an Indo-Malayan genus), has served as a focal point for discussion of the Greenland climate. Seward notes:

> Nathorst's discovery in the Cretaceous beds at Igdlokunguak (latitude 70° on Disco Island) of splendid leaves and pieces of the inflorescences exhibiting a remarkably close resemblance to the foliage and fertile shoots of *Artocarpus incisa* Forst. affords one of the most impressive illustrations of the contrast between Cretaceous and recent Arctic vegetation. (1926, p. 115)

The specimen referred to is preserved in the Natural History Museum in Stockholm and is undoubtedly the most fascinating plant that has ever come out of the Arctic and one of the most discussed of all fossils.

It seems certain that the Cretaceous flora could not have lived in a climate that approached the rigor of the present one in Greenland, yet many of the plants probably could have withstood some frost. The ginkgophytes were abundant here as in other Arctic Mesozoic floras—a short diversion from our main theme is inserted here for whatever bearing it may have on the general problem of the growing conditions of past Arctic floras. For a tree that apparently came close to extinction *Ginkgo biloba* does remarkably well under a variety of situations in the United States. There are specimens in the Missouri Botanical Garden in St. Louis that have thrived under the violent contrasts that that city can produce, fluctuating from a minimum of $-22°$ in winter to 110° F in summer. There are fine specimens on the campus of the University of Massachusetts at Amherst (latitude 42°) where the winter minimum is $-26°$. However, in the Bergianska Trädgården in Stockholm there is one tree which the Director, Dr. Rudolf Florin, has informed me is at least 40 or 50 years old, but it is small (under 20 feet tall), of stunted, atypical growth, and has never produced reproductive structures; it is clearly an unhappy ginkgo tree. The winter minimum in Stockholm, which lies at latitude 59° is $-22°$ F. Ginkgo does not survive outdoors in Leningrad (latitude 60°) where the winter minimum is $-35°$ F. It would seem as though some factor other than the cold is responsible for its failure in Stockholm and

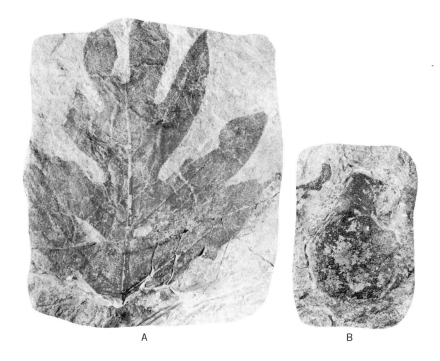

A   B

Fig. 14-1. Leaf (A) and fruit (B) of *Artocarpus dicksonii* from the Cretaceous of Greenland, about 0.5X. C. An *Artocarpus* growing on the island of Viti Levu, Fiji. (A, B Swedish Natural History Museum; C photograph by O. H. Selling.)

one cannot help wondering, in view of the great difference in latitude between that city and Amherst, whether light relations are responsible.

The dicots in the Greenland flora for the most part suggest a temperate climate, with *Artocarpus* and one or two others presenting notable exceptions. Seward concluded that the Cretaceous climate there was comparable with that of southern Europe today.

In his studies of the Cretaceous floras of Alaska Hollick recognized two distinct assemblages, one from the Yukon River region (latitude about 63 to 65°), considered as early Upper Cretaceous, and a second from the Alaska Peninsula. The Yukon River localities are rich in *Ginkgo* leaves and the cycadophyte *Nilssonia;* the dicots include:

> *Populus* (Salicaceae)
> *Quercus* (Fagaceae)
> *Macclintockia* (Proteaceae)
> *Aristolochia* (Aristolochiaceae)
> *Castalites, Nymphaeites, Paleonuphar* (Nymphaeaceae)
> *Menispermites* (Menispermaceae)
> *Magnolia* (Magnoliaceae)
> *Benzoin, Laurus, Daphnogene, Cinnamomum* (Lauraceae)
> *Platanus, Credneria* (Platanaceae)
> *Cotinus* (Anacardiaceae)
> *Sapindus* (Sapindaceae)
> *Vitis* (Vitaceae)
> *Sterculia* (Sterculiaceae)
> *Juglans* (Juglandaceae)

A preliminary report has been given by Dorf on a newly discovered late Cretaceous flora from the Knob Lake district of Labrador (latitude 55°). There are 36 species including 3 conifers and 27 angiosperms which are believed to have grown under humid, warm temperate conditions.

Jurassic Plants of Spitsbergen and King Karl's Land

Nathorst has described several florules from Spitsbergen (about 78.5° N latitude) of Jurassic age. The plants are predominantly leafy shoots and cones of conifers, cycadlike foliage, and fragments of fern fronds and ginkgophytes. Some are sufficiently scrappy as to suggest transport over some distance, but others such as *Elatides,* and a few nearly perfect cycadophyte leaves, indicate a nearby origin.

That the vegetation of Spitsbergen in Jurassic times was quite different from that of today is rather vividly brought out by Nathorst:

One of the coal seams at Cape Boheman furnishes a great abundance of *Podozamites* and *Pityophyllum;* sometimes the surface of the schists is as completely covered with the leaves of *Ginkgo digitata,* as the soil beneath a living *Ginkgo* tree may be in autumn. Since branches and seed of the same plant are also associated, it is natural to suppose that a *Ginkgo* forest occurred not far away from this spot. (1911, p. 221)

Late Jurassic silicified woods from King Karl's land (latitude 78°) have been reported up to 80 cm in diameter and showing 210 annual rings. In some of them the rings are more accentuated than in woods found at essentially the same horizon in continental Europe, which lead Nathorst to conclude that they grew in the region in which they were found and had not been transported from the south by marine currents. Five or six genera of coniferous woods have been described by Gothan from the Jurassic of Spitsbergen; the relationships of the woods with modern conifers is somewhat uncertain, but they at least reveal some diversity in the flora.

The Rhaeto-Liassic Flora of East Greenland

In a series of monumental works Harris (1926, 1931–1937) has given us the most informative account of an Arctic flora, as well as a very extensive and diverse one, that has been recorded to date. This is the Rhaeto-Liassic flora of Scoresby Sound (Fig. 14-2) on the coast of East Greenland at 71° N latitude. The plants come from two zones or horizons, the *Lepidopteris* zone which is Rhaetic and is regarded as transitional between the Triassic and Jurassic; and the *Thaumatopteris* zone which is considered to be lowermost Jurassic. Since it is the objective of this chapter to reveal the general floristics of the Arctic in the major periods rather than attempt a detailed stratigraphic analysis, the Scoresby Sound plants will be dealt with as a unit. Some have been described in other chapters and space here allows little more than a listing of the better known elements which display a striking contrast with the present vegetation of Greenland.

The liverworts are represented by both thalloid (*Thallites*) and leafy (*Hepaticites*) types. Several species of *Equisetites* have been recorded and one of these (*E. muensteri*) is described as being very close to certain living species. This is perhaps not surprising in view of the fact that the horsetails grow in Greenland today, but

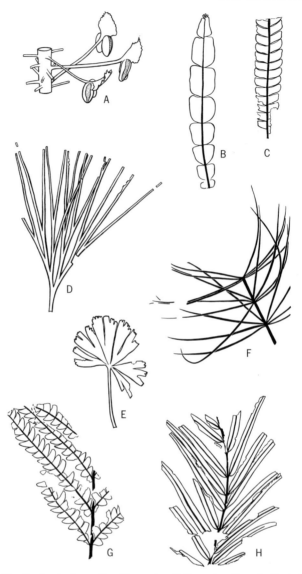

Fig. 14-2. Fossil plants from the Mesozoic of Scoresby Sound, East Greenland.
A. *Stenomischus athrous,* a male coniferous cone, possibly related to the extant *Cunninghamia.* B. *Anomozamites amdrupiana.* C. *A. nitida;* this bennettitalean genus is represented by several species. D. *Baiera spectabilis.* E. *Ginkgoites acosmia.* F. *Neocalamites hoerensis,* an articulate with especially long slender leaves. G. *Todites recurvatus;* both sterile and fertile fronds of several species of this osmundaceous genus are present. H. *Storgaardia spectabilis;* coniferous foliage twig possibly related to the podocarps. A is 6X, all others ⅓X. (From Harris, 1931–35.)

we encounter a contrast in the fossil *Neocalamites* with stems up to 10 cm in diameter and whorls of leaves which reach a length of 12 cm (Fig. 14-2F).

Several families of ferns are represented by abundant sterile and fertile foliage. There are five species of *Todites* (Osmundaceae) which bore dimorphic fronds apparently quite like certain living species. The Marattiaceae, a modern tropical family and one that we have traced back to the Carboniferous, is well represented. *Rhinipteris,* from Scoresby Sound, is compared with the Carboniferous *Ptychocarpus* and *Asterotheca. Marattiopsis,* which is very close if not identical with the living *Marattia,* is represented by fertile specimens with synangia which reach a length of 7 mm and include up to 17 sporangia on each side. Spore masses have been isolated intact and measure 900 μ by 350 μ and like the modern *Marattia* consists of several hundreds of spores.

Three other families of ferns are present, the Gleicheniaceae, Matoniaceae, and Dipteridaceae; since these are distinctive in their frond morphology and sporangial structure and well-preserved fertile specimens are present, there seems to be no question as to the identity of these fossils.

The variety of leaves assigned to the Ginkgoales is an especially distinctive feature of the flora; although they probably did not all live at one time in that area one gains the impression that it would have been an ideal place to have studied ginkgophyte evolution. There are several species of *Ginkgoites* which range from the slightly lobed *G. obovata* (Nathorst) Seward to species with deeply dissected leaves such as *G. acosmia* Harris, *G. minuta* (Nathorst) Harris, and *G. hermelini* (Hartz) Harris. There are also several species of *Baiera* as well as *Hartzia* and *Torellia.*

The coniferous elements of the Greenland flora were fairly numerous but they cannot be compared closely with modern ones. *Storgaardia* (Fig. 14-2H) is based on foliage shoots, the leaves being oppositely arranged, 10 cm long and 5 to 10 mm wide. The difficulty of classifying the Greenland conifers is typically illustrated here. Harris notes:

This striking fossil (*Storgaardia spectabilis* Harris) recalls *Podocarpus* subgenus *Eu-Podocarpus* in the form of its leaves, but subgenus *Nageia* in their insertion and the genus *Taxus* in the general appearance of its cuticle, although the stomata differ in detail in being monocyclic. (1935, p. 59)

Elatocladus is a genus with foliage superficially similar to *Taxus* and is represented by 16 species. *Podozamites* is a genus of broad-leafed conifers and Harris notes that it is probably not a natural one and may include species "with purely Cycad affinities." Other coniferous genera are *Araucarites* and several new ones erected for reproductive organs whose precise affinity is not known.

The Bennettitales is represented by 19 species of leaves and 12 of reproductive organs. Included in the former are *Anomozamites, Taeniozamites, Pterophyllum,* and *Otozamites.* Finally, the Scoresby Sound beds include several other fossils of particular interest which are not referable to the above mentioned groups and are considered elsewhere in this book: the *Caytoniales, Lepidopteris, Nilssonia, Hydropteridangium,* and others.

The Scoresby Sound flora is rich in its variety of plants and is a classic example of the wealth of information that can be obtained by the application of the proper techniques. As to the floristic contrast of this part of the Arctic between early Jurassic times and the present the following comment of Seward's seems especially apt:

> It is an impressive experience to stand on a flower-sprinkled Greenland heath and to see on a piece of shale split from the face of a ravine fragments of fronds closely allied to those of a *Gleichenia* which one had previously seen in very different circumstances growing on a tangled bank of ferns on the edge of a Malayan forest. The interval of millions of years separating the living from the dead is for the moment forgotten. In place of the bleak, treeless hills and the heath-covered slopes one sees a luxuriant undergrowth of ferns against a background of conifers and broad-leaved trees.

Carboniferous

Rather extensive Carboniferous floras have been found in Spitsbergen and northeast Greenland at points approximately 79° and 81° N respectively; the floras of both regions include lycopods, calamites, and sphenopterid foliage that is probably pteridospermous. *Lepidodendron* stems, from Spitsbergen, are described by Nathorst up to 40 cm in diameter and *Stigmaria* was observed with the rootlets radiating out through the sediments in which they were entombed; it is thus evident that the trees grew in the immediate vicinity and they were of respectable size. Silicified cordaitean (*Dadoxylon*) wood which lacks annual rings is also reported from Spitsbergen; this is admittedly an isolated bit of evidence which one would wish to confirm before arriving at any general conclusions, but if the long winter Arctic night prevailed at that time a pronounced indication of seasonal growth would be expected.

Lower Carboniferous collections from Traill Island, on the east coast of Greenland (latitude 72°), include *Calamites* and *Lepidodendron* stems.

Devonian

One of the most interesting Upper Devonian floras on record is from Bear Island, lying at approximately 74° N latitude between Spitsbergen and the North Cape of Norway. The early ferns were represented by magnificent fronds of *Archaeopteris roemeriana* (Goeppert) and *A. fimbriata* Nathorst, the latter being distinguished by delicately dissected pinnules which must have been magnificent in life. Another fern of exceptional interest is *Cephalotheca mirabilis* Nathorst, a genus that is closely related to, or identical with, *Rhacophyton* (see Chapter 3); massive clusters of sporangia were borne in pairs at the base of the frond and the general habit of the plant was probably quite like that of *Rhacophyton*. A fragmentary specimen of *Stigmaria* which measures about 15 cm in diameter is indicative of lycopods of arborescent dimensions; *Lepidodendron* stems are included in the flora and on some specimens isolated megaspores up to 2 mm in diameter are abundant. Spores of this size suggest lycopods of an advanced type. The diversity of Devonian heterosporous plants, including lycopods, is also indicated from Chaloner's study of megaspores from Ockse Bay, on the southwest coast of Ellesmere Island; these are mentioned in further detail in Chapter 13.

Most unusual of all the Bear Island plants is *Pseudobornia ursina*, an articulate that Nathorst regarded as constituting a distinct order, the Pseudoborniales. The shoots are about 0.5 cm in diameter and bear whorls of leaves, probably four at each node. Each leaf is several times dichotomized, the ultimate divisions being delicately fimbriate to the point of resembling small feathers. Associated articulate stems about 9 cm in diameter are believed to have borne the leafy shoots.

Professor Høeg has given us the most complete picture of Arctic Devonian vegetation for any single area in his study of several fossil floras from Spitsbergen. Some of the more interesting plants have been described in previous chapters—the problematical *Enigmophyton superbum*, the protopterid *Svalbardia polymorpha*, and the articulate *Hyenia vogtii* with its distinctive branch morphology. The lycopods are represented by stems of *Bergeria mimerensis* Høeg up to 10 cm in diameter; they are partially decorticated, making comparison with other genera rather difficult. Slender axes that are not more than 1 cm in diameter and characterized by a small, circular leaf scar located at the upper part of the slender, vertically elongate cushions are given the name *Protolepidodendropsis pulchra*. Both of these lycopods are from Middle Devonian or basal Upper Devonian horizons. *Actinopodium nathorstii* is a small, leafless axis with a star-shaped stele consisting of a central region of mixed parenchyma and multiseriate pitted tracheids surrounded by a narrow zone of radially aligned, apparently secondary tracheids. Unfortunately not enough of this plant is available to afford a very sound basis for speculation, but it may possibly represent an advanced psilophyte.

Summary Comments on the Arctic Climate

The evidence from fossil plants alone leaves no doubt that the frigid climate of the Arctic, which has prevailed through the Pleistocene to the present time, is very much the exception rather than the rule. The records from the Devonian on indicate conditions quite in contrast to what is known today. There have probably been significant fluctuations during the past few hundred millions of years and it is not implied that tropical forests ever forged their way into extreme northern latitudes. Yet, even though the present inhospitable climate may be temporary in terms of geologic time, it is with us and we are anxious to know not only what caused the change but what the future may hold in store. Are we now living in an interglacial period of the Pleistocene or are we on our way back to the generally warmer climates of pre-Pleistocene times?

There is a voluminous literature on the causes of climatic change in which one may find a sufficient range of explanations to suit almost any reader; a few have been especially popular such as changes in the position of the poles, drifting continents, and shifting ocean currents, but none thus far has met with anything approaching general agreement. At the risk of adding speculation on speculation it has seemed to me most appropriate to mention some of the approaches that have been undertaken by recent investigators.

Types of evidence are coming from sources that only a few years ago were quite unknown; the use of oxygen isotopes is one of these methods. It has been shown by H. C. Urey that when water evaporates oxygen 16, 17, and 18 are lost at different rates.

Slightly higher proportions of the common oxygen 16 are carried off in water vapor and as a result water becomes slightly enriched in the rare heavy isotopes 17 and 18. The ratios of the two will vary slightly in carbonates such as limestone, coral, or limy skeletons of other aquatic animals, partly according to the temperature of the water. Studies of bottom-dwelling foraminifera taken from deep sea drill cores in the Pacific indicate that the bottom temperatures, resulting from cooler surface waters of the Antarctic sinking and flowing northward, have dropped about 16° F in the past 32 million years.

A new and somewhat controversial approach to an understanding of fossil climates comes from paleomagnetic studies. Recent investigations of the magnetization of rock formations indicate that some sedimentary deposits may retain their direction of magnetization over long periods of geologic time and the direction perpetuates that of the earth's field at the time they were formed. For example, certain Carboniferous strata of Derbyshire and Yorkshire, England, are said to be magnetized with a low angle of dip which suggests an original magnetization in a field fairly close to the equatorial region. Paleomagnetic measurements in western United States also indicate a closer proximity to the equator in the late Paleozoic. In both western United States and in Great Britain studies of windblown sand deposits correlate with the paleomagnetic evidence in suggesting that these areas were within the belt of the northern trade winds, assuming that the present width of this belt is typical. Thus the evidence from this direction supports the view that the axis of rotation of the earth, also referred to as polar wanderings, may have undergone significant changes in the past.

I have mentioned elsewhere the possibility of physiologic changes in the plants themselves. This is summed up rather well in the following comment by Axelrod:

Cain (1944, p. 75) has pointed out that breadfruit (*Artocarpus*), which is now strictly tropical, and which has been recorded at higher latitudes, may have been represented in Cretaceous and Early Tertiary floras by one or more temperate to warm temperate species as in the case of certain genera today, such as persimmon (*Diospyros*), avocado (*Persea*) and magnolia (*Magnolia*). We may add in this connection that a number of genera which are frequently regarded as temperate, such as alder (*Alnus*), maple (*Acer*), birch (*Betula*), haw (*Crataegus*), katsura (*Cercidophyllum*), sweet gum (*Liquidambar*) and sycamore (*Platanus*), were widely represented in the warm temperate to subtropical zone during the Cretaceous and Early Tertiary. Pertinently, many of the widely distributed families in the northern hemisphere which are often considered as temperate, including the Aceraceae, Betulaceae, Fagaceae, Rosaceae, Ulmaceae and others, extend into the tropics. The most primitive living species of maple, poplar, birch, oak, cherry and hackberry are largely evergreen types, and are found in warm temperate to tropical regions. (1952, p. 51–52)

Some Fossil Plants of the Antarctic

As compared with the Arctic we have as yet very little information on the past floras of the South Polar region. During the course of the Swedish Southpolar Expedition of 1901 to 1903 a rather large collection of Jurassic plants was made at Hope Bay which is located near the tip of the Palmer Peninsula at latitude 63°. Included in the flora are: *Equisetites* stems up to 3 cm broad; well-preserved fertile and sterile fronds of *Todites,* an osmundaceous fern that is widely distributed through the Jurassic; and several species of *Cladophlebis* and *Coniopteis;* cycadophyte foliage is the most conspicuous element of the flora with large specimens of *Nilssonia taeniopteroides* Halle, *Pseudoctenis, Zamites, Otozamites,* and *Ptilophyllum;* judging from the long fragments preseved the leaves of *Nilssonia* reached a length of about 1 meter and those of *Pseudoctenis* are at least 14 cm broad; a few conifers are included but no angiosperms or lycopods.

Several coniferous and dicot woods have been found in lower Tertiary deposits on Seymour and Snow Islands, also located on Palmer peninsula, at 64° S latitude.

A Tertiary leaf flora from Seymour Island includes *Lauriphyllum, Drimys, Iliciphyllum, Lomatia, Knightia, Fagus, Nothofagus, Myrica,* several ferns, and two species of *Taeniopteris.* The preservation of these plants is for the most part rather poor and many of the identifications perhaps questionable. A preliminary study by Cranwell of spores and pollen found in a rock sample from Seymour Island offers the hope that this approach will greatly enhance our knowledge of Antarctic floras. The age of the sample is somewhat uncertain and may include mixed Cretaceous and mid-Tertiary fossils. The dominant elements are conifers, including *Araucaria, Agathis* and *Podocarpus,* and several southern beeches (*Nothofagus*). The fern families Cyatheaceae and Schizaeaceae are represented and pollen of the following flowering plant families are tentatively identified: Cruciferae, Myrtaceae, Proteaceae, Loranthaceae, Oenotheraceae, and possibly the Winteraceae.

It seems evident from the little that is known at present that the Antarctic, like the Arctic, has enjoyed a much milder climate in the past.

REFERENCES

Cranwell, Lucy M. 1959. Fossil pollen from Seymour Island, Antarctica. *Nature,* 184: 1782–1785.

Dorf, Erling. 1959. Cretaceous flora from beds associated with iron-ore deposits in the Labrador trough. (abstract). Geol. Soc. Amer. 1959 program of meetings, p. 33A.

Dusén, Carl H. 1908. Über die Tertiäre flora der Seymour-Insel. Wissensch. *Schwedischen Sudpolar-Exped.,* 1901–1903. vol. III, lief 3: 1–27.

Emiliani, Cesare. 1958. Ancient temperatures. *Scientific Monthly,* 198 (2): 54–63.

Gothan, Walther. 1908. Die Fossilen Holzer von der Seymour und Snow Hill Insel. Wissensch. *Schwedischen Sudpolar-Exped.,* 1901–1903. vol. III, lief 8: 1–31.

————. 1908. Die fossilen Hölzer von König Karls Land. *Kungl. Svenska Vetensk. Akad. Handl.,* 42 (10): 1–44.

————. 1910. Die fossilen Holzreste von Spitzbergen. *Kungl. Svenska Vetensk. Akad. Handl.,* 45 (8): 1–56.

Halle, Thore G. 1913. The Mesozoic flora of Graham Land. Wissensch. *Schwedischen Sudpolar-Exped.,* 1901–1003. Vol. III, lief 14: 1–123.

————. 1931. Younger Palaeozoic plants from East Greenland collected by the Danish expeditions 1929 and 1930. *Meddel. om Grönland,* 85 (No. 1): 1–26.

Harris, Thomas M. 1926. The Rhaetic floras of Scoresby Sound, East Greenland. *Meddel. om Grönland,* 68: 45–147.

————. 1931–37. The Fossil flora of Scoresby Sound, East Greenland. *Meddel. om Grönland.* Pt. 1, vol. 85, No. 2; pt. 2, vol. 85, No. 3; pt. 3, vol. 85, No. 1; pt. 4, vol. 112, No. 2; pt. 5, vol. 112, No. 3.

Heer, Oswald. 1868. Die fossile Flora der Polarländer, enhaltend die in Nordgrönland, auf der Melville-Insel, im Banksland, am Mackenzie, in Island und in Spetzbergen entdeckten fossilen Pflanzen. pp. 1–192. Zürich.

Høeg, Ove Arbo. 1942. The Downtonian and Devonian flora of Spitsbergen. *Norges Svalbard-og Ishavs-Undersökelser,* 83: 1–228.

Hollick, Arthur. 1930. The Upper Cretaceous floras of Alaska. U. S. Geol. Survey Prof. Paper, 159: 1–116.

————. 1936. The Tertiary floras of Alaska. U. S. Geol. Survey Prof. Paper, 182: 1–185.

Manum, Svein. 1954. Pollen og sporer I Tertiaere Kull fra Vestspitsbergen. *Blyttia (Oslo),* 12: 1–9.

Nathorst, Alfred G. 1890. Ueber die reste eines Brotfrucktbaums, *Artocarpus dicksoni* n. sp., aus den Cenomanen Kreideablagerungen Grönlands. *Kongl. Svenska Vetensk. Akad. Handl.,* 24I 1–10.

————. 1894. Zur Palaozoischen flora der Arctischen Zone. *Kongl. Svensk. Vet. Akad. Handl.,* 26, No. 4.

————. 1897. Zur Fossilen Flora der Polarländer. *Kongl. Svenska Vets. Akad. Handl.,* 30 (1): 1–77.

————. 1911. Contribution to the Carboniferous flora of Northeastern Greenland. *Meddel. on Grönland,* 43: 339–346.

————. 1911. On the value of the fossil floras of the Arctic regions as evidence of geological climates. *Geol. Mag.,* 48, 8: 217–225.

————. 1907. Über Trias und Jurapflanzen von der Insel Kotelny. *Mem. Akad. Imp. Sci. St. Petersbourg,* ser. III, vol. 21, No. 2.

Opdyke, N. D. and S. K. Runcorn. 1959. Palaeomagnetism and ancient wind directions. *Endeavour,* 18: 26–34.

Runcorn, S. K. 1955. The permanent magnetization of rocks. Endeavour 14: 152–159.

Schloemer-Jäger, Anna. 1958. Altertiäre Pflanzen aus Flözen der Brögger-Halbinsel Spitsbergens. *Palaeontographica,* 104B: 39–103.

Seward, Albert C. 1922. *A Summer in Greenland.* Cambridge Univ. Press. Pp. 1–100.

Seward, A. C. 1925. Arctic vegetation past and present. *Journ. Roy. Horticultural Soc.,* 50: 1–18.

————. 1926. The Cretaceous plant-bearing rocks of western Greenland. *Phil. Trans. Roy. Soc. London,* 215B: 57–175.

15

HEPATOPHYTA and BRYOPHYTA
— the bryophytic plants

The mosses and liverworts are ancient groups and their position in the plant kingdom is controversial; they are all small by comparison with the vascular plants although very common and widely distributed over the surface of the earth. Quite a few favor moist shady places, many are epiphytes on large plants in both temperate regions and the tropics, and others are able to thrive in the most exacting habitats. Although they are not often preserved as fossils we have a few records that are of exceptional interest.

In order that the use of terms may be clear a brief consideration of the classification of the plants under consideration seems essential. It is, however, recommended that the student who is not familiar with representative mosses and liverworts should consult one of the many good elementary botany texts, virtually all of which devote a chapter to these plants. For the purpose of the following discussion it will be helpful to become familiar with the essential features of such plants as *Marchantia, Porella, Anthoceros, Polytrichum,* and *Sphagnum.* In all of them the sexual phase is the dominant one in the life cycle, the opposite being the case in the pteridophytic groups and seed plants. The first three genera cited are liverworts (Hepatophyta) although they differ from one another in no small degree. In *Marchantia* the main body of the plant is a flat, ribbon-shaped, dichotomously branching structure with a complex chambered internal anatomy. *Porella* is one of the many leafy liverworts and superficially resembles certain mosses, although the lack of a midrib in the leaf and other important characters distinguish it from the group. *Anthoceros* (the horned liverwort) consists of a thin, irregularly dichotomizing thallus which lacks the chambered organization of *Marchantia,* but it has a sporophyte that is larger and more conspicuous than that of other liverworts; it is, moreover, photosynthetic and thus partially independent and it possesses a basal meristem which allows continued growth for several months. *Polytrichum* (hair-cap moss) is a very common moss with upright, radially symmetrical leafy stems with conspicuous sporophytes that develop at the top of the female plant. *Sphagnum* is the peat moss, a distinctive feature of the northern bog vegetation. Perhaps its most striking feature lies in the leaves which are composed of two types of cells, small photosynthetic ones which give the plant its pale green color, and large empty ones that serve as minute water storage tanks and which render the plant valuable in gardening as a conditioner for heavy soils.

Living species and genera of mosses and liverworts are often delimited by structural details of microscopic magnitude. Consequently it is rarely possible with fossils to compare them with modern species, genera, or even families. Thus we need not be concerned with a detailed classification although a few notations on the major categories may be in order. The mosses and liverworts have long been treated as two subgroups under the major division Bryophyta; the differences between the two are, however, great and it seems to me that the classification given by Bold in his recent book *Morphology of Plants* is more realistic in regarding them as two distinct divisions. Since it is, nevertheless, convenient to have an inclusive designation for the entire assemblage, the term *bryophytes* or bryophytic plants will be employed here in that way. A few other comments on the classification and evolution of these plants will be more appropriate at the close of the chapter.

Standard classification	*Classification of Bold*
BRYOPHYTA	HEPATOPHYTA (liverworts)
Hepaticae (liverworts)	BRYOPHYTA (mosses)
Musci (mosses)	

Fossil Hepatophyta

The oldest authentic liverwort is *Hepaticites kidstoni* Walton from the Yorkian series of the Upper Carboniferous of Shropshire, England. Although its exact affinities are not known it is remarkably similar to some of the modern leafy liverworts and indicates that the group had already enjoyed a long period of evolution. It was a small plant, the specimen shown (Fig. 15-1A) being but a few millimeters long. The cellular structure is quite well preserved; in addition to two lateral rows of large conspicuous leaves there are two alternating series of small semicircular flaps on the under (or upper?) surface.

A thalloid liverwort, *Hepaticites willsi* Walton, is also known from the Upper Carboniferous (Staffordian series of Staffordshire). It is a dichotomously branching plant averaging about 1 mm in breadth. In another thalloid type, *Hepaticites langi* Walton from the Upper Carboniferous of Shropshire, the rhizoids were found to be preserved.

The fossils described above indicate that both thalloid and leafy types existed in the Upper Carboniferous, but little more can be said about their affinities within the Hepatophyta.

The naming of fossil liverworts or presumed liverworts has tended to be somewhat haphazard; owing in part to considerable variation in the quality of preservation, a completely acceptable system is probably not possible, but the following use of terms seems reasonably satisfactory:

Thallites Walton is employed for thalloid plant remains which may be liverworts but which cannot positively be distinguished from certain other groups of plants.

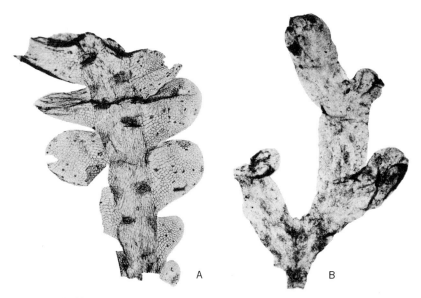

Fig. 15-1. A. *Hepaticites kidstoni;* Upper Carboniferous, Shropshire, 30X. B. *Hepaticites willsi;* Upper Carboniferous, Staffordshire, 16X. (From Walton, 1925.)

Hepaticites Walton is employed for fossils that can with some assurance be determined as liverworts and in which characters are present that enable one to distinguish them from algae, pteridophytes, and other groups of plants, yet their structure is not sufficiently complete to warrant the use of a distinctive generic name or one that implies close affinity with living genera.

Marchantites Brongniart, according to Harris "is restricted to hepatics agreeing with *M. sezannensis* Brongniart, the type species, in which marchantiaceous air chambers, ventral scales and reproductive organs have been demonstrated." (1942, p. 394)

There are many reports of thalloid liverworts from Mesozoic and Tertiary horizons, but for the most part they only inform us that these plants existed, probably in about the same form that we find them today; the preservation is such that one can usually do little more than assign them to the noncommittal *Hepaticites,* although a few offer more precise information.

A rather rich liverwort assemblage has been described by Miss Lundblad from the Rhaetic-Liassic coal mines of Skromberga, Sweden, which includes *Ricciopsis florinii,* a small dichotomously branching thallus about 1 to 3 mm wide which developed a rosette-like growth 2.5 cm in diameter. It is considered to be closely related to certain modern members of the genus *Riccia.*

Marchantites hallei Lundblad from the Lower Cretaceous of Patagonia shows air pores and two rows of ventral scales plus numerous rhizoids, but the internal structure is not known.

Harris has described several species of *Hepaticites* from Scoresby Sound, Greenland: *H. rosenkrantzi* is a thalloid plant up to 2 cm long and 3 mm wide; no midrib is evident but round or oval cups 1.5 to 3 mm in diameter were borne on the surface which may represent gemma cups such as are found in *Marchantia.*

A few fossil liverworts are well enough preserved to allow a somewhat more precise identification. *Metzgeriites glebosus* (Harris) Steere, also from Scoresby Sound, is a dichotomously forking thalloid plant with a lamina that is divided into irregular lobes, and numerous unicellular unbranched rhizoids are attached to the underside of the midrib. It has been assigned this generic name because of undoubted affinities with the anacrogynous Jungermanniales.

Fossil Bryophyta

Fossil mosses were described by Renault and Zeiller in 1888 from the Upper Carboniferous of Commentry in central France under the name *Muscites polytrichaceus*. Professor Walton has reexamined the specimens and concludes that they have been correctly identified. It is, however, of considerable interest to note the discovery of beautifully preserved mosses in Upper and Lower Permian horizons in the Kuznetsk basin and in Vorkouta, respectively, of the USSR. Professor Neuburg has described some remarkably well-preserved plants (Fig. 15-2) representing several species; the preservation is such that perfect leaf compressions can be removed from the shale and mounted on a slide for study in much the same way that a modern moss leaf would be handled. They show the characteristic leaf structure of the Bryales, a subdivision of the Bryophyta; the central vein is conspicuous and in at least one species there seems to be a tendency toward the development of side veins. During a visit to her laboratory in October of 1958 Mrs. Neuburg also showed me small rosette-shaped fossils composing a branching moss plant, or a portion of one, in which the leaves have two distinct cell types which are strongly suggestive of the Sphagnales (peat mosses). The spore-bearing

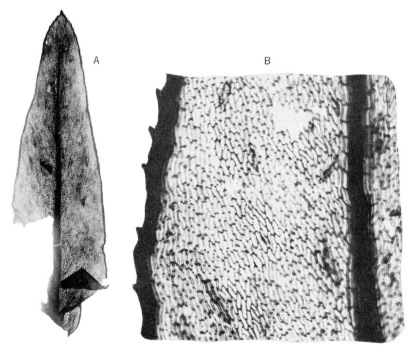

Fig. 15-2. *Intia vermicularis,* a moss from the Permian of the USSR. A. a nearly complete leaf; B. a portion more highly enlarged. (Photographs courtesy of Maria F. Neuburg.)

organs of these Permian mosses have not been found, so that more precise comparisons with living forms are not possible. Their importance lies in the fact that they are well preserved, they are unquestionably mosses, and sufficiently similar to modern ones in their vegetative organization as to suggest no major changes in moss evolution since that time.

Quite a few Tertiary mosses have been described but, as with the liverworts, close comparisons with modern genera are usually not possible.

Polytrichites spokanensis Britton from the Upper Miocene of Washington is an upright or slightly decumbent plant with stems up to 2 cm long; it is considered to bear a close resemblance to *Polytrichum*.

The genus *Plagiopodopsis* has been proposed for acrocarpous fossil mosses (with upright stems and terminal sporophyte). Mac-Ginitie has figured a fine specimen of *P. cockerelliae* (Britton and Hollick) Steere from the Florissant beds in Colorado; sporophytes are present but, judging from the presence of the calyptra, they were immature at the time of fossilization.

Palaeohypnum has been established by Steere for fossils that are referable to the Bryales Pleurocarpi—mosses with creeping stems and laterally attached sporophytes. Several species have been reported from various Tertiary horizons of western United States.

Although most fossil specimens are sterile (the gametophyte only is preserved), a moss theca has been reported from Oligocene brown coal in Victoria and given the name *Muscites yallournensis* Clifford and Cookson. It is quite small, being 0.8 mm long and 0.2 mm in diameter, and only the epidermis of the theca and its operculum were preserved.

Some Problematical Bryophytic Plants

Naiadita lanceolata Buckman

There are a few fossil plants which in themselves are expressive of the whole history of paleobotanical investigation—confused nomenclature, unique preservation, occasional abundance of specimens, mistaken identity, and the wealth of information that may be wrested from the rocks with the proper techniques and a competent investigator. Of such landmarks in the science *Naiadita* deserves a place of some prominence; the information given here is taken from a comprehensive study by T. M. Harris.

Naiadita comes from an English Rhaetic outcrop extending from Somerset east of the Mendip Hills to Worcestershire and Warwickshire. Although specimens are fragmentary they occur in abundance, as many as 33 leaves, leaf fragments, and other parts being found per square centimeter of rock surface, and this is said to be typical over a stretch of about 90 miles of the outcrop.

It was a smallish plant (Fig. 15-3) consisting of stems usually unbranched and rarely over 2 cm high which bore lanceolate, spirally arranged leaves, mostly under 5 mm long. Large numbers of gemmae (small, specialized vegetative reproductive structures) are found among the leaf and stem remains; they are oval in shape and range from $500 \times 300 \ \mu$ to $200 \times 120 \ \mu$ and a few have been found in terminal gemmae cups. Archegonia have been found in some abundance on the stems and their position relative to other organs is problematical; they do not occupy a position in the leaf axils and it has been suggested that they may take the place of a leaf. Some of the archegonia are enclosed in a "perianth" of several leafy lobes. Following fertilization (the antheridia have not been found) a pedicel developed resulting in a short branch which carried the developing sporophyte at its tip. The sporophyte had a short bulbous foot and a globose sporangium varying from 0.3 to 1.2 mm wide and with 100 to 400 spores which are lens-shaped with a characteristic equatorial flange.

In its general habit, notably the arrangement of the leaves, *Naiadita* is typically mosslike, but the lack of a midrib in the leaves suggests a liverwort. There are other liverwort characters such as unicellular rhizoids, an archegonium with a one-layered venter and lack of tissue differentiation in the stem. The general form of the sporophyte is comparable with that organ in the Marchantiaceae (a family of Hepatophyta), but elater cells are not found in *Naiadita*.

Sporogonites exuberans Halle

In 1916 Halle described a fossil from the Lower Devonian of Norway which consisted of an oval-shaped sporangium borne at the end of a slender unbranched stalk at least 4 cm long; the remainder of the plant was not found. The sporangium was sufficiently well preserved to reveal a multilayered wall with a dome-shaped mass of spores within; it was assumed that this indicated the presence of a columella but lack of preservation in the central portion left the matter in doubt. Columellate sporangia are found in the modern *Sphagnum* (peat moss) and in *Horneophyton.*

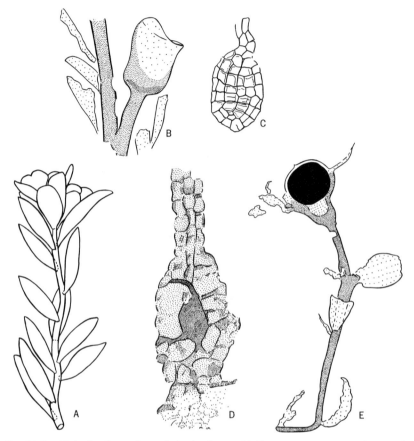

Fig. 15-3. *Naiadita lanceolata;* A. leafy shoot, 6X; B. gemma cup, 55X; C, a single gemma, 50X; D. archegonium, 200X; E. plant with a mature sporophyte, 8X. (From Harris, 1938.)

When *Sporogonites* was first discovered, Halle favored a bryophytic alliance but when the Rhynie plants were made known a comparison with *Horneophyton* seemed more likely and in most accounts dealing with early land plants *Sporogonites* is listed under the psilophytes.

A short time ago this writer had occasion to examine abundant collections of *S. exuberans* in the Brussels Museum which have been collected by Dr. Stockmans from the Lower Devonian of Belgium. A notable feature of the Belgian specimens is the consistent parallel alignment of the long sporangiophores; this is true for literally hundreds of specimens and it did not seem likely that it could be the result of chance deposition. A careful survey of the collection revealed several attached to an irregularly shaped carbonaceous film which is interpreted as a thalloid structure (Fig. 15-4). In one specimen this "thallus" measured about 15 × 5 cm and was clearly only a fragment of the whole. No evidence of vascular tissue was detected in this structure nor has vascular tissue ever been found in the stalks. The plant is therefore tentatively considered to be a bryophyte.

Relationships of the Bryophytes to Other Plant Groups

It is necessary to reflect back for a moment to the classification of the plants considered in this chapter. In dividing them into two distinct divisions Professor Bold noted that:

When one reviews the structure and reproduction of the members of these several groups, the diversities among them appear more striking than the resemblances. (1957, p. 312)

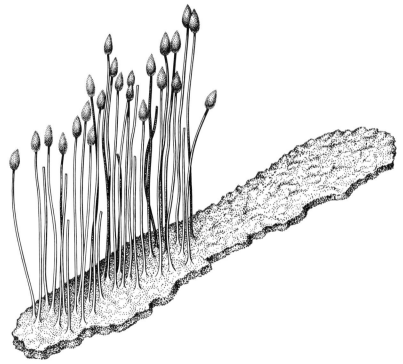

Fig. 15-4. A tentative restoration of *Sporogonites exuberans* based on specimens in the Natural History Museum, Brussels. (From Andrews, 1960.)

Probably one of the primary reasons for retaining the mosses and liverworts within one group has been the general similarity in their life cycle; that is, both have a dominant sexual phase and a much smaller sporophyte one that is either partially or wholly dependent on the former. This is quite the opposite of what is found in the pteridophytic groups. This in itself does not necessarily imply close relationship; some of the algae, for example, display essentially the same type of life cycle. It is precarious to depend on such comparisons, but I hazard the suggestion that the mosses and liverworts, so different in their gametophytic structure, are no more closely related than the various groups of pteridophytes are to each other.

The fossil record reveals both thalloid and leafy liverworts of essentially modern aspect in the Upper Carboniferous; mosses are also known from that age although not well preserved. However, the Permian specimens seem quite modern in their general organization. It seems reasonable to assume that the origins of the two groups will be found at a much lower horizon. All of the bryophytic plants are small and there has been very little systematic searching for fossil remains as yet.

Plants like *Naiadita* complicate the evolutionary picture in that they suggest a greater diversity among the bryophytes than is indicated by the modern flora; it may well be that many lines will ultimately be found in the Paleozoic.

We do not know for sure where *Sporogonites* fits into the scheme of things, but enough information is available to suggest that it may represent a group of plants equal in status with the Hepatophyta and Bryophyta. The size and shape of the structure that is interpreted as thallus and the densely borne, unbranched sporangiate stalks are distinctive characters.

The bryophytes are of more than passing interest, for botanists have long toyed with the notion that some of them may have been the progenitors of vascular plants; *Anthoceros* has been pointed to as a plant in which the sporophyte shows a tendency towards dominance over the gametophyte. Others have chosen to regard the group as reduced, and as a third possibility they may be wholly independent of any relationship with the pteridophytes.

REFERENCES

Andrews, Henry N. Jr. 1960. Notes on Belgian specimens of *sporogonites*. *The Palaeobotanist,* 7: 85–89.

Bold, Harold C. 1957. *Morphology of Plants.* Harper and Bros. 669 pp.

Clifford, H. T. and Isabel C. Cookson. 1953. *Muscites yallournensis,* a fossil moss capsule from Yallourn, Victoria. *The Bryologist,* 56: 53–55.

Halle, Thore G. 1916. A fossil sporogonium from the Lower Devonian of Röragen in Norway. *Bot. Notiser, Lund (1916):* 79–81.

————. 1936. Notes on the Devonian genus *Sporogonites. Svensk. Bot. Tidskrift,* 30: 613–623.

Harris, Thomas M. 1938. *The British Rhaetic Flora.* Brit. Mus. Nat. Hist. pp. 1–84.

————. 1942. On two species of hepaticae of the Yorkshire Jurassic flora. *Ann. Mag. Nat. Hist.,* ser. 11, 9: 393–401.

Lundblad, Britta. 1954. Contributions to the geological history of the hepaticae. *Svensk. Bot. Tidsk.,* 48: 381–417.

————. 1955. Contributions to the geological history of the hepaticae II. *Bot. Notiser,* 108: 22–39.

MacGinitie, Harry D. 1953. Fossil plants of the Florissant beds, Colorado. *Carnegie Inst. Washington Pub.,* 599.

Neuburg, Maria F. 1955. Bryophytes from Permian sediments of the USSR. *Dokl. Akad. Nauk. SSSR,* 107 (2): 321–324.

Renault, Bernard and Zeiller, Rene. 1888–1890. Études sur le terrain houiller de Commentry. Flore fossile. Pts. 1, 2. *Soc. Industrie Min. St. Étienne Bull.*

Steere, William C. 1946. Cenozoic and Mesozoic bryophyta of North America. *Amer. Mid. Nat.,* 36: 257–380.

Walton, John. 1925. Carboniferous Bryophyta. I. Hepaticae. *Ann. Bot.,* 39: 563–572.

————. 1928. Carboniferous Bryophyta. II. Hepaticae and musci. *Ann. Bot.,* 42: 707–716.

————. 1949. A thalloid plant showing evidence of growth, etc. *Trans. Geol. Soc. Glasgow,* 21: 278–280.

16

some Paleozoic and
Mesozoic floras

In this chapter floristic surveys are made of certain ages
and areas that have been omitted thus far or touched on
only incidentally. The coverage is by no means complete, the approach being similar to that used in presenting the Tertiary angiosperm floras; that is, plant assemblages of particular interest have
been selected from Paleozoic and Mesozoic horizons with the intent
of revealing their distinctive features as well as the problems that
remain unsolved. In large part the floras dealt with are from sections of the world that have not been emphasized in previous
chapters and from which we may expect to learn a great deal in
the future.

In certain of the older works on fossil plants, as well as some
recent ones, there is a tendency to regard floras of certain ages as
world-wide in extent. As our knowledge of them has increased,
through more extensive collecting and improved techniques for
extracting information, the floras from various parts of the world
for a particular period, say the Jurassic or Permian, show marked
differences in composition. It is my impression that future study
will continue to emphasize the diversity of floras in different
regions for any one point in time. In the following pages a few
vistas are opened which reveal a little of what has gone on in the
past, but to attempt a generalization in the nature of world-wide
floristics would be precarious if not quite misleading.

A tremendous amount of floristic information is gathered together in Seward's *Plant Life through the Ages* and, although
much has been added to our fund of knowledge since its date of
publication, it will continue to be a valuable reference. It is particularly recommended to students who have already acquired
some understanding of fossil plants.

408

The Distribution of Vascular Plants in the Lower and Middle Devonian

The statement has been made by several competent students of early land vascular plants that the floras of the Lower and Middle Devonian were essentially uniform throughout the world; it seems to me that this viewpoint has been somewhat overemphasized. In the first place the localities from which these early evidences of plant life on the land have been obtained are somewhat short of being world-wide in coverage. Second, in view of the relative simplicity of many of the early Devonian plants and the fact that they are often incomplete or poorly preserved, it is my feeling that their interrelationships may not be as close as has been supposed.

The greater part of our knowledge of the earlier Devonian floras has come from studies of fossils found in northwestern Europe, eastern North America, Australia, and more recently, Siberia. The earlier chapters of this book have drawn chiefly from these geographical sources only because they have yielded the most significant plant remains. The more important ones found in other parts of the world may be summarized briefly.

A Middle or Lower Devonian flora has been reported recently by Teichert and Schopf from the canyon of the Salt River in central Arizona. Initial studies reveal plants similar to *Cooksonia, Rhynia,* and *Hicklingia* as the more common elements; it may be noted that *Hicklingia edwardi,* as originally described by Kidston and Lang from the Middle Old Red Sandstone of Scotland, was a densely tufted, dichotomously forking, leafless plant with terminal sporangia and is possibly quite close to *Rhynia.* Other plants of the Salt River flora suggest *Hedeia* and *Dawsonites;* isolated spores have been found and although poorly preserved it is interesting to note that some of them are regarded as being the spores of heterosporous plants. The only other early Devonian flora from the western states is the one from Beartooth Butte, Wyoming, which has been mentioned previously in connection with *Psilophyton wyomingense* and *Bucheria ovata.* It also includes *Hostimella* and *Bröggeria strobiliformis* Dorf, the latter being a problematical strobiluslike body bearing some resemblance to an *Equisetum* cone. The type species, *B. norvegica* Nathorst, was found in the lower Middle Devonian of Norway and appears to be a unique cone of some sort, but nothing more precise can be said about it.

Sahni has reported very fragmentary small, naked, forking branches and one spiny specimen from Spiti province in the northeast Himalayas; although the preservation is not good, these offer evidence of early vascular plants. From the Lower or Middle Devonian of South Africa Høeg has reported *Dutoitia pulchra,* consisting of small dichotomous fragments with terminal, flat-topped sporangia about 3 mm broad; no spores have been found in them. Kryshtofovich has recorded *Psilophyton, Sporogonites exuberans,* and *Dawsonites arcuatus* from Turkestan in Central Asia.

Hsü has described several plants of particular interest from rocks of Lower or Middle Devonian age of central Yunnan in southwestern China. Plants tentatively referred to *Taeniocrada dubia* are preserved in the form of thin, flattened incrustations up to 1 mm thick and he suggests that it was not a ribbon-shaped axis; the tracheids of the central strand have small, round, uniseriate bordered pits with elliptic pores. Other fossils are tentatively assigned to *Asteroxylon* and *Protolepidodendron;* additional lycopods are present which indicate a horizon somewhat higher than the Lower Devonian.

Halle discovered some rather poorly preserved plants from the Devonian (presumably Lower) in the Falkland Islands. These include slender, leafless, dichotomously forking shoots, some with globose bodies at the tip which may be sporangia; there are also doubtful fragments which may represent the terminal spike of a *Zosterophyllum.* Some problematical plants have been described by Termier from central Morocco; they are of lower Middle Devonian age and suggest species of *Psilophyton, Asteroxylon,* and *Aneurophyton.*

So far as I am aware the records cited above are the chief ones that we have for the earliest land vascular plants aside from the North American, northwest European, and Siberian and Australian localities. It should be noted that very few are positively known to be Lower Devonian and the preservation for the most part is such that identification to the genus is precarious. The conclusion is unavoidable that our knowledge of the earliest land vascular plants is wholly inadequate to allow any generalization on a worldwide basis.

By Middle Devonian times a rather diverse assemblage of plants was established on the land; they were generally larger and more complex than those from the Silurian and Lower Devonian. One of the most significant insights that we have into the landscapes of this age comes from Professor Høeg's studies of the Spitzbergen flora; a few of the plants have been mentioned previously and some others may be introduced here. It has not been possible to determine the exact stratigraphic position of all of the Spitsbergen fossils; however, those cited below fall within the range of the upper part of the Lower Devonian and the lower part of the Upper Devonian.

Psilophyton arcticum Høeg seems to have borne a fairly close resemblance to *P. princeps* and *P. wyomingense;* the stem fragments attain a breadth of 5 mm and larger specimens branch by unequal dichotomies; thus the plant probably had an upright habit similar to that of *Asteroxylon.* The spines measure 5 to 6 mm long with some as much as 8 mm. *Psilodendrion spinulosum* Høeg is another probable psilophyte; the largest stems reach a diameter of 1 cm and the main branches are arranged oppositely or nearly so. These primary order branches are monopodial, whereas those of higher orders are dichotomous. Spines are present but rather sparsely distributed and somewhat under 2 mm long. *Svalbardia polymorpha* has been described previously as a plant that is seemingly transitional between the psilophytes and protopterid ferns. Lycopod stem casts described under the name *Bergeria mimerensis* Høeg range up to 10 cm in diameter and it is quite clear that they represent small trees. Other stem impressions have been found, as much as 15.5 cm broad, which date to the lower Middle Devonian or older; although unidentifiable they present drammatic evidence of trees of respectable size in the early part of the Devonian. It might be added that the specimens show distinct longitudinal striations, with ridges several millimeters apart, which suggest cortical fibers; it is thus certain that they do not represent the presumed alga, *Prototaxites.* A few petrified plants are present in the Spitsbergen flora, one of the most unusual being *Actinopodium nathorstii* Høeg. It is a stem about 16 mm in diameter containing a seven-rayed stele with a central mixed pith of parenchyma and tracheids; the protoxylem is probably mesarch, and of special interest are the radially aligned, presumably secondary, tracheids with multiseriate pits in their radial walls. In the mesarch protoxylem, fairly well-developed pith, and apparent secondary wood there is a suggestion of pteridosperm affinities. Adding to the variety of this mid-Devonian flora is *Enigmophyton superbum* and a few other problematical plants.

Several years ago Professor Suzanne Leclercq discovered a remarkable Middle Devonian plant horizon near the village of Goé in eastern Belgium, the specimens of *Calamophyton bicephalum* described in Chapter 9 having been obtained from there. The preservation is generally excellent; the flora includes several distinctly new plants and it may be expected to shed a great deal of light on others that are inadequately known at present. It is safe to assert that the Goé deposit, when fully studied, will yield a more diverse assemblage of plants than has yet been found in the Middle Devonian and it will reveal vascular plants of greater complexity than was formerly supposed existed at that time. I am indebted to Professor Leclercq for allowing me to insert this comment on the Goé plants since it will be some few years before they are fully studied.

Upper Devonian floras are known from eastern Canada and the United States, Ireland, Great Britain, Germany, the USSR, Bear Island, Spitsbergen, and Australia. Among the widely distributed elements of them are several species each of *Archaeopteris* and *Callixylon;** many others occur in one or more localities and it is evident that the vegetation of this age was a highly diversified one. To present, briefly, representative views of Upper Devonian landscapes two areas will be considered, Belgium and the east central United States.

The Upper Devonian Plants of Belgium

As a result of the investigations of Stockmans, Leclercq, and others the late Devonian rocks of Belgium have yielded a considerable number of interesting plants. *Archaeopteris, Aneurophyton,* and *Rhacophyton* have been mentioned in Chapter 3. Six species of *Sphenopteris* and one of *Diplotmema* have been described; among the fernlike fronds *Sphenocyclopteridium begicum* Stockmans is particularly distinctive, the fronds being at least twice pinnate and with pinnules that are nearly circular in outline. In Chapter 13 note was made of several fossils that are suggestive of being seeds or cupules, such as *Moresnetia, Condrusia,* and *Xenotheca;* the association of sphenopterid foliage adds to the probability that pteridosperms, or plants close to that level of organization, were already in existence. The high level that the articulates had attained is evinced by the *Eviostachya hφegi* cones with their complex sporagiophores. Stem impressions nearly 10 cm in diameter

---

* As this was going to press the important discovery has been announced (Beck, 1960) of a pyritized stem with *Callixylon* wood structure bearing several fragments of *Archaeopteris* fronds.

have been found and, while not identifiable, they indicate here, as elsewhere in the Devonian, plants of arborescent dimensions.

Upper Devonian Plants from East Central United States

In view of what is already known and what may be expected within the next few decades the fossil floras of the Black Shales of the east central states, especially central Kentucky, are beginning to assume a niche of great importance. There is some disagreement concerning the exact stratigraphic position of the plant horizons, whether uppermost Devonian or basal Mississippian, the latter being most probable for the majority of the 28 or more genera discovered thus far. Unlike most Upper Devonian floras the Kentucky one is based entirely on petrified remains; it is of special interest by virtue of the variety of plant form that is represented, and by the fact that many of the fossils are quite unlike any discovered elsewhere.

Among the more unusual fossils are several that are best regarded as problematical pteridophytic plants.

Stenokoleos simplex (Read and Campbell) is known from stems about 3.5 mm in diameter with a more or less cruciform vascular strand that occupies a considerable portion of the stem (Fig. 16-1). There is an apparent resemblance to *Asteroxylon,* but the two

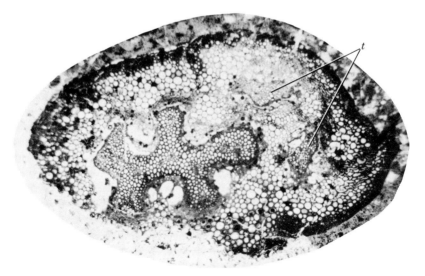

Fig. 16-1. *Stenokoleos simplex,* a transverse view showing the conspicuous stele and two traces (*t*) departing at the left, about 25X. (From Beck, 1960.)

differ fundamentally. In *Stenokoleos* the stele occupies a larger portion of the stem, it lacks the middle trabecular cortex of *Asteroxylon,* possesses numerous (about 14) mesarch protoxylem points, but most striking is the organization of the appendages. Rather large triangular-shaped traces are given off in pairs from opposite arms of the stele; the diameter of these appendages somewhat exceeds that of the main axis at the point of departure, but they rapidly taper to about 2 mm in diameter. The plant is regarded by Beck as possibly referable to the Zygopteridaceae, but this is admittedly uncertain.

Guycampbellia microphylla (Read and Campbell) is another actinostelic type in which the stems are known up to 5 mm in diameter. It apparently bore closely appressed microphyllous leaves 6 to 10 mm long which were twice forked. There is some resemblance to *Protolepidodendron* in the form of the foliage but the stele, with deeply embedded mesarch protoxylem, is certainly not typically lycopodiaceous.

Polyxylon elegans (Read and Campbell) is known from stems 5 to 8 mm in diameter in which the stelar system consists of a ring of U-shaped strands; the tracheids are large and scalariform. Nothing is known of the appendages of this interesting fossil.

Pietzschia polyupsilon (Read and Campbell) is based on stems of about 25 mm in diameter with a single peripheral ring of about 54 radially aligned primary strands, some of which appear Y-shaped as a result of frequent branching and anastomosing of the meristeles. It has been suggested that *Pietzschia,* which is also known from the Upper Devonian of Saxony, is related to *Cladoxylon.*

Siderella scotti (Read) is especially perplexing. The stem is a little under 1 cm in diameter and the star-shaped stele, with 8 to 10 rays or lobes, gives off vascular strands of two types: there are small ones that depart in whorls and each trace divides once or twice in its outward course; the others are relatively large, H-shaped "zygopterid" strands and these depart in opposite pairs. This distinctive vascular nodal anatomy has led to the supposition that the plant may be allied to both the articulates and the zygopterid ferns. It seems most unlikely, however, that there could be "intermediate" forms between two such divergent groups; it may be recalled that whorled appendages are not exclusive to the articulate group. *Siderella* has received only a cursory study as of this writing, but it seems very possible that it may represent a group of plants quite distinct from any known at present.

Small lycopod stems have been attributed to *Lepidodendron,* but these are for the most part rather poorly preserved. A cone, *Lepidostrobus fischeri,* described some years ago by Scott and Jeffrey, stands in contrast to the small and scanty vegetative remains. It is known from a fragment only 8 cm long but is sufficiently intact to indicate an over-all diameter of 4 cm; thus it compares favorably with the large cones of the arborescent Upper Carboniferous lycopods. The opinion has also been expressed that the bisporangiate cone *Lepidostrobus noei* (mentioned in Chapter 8) was derived from a horizon close to that from which the other Black Shale plants originated.

Certain stems and petioles have been assigned to several species of the *Calamopityeae,* a group of presumed pteridosperms. *Diichnia kentuckiensis* Read may be taken as an example; it is a stem with an angular pith about 7 mm broad and although predominantly parenchymatous there are tracheids or clusters of tracheids scattered through it. Excentrically mesarch primary vascular strands are present in the angles of the pith which is enclosed by a broad zone of secondary wood; the latter is composed of multiseriate pitted tracheids and rays which vary considerably in width, but many are eight cells broad and thus lend a conspicuous pattern in transverse sections. The outer cortex consists of alternating bands of parenchymatous and sclerotic (apparently fibrous) cells. The petiole vascular supply originates from branches of two primary strands; as these enter the cortex they tend to elongate and divide to form a semicircle of bundles in the base of the petiole. The genus *Calamopitys,* which is known from several species both in this country and Europe, is basically similar to *Diichnia;* perhaps the most notable difference lies in the development of the petiole vascular supply from one peripheral primary bundle of the stem rather than two.

Callixylon is represented by numerous small fragments and occasional logs in excess of 1 foot in diameter.

In summary, the 45 or more species now known from the Black Shales are an especially diverse aggregation of plants; the Calamopityeae are probably seed plants and several pteridophytic groups are included, some of which are very likely new. Not a few are especially problematical, such as *Steloxylon* and the *Foerstia-Protosalvinia* complex, and investigations which are now in progress will undoubtedly bring others to light. Although the petrifactions are not especially abundant it seems to me that this is a most promising source of information concerning a very critical age in the evolution of pteridophytic and early seed-plant groups.

Asian and Southern Floras of the Late Paleozoic and Early Mesozoic

According to our present knowledge the floras of the late Carboniferous to the early Triassic fall into four groups as follows:

1. The Euramerican flora extends from the Ural Mountains through western Europe and takes in the United States and Canada to eastern Kansas, with western outliers in Colorado and northern New Mexico. Fossils found in this general area form a large part of the subject material considered in Chapters 3 to 5 and 8 to 9.

2. The Angara (also known as the Siberian or Kuznetzk) flora occupies the territory known as Angaraland which is bounded by the Urals in the west and extends east through Siberia to the Pacific, and from the Arctic ocean south to outer Mongolia.

3. The Cathaysia (or *Gigantopteris*) flora extends from Shansi, China and Korea south through Sumatra, New Guinea, and with what appear to be outliers in Colorado, Oklahoma, and Texas. It may be noted on Fig. 16-2 that this overlaps a corner of the Angara province.

4. The *Glossopteris* flora occupied an area referred to as Gondwanaland which includes India, the central and southern part of Africa, the southern half of South America, Australia, and Antarctica.

It is convenient but not quite correct to apply the phrase "the flora" to the fossil assemblage to any of these areas. We are dealing with large geographic areas and long time spans in each one; the Euramerican and Angara provinces are best known and it is well established that the plant assemblages vary considerably, both geographically and stratigraphically. We must of necessity deal with them here in rather general terms.

One of the major problems involved in understanding these floras is the matter of their distribution, this being especially true of the Euramerican and *Glossopteris* provinces. The most favored explanations have been the former presence of *land bridges* or *continental drift*. The latter has been pronounced both impossible and indispensable; a rather tremendous literature has been developed by its proponents and opponents and since it is still a very live issue in connection with the distribution of the *Glossopteris* plants a brief consideration of it seems appropriate. For further

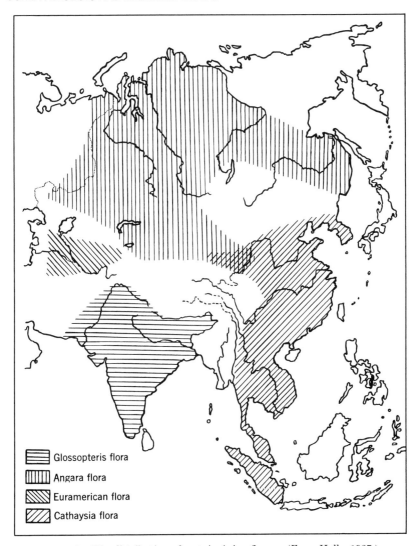

Fig. 16-2. The distribution of certain Asian floras. (From Halle, 1937.)

information the student is referred to Chaloner's well-formulated summary of the problem cited at the end of the chapter.

In 1915 Alfred Wegener published his detailed treatise of the proposal that the continents were once aggregated together into a single huge land mass; this concept was later modified by A. L. DuToit who proposed two primitive land masses, a northern one including what we now call North America, Europe, and Asia, and a southern one composed of South America, Africa, Antarctica, Australia, and India. The separation of this great southern land mass is thought, by many supporters of the idea, to have started early in Tertiary times.

Some of the points which have been presented in support of continental drift are the general similarity in outline of the west coast of Africa and the east coast of South America and a close comparison in composition and grain size of the rocks along the respective opposing coasts. There is also agreement in a line of ancient mountain chain systems along the east side of North America and continuing through parts of Britain and northern Scandinavia; a comparable structural line is found extending through the tip of the Cape and the opposite coast of South America.

Next, there is evidence of widespread glaciation in late Paleozoic times in southern Australia, extreme southeastern South America, the southern tip of Africa and in India. The latter area is especially perplexing since the direction of ice movement seems to have been north from a point near the equator. The great difference in both latitude and longitude of these presently remote areas is difficult to explain other than on the basis of a former close association; correlated with this is a presumed change in the position of the poles, a point that is anathematic to many physical scientists.

A considerable fund of biological evidence has been marshalled in defense of the idea from distribution patterns of both modern and fossil plants. Several genera and families of flowering plants seem to afford cogent support: the Proteaceae is distributed through most of South America, the southern half of Africa, Australia, India, and Madagascar; the Restionaceae is found in the extreme southern tips of South America and Africa, in Madagascar, Malaya, and Australia; the genus *Symphonia* (Guttiferae) is distributed through the northern half of South America and in west central Africa. An interesting case of an animal is found in the small Triassic reptile, *Mesosaurus,* which is known only from the east coast of South America and along the African coast on the opposite side of the ocean; having been a small animal that is thought to have lived in a delta swamp environment it does not seem possible that it could have crossed the south Atlantic with its present breadth. Most impressive of all the positive evidence is the distribution of the *Glossopteris* flora, which is presented in some detail below.

Proponents of former land bridges have frequently voiced their case with enthusiasm and not entirely without evidence that commands some consideration. The mid-Atlantic Ridge presents an interesting case in point; one of the most intriguing bits of evidence that has turned up recently supporting the belief that islands, perhaps of considerable magnitude, once existed comes from Kolbe's report of abundant fresh-water diatoms as well as the distinctive silica "casts" of the epidermal cells of grasses and sedges, found in deep sea cores taken about 1000 kilometers off the west coast of Africa between the equator and approximately 10° N latitude. It is possible that the diatoms may have been transported by sea or air currents from the African mainland, but Kolbe regards it as much more likely that the cores, in which the diatoms are most abundant, came from a lake bed of an island of the mid-Atlantic Ridge.

The Angara or Kuznetzk Flora

The Angara flora comes chiefly from the Kuznetzk and Minussinsk depressions in central Asia within the Altai-Sayany mountain regions. In the Kuznetzk basin the succession of sediments bearing the fossil plants begins with the Lower Carboniferous, but it is not until the Lower Permian that the flora, with all of its unique elements, reaches its full development. There is a strong lycopod component in the Lower Carboniferous and this gives way to what has been called a cordaite flora, although many other groups are present; there is, in fact, a total of some 600 species according to a recent summary by Neuburg (1958), although full descriptions of all of them have not been given.

The following summary refers chiefly to late Carboniferous and Permian plants of the floral succession. The cordaites are represented by the leaf genus *Noeggerathiopsis,* 17 species having been recognized; in general organization they are comparable with the leaves of *Cordaites* and may be as large, but in most species they are smaller, some being only a few centimeters long, and they vary from lanceolate to spatulate in outline.

Fernlike fronds referable to *Sphenopteris, Pecopteris, Callipteris,* and *Neuropteris* are represented by several species each, whereas there are many other genera that indicate plants of a rather different nature from those found in the Euramerican upper Paleozoic, such as *Gondwanidium, Tschirkoviella, Angaropteriduim,* and *Angaridium* (Fig. 16-3); reproductive organs are known for

Fig. 16-3. Some plants of the Angara flora. A. *Ginkgophyllum vsevolodi;* B. *Angaridium mongolicum;* C. *Gondwanidium petiolatum;* D. *G. sibiricum;* E. *Tschirkoviella sibirica;* F. *Angaropteridium cardiopteroides.* (All from Neuburg, 1948.)

very few and future studies will almost certainly reveal many new ferns and pteridosperms. The articulates are represented by *Sphenophyllum, Annularia, Phyllotheca* (with exceptionally long slender leaves), and *Paracalamites robustus* with pith casts at least 8 cm in diameter. Among a few apparent ginkgophytes

Ginkgophyllum vsevolodi Zalessky is suggestive of *Dichophyllum moorei,* a Kansas fossil from the top of the Carboniferous mentioned in Chapter 11.

Several true mosses (Bryales) as well as early peat mosses (Protosphagnales) have been found which are considered to indicate a trend toward a more temperate climate in Angaraland.

Conifers are very scarce in contrast to their relative abundance in the late Paleozoic floras of the Euramerican province. Finally, as suggestive of the unique elements that will probably continue to turn up in the Angara flora, reference may be made to *Vojnovskya paradoxa* which was described in Chapter 12.

The Cathaysia Flora

Much of our knowledge of this flora comes from Halle's studies of uppermost Carboniferous and Lower Permian plants from the province of Shansi in east central China. His original collections, made during several months of field work in 1917, were lost when the ship carrying them to Stockholm went down in a typhoon; later collections formed the basis of his classic work (1927). In view of the many interesting plants that it contains one cannot help wondering what was lost and what may turn up in the future. The following plants are representative of the flora:

Annularia seems to have been quite abundant and Halle notes that specimens of *A. stellata* agree closely with European ones. A distinct articulate genus, *Lobatannularia,* is characterized by markedly excentric whorls of leaves; that is, some are notably smaller than others (Fig. 16-4D); it also differs from *Annularia* in that there are four branches at a node. Several species of *Sphenophyllum* are present; the leaves of *S. thoni* Mahr are quite large, being nearly 5 cm long and the vascular system is fan-shaped with vein endings distributed along the sides as well as the distal edge. In some specimens of *S. rotundatum* Halle the leaves are almost round.

The Gleicheniaceae was represented by *Oligocarpia gothanii* Halle which apparently was a small plant. Halle figures a specimen consisting of several leaves not much more than 10 cm long attached to what is probably a rhizome; it is suggested that it was a creeping plant occupying mud flats or was possibly floating. A closely allied genus, *Chansitheca,* is distinguished by elongate sori and more sporangia per sorus (as many as 16 in *C. kidstoni* Halle) than in *Oligocarpia.*

A species of *Cladophlebis* is included in the flora; this is a com-

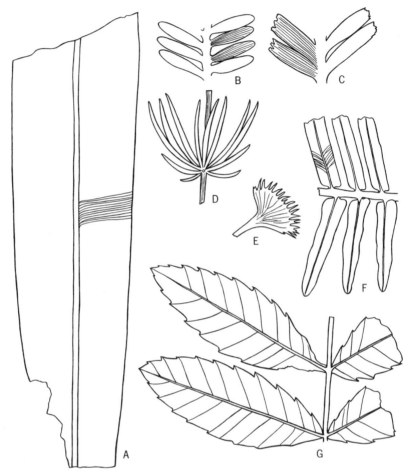

Fig. 16-4. Some plants of the Shansi flora. A. *Taeniopteris latecostata;* B. *Plagio-zamites oblongifolius;* C. *Tingia carbonica;* D. *Lobatannularia ensifolius;* E. *Norinia cucullata;* F. *Protoblechnum wongii;* G. *Gigantopteris nicotianaefolia.* (Drawn from specimens in the Swedish Natural History Museum and from Halle, 1927.)

mon Mesozoic genus and is usually regarded as representing the Osmundaceae.

The presence of the pteridosperms is implied by quite a number of the genera and there can be no doubt that there were many members of the group in the flora. A few have been found with seeds attached, namely, *Pecopteris wongi, Alethopteris norinii,* and *Emplectopteris triangularis* (see Chapter 5).

The form-genus *Taeniopteris* is an abundant element of the flora, but Halle himself recognized the uncertainty of clearly defined limits for the 11 species that he named. *Taeniopteris* is widely distributed and ranges from the uppermost Carboniferous to basal Cretaceous; it may be recalled that one species was cited in Chapter 10 as the foliage of *Williamsoniella coronata*. The entire, lanceolate, leaves have a strong midrib and side veins that depart at an angle of 70 to 90°; these usually divide but once immediately after leaving the midvein and thus appear parallel. In his study of the Scoresby Sound plants (Chapter 7) Harris proposed the segregation of certain species into three natural groups: those that are known to be the fronds of marattiaceous ferns; those with bennettitalean type stomata (as *Taeniozamites*); those with cycadean type stomata (as *Doratophyllum*); and finally, species that are otherwise indeterminate are retained in *Taeniopteris*. Some of the Shansi taeniopterids attained a respectable size; there are fragmentary specimens of *T. nystroemii* Halle in the Swedish Natural History Museum that are 20 cm broad; very likely they appeared similar to a fair-sized banana leaf in life.

The "feature element" of this flora is *Gigantopteris nicotianaefolia* Schenk, and lends the frequently used name "*Gigantopteris* flora." The fronds were large although the total size and form are not fully known. They were at least once pinnate, 30 cm or more broad, and the main rachis, which was 13 mm in diameter, bore opposite, slightly overlapping pinnae. The latter attained a length of 15 cm and were dentate to almost entire; a conspicuous midvein produced nearly parallel laterals that terminate in the teeth apices, and the finer venation in between these is netted. Towards the apex of the frond the pinnae merge into each other forming a massive terminal portion. Halle established another species, *G. lagrelii,* which he regarded as intermediate between *G. nicotianaefolia* and *Emplectopteris triangularis*. In a later study he described tendrillike structures which have the same general branching pattern as the normal leaves of *G. nicotianaefolia,* indicating the possibility of a vinelike habit for the plant.

A few of the other plants of the flora may be mentioned briefly. *Cordaites* is represented by two species and in one the leaves attain a breadth of 45 mm. In a more recent study Sen and Bose have described a petrified stem, *Cordaites sahnii,* from the Upper Carboniferous or Lower Permian of central Shansi; it is a fragment with a radial dimension of about 4 cm and shows no evidence of growth rings. *Saportaea nervosa* has been mentioned in the chapter on ginkgophytes as a possible early member of that group; there are also two species of *Baiera* recorded and one of these, *B. tenuistriata,* is based on exceptionally large "leaves" in which the segments attain a breadth of 1.5 cm. There is a fragmentary specimen in the Museum in Stockholm that is 15 cm long and obviously only a small portion of the leaf; this is a most interesting fossil but must be regarded as a doubtful ginkgophyte. There are several species of the problematical genus *Tingia;* as an example, *T. carbonica* (Schenk) Halle consists of an axis 3 to 8 mm thick which bears a strong resemblance to a cycadophyte frond, but in addition to two lateral rows of "leaflets" there are two more on the under side although they are somewhat smaller.

In summary the following features of the flora seem most significant. Nearly 50% of the species are attributed to the ferns or pteridosperms. It seems safe to assume that the latter were abundantly represented and included some unique forms; of special interest are the three species with seeds attached to the foliage. The Equisetales and Sphenophyllales are characterized by anisophyllous species, that is, ones in which leaves of varying size are found in the same whorl. The lycopods compose a very minor element of the flora, and the cordaites, although fairly abundant in specimens, are represented by only two species. The extreme scarcity of the conifers is also noteworthy, especially the absence of *Walchia* which is a conspicuous feature of Lower Permian floras in the Euramerican province. Several plants are attributed to the ginkgophytes and although of considerable interest it seems to me their classification must be regarded as tentative.

The Glossopteris Flora

The literature dealing with the *Glossopteris* flora is now extensive and quite scattered and having accumulated over a period of about one century it is encumbered by some error and no small amount of divergence of opinion. Krishnan has recently given a useful stratigraphical analysis of the entire Gondwana era and it includes charts for the beds in several parts of India as well as correlations for South Africa, Australia, Brazil, and Argentina. In India the Gondwana era is regarded as beginning with the Talchir tillite in the Upper Carboniferous and continuing into the middle Cretaceous. As might be expected this range includes a great

variety of plant forms. *Glossopteris* and its characteristic associates appear in the Upper Carboniferous and extend into the early
part of the Triassic but are rare thereafter; this constitutes the so-
called Lower Gondwana flora with which we shall be concerned in
the following pages. Although this extends through southern
South America, the Falkland Islands, Antarctica, South Africa,
Australia, and India, the discussion will deal largely with the flora
as it occurs in India. The geology of the Gondwana system is best
known in that country and especially active studies of the flora
have been conducted there in recent years.

  Glossopteris and *Gangamopteris* are the most notable elements
of the flora; some 40 species of the former have been described and
about 18 of *Gangamopteris*. Recent studies of the epidermal
structure confirm the fact that numerous species existed, but the
number may not be quite as high as the figures cited. In both
genera the leaves are net-veined, *Gangamopteris* being separated
by the supposed lack of a distinct midrib. A third leaf type may
be appropriately introduced here, *Palaeovittaria kurtzi* Feistmantel, which is of the same general shape but is usually described as
having dichotomous, open (not netted) venation. However, according to recent studies the three cannot be sharply delimited by
gross macroscopic characters or cuticular structure. *Vertebraria
indica* Royle is a name that was given to axes 2 or 3 cm in diameter, with characteristic radially aligned wood sectors, that are
thought by some workers to have been the rhizome of *Glossopteris*.
More recently *Glossopteris* leaves have been found attached in
whorls to slender axes little more than 1 mm broad. The reproductive organs found attached to *Glossopteris* and *Gangamopteris*
leaves have been mentioned in Chapter 12. Mrs. Plumstead has
summarized the evidence bearing on the habit of the glossopterids
and suggests that they may have been large deciduous trees with
the foliage arranged on both long and short shoots.

  Noeggerathiopsis, although present, apparently was not a dominant element as in Angaraland; several species of cordaitean wood
(*Dadoxylon*) have been found in India and other parts of Gondwanaland, most of which display distinct annual rings. There are
only three genera of conifers known, *Buriardia, Paranocladus,* and
Walkomiella; W. australis (Feistmantel) Florin is regarded as having been a large forest tree probably similar in general appearance
to the modern *Araucaria cunninghamii* which is found in Queensland and New South Wales. In another species, *W. indica* Surange

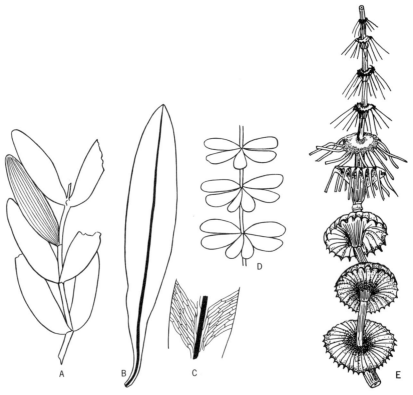

Fig. 16-5. Some plants of the *Glossopteris* flora. A. *Schizoneura gondwanensis;* B. *Glossopteris* sp.; C. a portion of the leaf enlarged to show venation; D. *Sphenophyllum speciosum;* E. *Phyllotheca etheridgei.* (A from Feistmantel, 1881; D from a specimen in the British Natural History Museum; E from Saksena, 1954.)

and Singh, small seed-bearing shoots have been identified which compare with those of *Lebachia;* very possibly they were arranged in a cone like the dwarf-shoots of the latter genus.

There are a few distinctive articulate types. *Schizoneura gondwanensis* Feistmantel has been described as having leaf sheaths that are usually divided into two parts which are ovate-lanceolate with a rounded tip; occasionally the sheath (or paired leaves) may be dissected into several segments each. The genus also has been reported from several widely scattered Triassic-Rhaetic localities in Europe and China. Srivastava has described a cone from the Raniganj coalfield of Bengal which is believed to have been borne on *S. gondwanensis;* it is 30 mm long and bore appendages that seem to have been similar to those of *Equisetum,* there being no evidence of associated bracts. Recent restorations of two species of *Phyllotheca* show them as plants with foliage sheaths that were deeply dissected in the younger parts of the branchlets and as they matured the sheath developed an umbrellalike form. *Sphenophyllum* was fairly abundant as an undergrowth plant.

The lycopods are not common in the *Glossopteris* flora and the few specimens available are poorly preserved; as a consequence there has been considerable disagreement over their identification. *Lycopodiopsis derbyi* Renault is known from petrified stems in which the primary xylem cylinder is discontinuous, that is, dissected; otherwise it is said to be typically lepidodendraceous. *Lepidodendron* and *Sigillaria* have both been reported, but in a recent study of southern hemisphere lycopods Edwards concludes that the *Glossopteris* flora does not include any *species* that are common to both southern and northern floras, and that there is no satisfactory evidence to indicate that any *genera* are common to both.

A few other plants of the flora may be cited briefly: there are several species of sterile *Sphenopteris,* the affinities of which are not known. A fertile pecopterid frond fragment from the Raniganj coalfield of India has been tentatively assigned to the marattiaceous genus *Ptychocarpus;* the discovery of a rather well-preserved *Psaronius* stem in Brazil many years ago tends to confirm the presence of the Marattiaceae in the flora. Fronds originally described under the name *Neuropteridium* appear to be quite closely related to some of the Angara species of *Gondwanidium* and have been transferred by certain authorities to that genus. *Belemnopteris woodmasoniana* Feistmantel is a distinctive arrow-shaped frond with net venation which has been found only in the Upper Permian of India. Surange has recently described a fossil from the Raniganj coalfield consisting of a cylindrical axis about 1 cm broad and 5 cm long which was borne on a slender stalk 1 mm in diameter; numerous sessile, exannulate sporangia, each containing many spores, were borne on, or slightly embedded in, the axis; it is regarded as being a fern or pteridosperm but is a unique spore-bearing organ.

Relatively little has been done as yet with the palynology of the Lower Gondwanas, but in two recent studies more than 40 distinctive microspore types were found in coal from Bihar and 20 species of megaspores have been identified. The megaspores are classified in the genus *Triletes* which is regarded as being lycopodiaceous, but it is by no means certain that all the species are correctly assigned to that group. There is, however, a strong suggestion that the lycopods were more abundant than is indicated by the macrofossil record.

The evidence available at present indicates that the *Glossopteris* assemblage was not nearly as rich a one as is found in the northern floras. In view of the widespread glaciation, as evidenced by extensive tillites or boulder beds, there has been considerable discussion over the matter of the Gondwana climate. *Gangamopteris* and *Phyllotheca* have been found as low as the Dwyka series (Upper Carboniferous of South Africa) and spores have been found in some abundance in the Dwyka and in the Bacchus Marsh beds (Upper Carboniferous of Victoria). It thus seems evident that the flora existed during times of extensive glaciation and the conclusion that it was a cool temperate one seems reasonable.

The effect of a glacier on the vegetation of the immediate vicinity depends in part at least on the size of the ice sheet and the latitude. Tree ferns grow within a mile of the terminal face of the Franz Josef Glacier in New Zealand, a dramatic photo of this contrast being given in Seward's *Plant Life through the Ages*. On the other side of the world, glaciers in western Norway come down very nearly to sea level and exist in close proximity to forests of fir, birch, and alder, as well as fine apple orchards. One of the most memorable sights I have ever encountered for geological and botanical interest, as well as sheer beauty, is the view across Sogne Fjord in western Norway. Looking north from the highland above the town of Vik one can see part of the great Jostedals Glacier, covering some 1076 square kilometers, and in autumn the apple harvest will be in full swing in the many small farms bordering the fjord—an impressive contrast to the nearby icefield.

Triassic Floras

The Middle Triassic Ipswich flora of Queensland

The early Mesozoic floras of the southern hemisphere are frequently referred to as *Thinnfeldia* floras because of the numerous species of this widely distributed genus. The fronds are once or twice pinnate with a stout rachis that may display a single dichotomy and the pinnules vary, in different species, from ovate to narrowly elongate. The presence of a rather heavy cuticle has led most authors to regard them as pteridosperms, but in view of the morphological variation it is possible that the natural relationships may be correspondingly diverse. In the instance of *T. feistmanteli* Johnson the pinnules are more or less rhombic and attached along their entire base; exannulate sporangia are grouped three or four to a sorus, and there are two or three rows of the latter on each pinnule. In another species, *T. lancifolia* (Morris), in which the pinnules are variable in shape but more or less lanceolate, sporangia have also been reported and seem to be nearly identical with those of *T. feistmanteli.*

Stenopteris is another difficult genus that has been considered as closely related to *Thinnfeldia* and is reported from northern as well as the southern floras. The rachis may or may not be dichotomized and the lamina of the pinnules is almost nonexistent; the strongly developed cuticle is regarded as indicating gymnosperm affinities. Of particular importance in this problematical assemblage is a series of specimens (Fig. 16-6C) reported by Jones and de Jersey from the Ipswich coal measures that reveal an evolutionary sequence of increasing complexity; it is rarely possible to clearly demonstrate such trends through a known stratigraphic sequence. All of the forms that occur in the Ipswich flora were formerly assigned the binomial *S. elongata* (Carruthers) Seward but this no longer seems tenable. Specimens taken from the base of the series bear "pinnules" in which essentially no lamina is present (Fig. 16-6C1); at a higher horizon some lobing appears in the base of the pinnules and these are supplied with branch veins (Fig. 16-6C3); still higher this lobing becomes constant throughout the length of the pinnule and finally the lobes in turn begin to proliferate (Fig. 16-6C6). The transition indicated in the figures takes place through approximately 3000 feet of sediment, and it is also interesting to note that the simpler forms do not immediately disappear but with the advent of the more complex ones they do show a marked decrease.

Plants that can be positively identified as true ferns are not abundant in the Ipswich flora. Three or four species of *Cladophlebis* have been recorded, but the few fertile specimens are not well preserved. Several species of *Dictyophyllum* are known and these are assigned to the Dipteridaceae.

In one of several contributions on the Ipswich flora Walkom (1917) described a considerable variety of ginkgophyte foliage under the names *Ginkgo* and *Baiera*. Most of the leaves are quite deeply dissected and petiolate, thus falling within the range of *Ginkgoites* as defined by Harris (see Chapter 11). It is a little disturbing, however, to find *Ginkgo antarctica* Saporta at this low

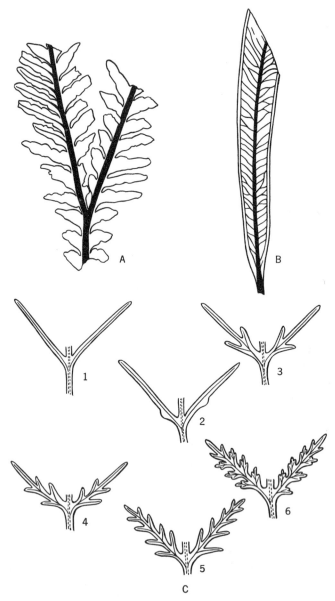

Fig. 16-6. Some plants from the Triassic Ipswich flora of Queensland. A. *Thinnfeldia talbragarensis;* B. *Yabeiella brackebushiana;* C. a series showing evolution in *Stenopteris* leaves; 1, 2, *S. elongata;* 3–5, *S. spinifolia;* 6, *S. tripinnata.* (All from Jones and de Jersey, 1947.)

horizon; it is a species in which the leaf is entire or nearly so and closely resembles the foliar organs of the modern *G. biloba.* The variety of leaf types figured by Walkom suggests that the group was quite diversified and very likely had its origin not later than the Permian.

The flora includes a few cycadophytes as well as some leaf types that are of questionable affinities. The Bennettitales is represented by very fragmentary remains of *Williamsonia* reproductive organs, and leaves of *Pterophyllum.* The Cycadales seem to have been slightly more abundant with *Ctenis, Nilssonia,* and *Doratophyllum* present, the last being a taeniopterid leaf with the cycadean type stomata. Another interesting and problematical leaf genus, *Yabeiella,* is represented by several species; the leaves are entire, lanceolate to oval-lanceolate, and up to 15 cm long. The venation is quite distinctive; two lateral veins depart near the base of the blade and assume a near-marginal position; the numerous other minor laterals occasionally anastomose and terminate in these marginals.

The Triassic Flora of Arizona

In conclusion it seems of interest to compare the Ipswich flora with one on the other side of the globe, namely, the Upper Triassic Chinle flora of Arizona. The two were not exactly contemporaneous, the latter being somewhat younger, but the differences in composition are quite striking.

The Chinle flora is known from western Texas to Nevada although the most productive localities are in Arizona. The extensive accumulations of silicified logs composing the "petrified forest" of Arizona have long been famous; however, a more significant contribution to our understanding of the plants of that area was made by Daugherty a few years ago following the discovery of impression fossils from several localities. The logs are colorful and impressive but are often poorly preserved and comprise only a very few species.

The cycadophytes and certain ferns are the dominant elements; the Osmundaceae is represented by a petrified stem, *Osmundites walkeri* Daugherty, which is of quite modern aspect; another stem has been referred to the family, *Chinlea campii* Daugherty, which is rather poorly preserved but appears to have scalariform tracheids scattered through the central portion and is possibly intermediate between *Thamnopteris* and *Osmundites*. Leaf impressions of *Todites* and *Cladophlebis* are also referred to this family.

Some fine frond specimens of *Phlebopteris smithii* testify to the presence of the Matoniaceae and a leaf specimen, regarded as sufficiently distinct to merit the name *Apachea arizonica* Daugherty, is assigned to the Dipteridaceae.

There are several cycadophytes in the Arizona flora, but the absence of preserved cuticular structure renders their identification unsatisfactory. *Otozamites powelli* (Fontaine) Berry is a typical once pinnate cycadophyte frond with oblong pinnules having a truncated tip. Leaves of *Macrotaeniopteris* are described (*M. magnifolia* (Rogers) Schimper) which are simple and up to 17 cm broad and a meter long; poorly preserved fructifications were observed which may be fern sori but this is uncertain. Whatever the plant's affinities were it was abundant; although recorded at only one locality it occurs as a mass of leaves more than a foot thick. *Macrotaeniopteris* is associated with *Equisetum*-like stems and largely on this account the plant is regarded as having grown in low, marshy places.

A few additional notations may be inserted here concerning the articulates; their record consists of very fragmentary specimens but several are worthy of note: *Neocalamites virginiensis* (Fontaine) Berry is known from stem casts that exceed 10 cm in diameter; these are small by comparison with the Carboniferous calamites but considerably larger than our modern equisetums. The stems of *Equisetites bradyi* Daugherty attain a diameter of only 14 mm, but have leaf sheathes that appear very similar to those of a modern *Equisetum*. The distinctive spores of *Equisetosporites chinleana*, with their two elaters, have been mentioned in Chapter 9.

Many of the petrified logs are those of a conifer, *Araucarioxylon arizonicum* Knowlton; specimens have been found up to 7 feet in diameter and 120 feet long and they may, therefore, be properly referred to as a dominating as well as a dominant element of the flora! Growth rings are faintly discernible. *Schilderia adamanica* Daugherty is a wood with unique ray organization; it is thought to have been a small tree since the diameter of the trunks does not exceed 30 cm shortly above the fluted base. The wood consists of tracheids, with some parenchyma cells toward the end of each faintly defined growth ring, and small uniseriate rays; it is thus essentially coniferous with the exception of rather regularly spaced multiseriate rays. These are 0.3 mm broad and 10 mm high and form a very conspicuous feature; they are composed of irregularly shaped parenchyma cells as well as some tracheids that bend into them. The affinities of this curious wood are not known.

It is precarious to speculate on the climatic conditions under which these plants grew. In all probability several habitats are represented including a low, swampy one as well as an upland environment. The ferns of the Matoniaceae and Dipteridaceae imply a warm climate, but it seems to me that the cycadophytes or, more precisely, the cycadlike foliage, is not classified precisely enough to give any dependable clues. Growth rings of several species are sufficiently well defined to indicate some seasonal fluctuation.

REFERENCES

Arber, E. A. Newell. 1905. Catalogue of the fossil plants of the *Glossopteris* flora in the Department of Geology, British Museum (Natural History), London. 255 pp.

Beck, Charles B. 1960. Studies of New Albany shale plants. I. *Stenokoleos simplex* comb. nov. *Amer. Journ. Bot.,* 47: 115–124.

————. 1960. Connection between *Archaeopteris* and *Callixylon. Science,* 131: 1524–1525.

Chaloner, William G. 1959. *Continental Drift.* New Biology (Penguin Books), No. 29: 7–30.

Coleman, A. P. 1926. *Ice Ages Recent and Ancient.* Macmillan and Co., London. 296 pp.

Daugherty, Lyman H., and Howard R. Stagner. 1941. The Upper Triassic flora of Arizona. Carnegie Instit. Washington Pub., 526, 108 pp.

DuToit, A. L. 1937. *Our Wandering Continents, an Hypothesis of Continental Drifting.* Oliver and Boyd, London. 366 pp.

Florin, Rudolf. 1940. On *Walkomia* n. gen., a genus of Upper Palaeozoic conifers from Gondwanaland. *Kungl. Svenska Vetenskapsakad. Handl.,* ser 3, 18 (5): 1–23.

Frenguelli, Joaquain. 1952. The Lower Gondwana in Argentina. *The Palaeobotanist,* 1: 183–188.

Halle, Thore G. 1911. On the geological structure and history of the Falkland Islands. *Bull. Geol. Inst. Univ. Uppsala,* 11: 115–229.

————. 1927. Palaeozoic plants from central Shansi. *Geol. Survey China, Palaeontologia Sinica,* ser. a, 2 (1): 1–316.

————. 1929. On the habit of *Gigantopteris. Geol. Fören. Stockholm Förhandl.,* 51: 236–242.

————. 1937. On the relation between the late Palaeozoic floras of eastern and northern Asia. *Deux. Congrés. Strat. Carbonifere Heerlen,* 1: 237–245.

Høeg, Ove A. 1930. A psilophyte in South Africa. *Konelige Norske Videnskabers Selskab Forhand.,* 3 (24): 92–94.

————. 1937. Plant fossils and paleogeographical problems. *Deux. Congrés Strat. Carbonifére, Heerlen,* 1: 291–311.

Holden, Ruth. 1917. On the anatomy of two Palaeozoic stems from India. *Ann. Bot.*, 31: 315–326.

Hoskins, John H., and Aureal T. Cross. 1951. The structure and classification of four plants from the New Albany shale. *Amer. Mid. Naturalist*, 46: 684–716.

————. 1952. The petrifaction flora of the Devonian Mississippian Black Shale. *The Palaeobotanist*, 1: 215–238.

Hsü, J. 1947. Plant fragments from Devonian beds in Central Yunnan, China. *Journ. Indian Bot. Soc.*, (Iyengar Commem. Vol., 1946): 339–360.

Jones, O. A. and N. J. de Jersey. 1947. The flora of the Ipswich coal measures-morphology and floral succession. Univ. Queensland Papers, Dept. Geol., 3: 1–88.

Just, Theodor. 1952. Fossil floras of the southern hemisphere and their phytogeographical significance. *Bull. Amer. Mus. Nat. Hist.*, 99: 189–203.

Kidston, Robert and William H. Lang. 1923. Notes on fossil plants from the Old Red Sandstone of Scotland. I. *Hicklingia edwardi* K. and L. *Trans. Roy. Soc. Edinburgh*, 53: 405–407.

Kolbe, R. W. 1957. Fresh-water diatoms from Atlantic deep-sea sediments. *Science*, 126: 1053–1056.

Krishnan, M. S. 1954. History of the Gondwana era in relation to the distribution and development of flora. Seward Memorial lecture, Sahni Institute of Palaeobot., Lucknow, pp. 1–15.

Mendes, Josué C. 1952. The Gondwana formations of southern Brazil: some of their stratigraphic problems, with emphasis on the fossil flora. *The Palaeobotanist*, 1: 335–345.

Neuburg, Maria F. 1948. *Upper Paleozoic Flora of the Kuznetsk Basin.* Paleontology of the USSR, Acad. Nauk USSR, vol. 12, pt. 3, fasc. 2, pp. 1–342. Moscow.

————. 1958. Present state of the question on the origin, stratigraphic significance and age of Paleozoic floras of Angaraland. (Rept. for 4th Internat. Congr. Carbonif. Stratigr. and Geol., Heerlen, 1958), *USSR Acad. Sci., Geol. Institute*, Moscow, pp. 1–27.

Plumstead, Edna P. 1958. The habit of growth of Glossopteridae. *Trans. Geol. Soc. South Africa*, 61: 81–94.

Rao, H. S. 1940. On the anatomy of *Lycopodiopsis derbyi* Renault with remarks on the southern Palaeozoic lycopods. *Proc. Indian Acad. Sci.*, 11B: 197–217.

Read, Charles B. 1936. The flora of the New Albany shale, Part 1. *Diichnia kentuckiensis*, a new representative of the Calamopityeae. U. S. Geol. Survey Prof. Paper, 185-H: 149–155.

———— and Guy Campbell. 1939. Preliminary account of the New Albany shale flora. *Amer. Mid. Naturalist*, 21: 435–453.

Sahni, Birbal. 1953. (Posthumous work edited by T. M. Harris). Note on some possible psilophyte remains from Spiti, north-west Himalayas. *The Palaeobotanist*, 2: 1–3.

Saksena, Shivdayal. 1954. Reconstruction of the vegetative branches of *Phyllotheca etheridgei* Arber and *P. sahni* Saksena. *The Palaeobotanist*, 3: 51–53.

Scott, Dukinfield H., and Edward C. Jeffrey. 1914. On fossil plants, showing structure, from the base of the Waverley shale of Kentucky. *Phil Trans. Roy. Soc. London*, 205B: 315–373.

Seward, Albert C. 1931. *Plant Life through the Ages.* Cambridge Univ. press. 601 pp.

————, and John Walton. 1923. On a collection of fossil plants from the Falkland Islands. *Quart. Journ. Geol. Soc. London*, 79: 313–332.

Srivastava, P. N. 1954. Studies in the *Glossopteris* flora of India: 1. Some new fossil plants from the Lower Gondwanas of the Raniganj coalfield, India. *The Palaeobotanist,* 3: 70–78. (And several other contributions under this title in the same journal.)

Surange, K. R., and Prem Singh. 1953. The female dwarf shoot of *Walkomiella indica*—a conifer from the Lower Gondwana of India. *The Palaeobotanist,* 2: 5–8.

————, and P. N. Srivastava. 1953. Megaspores from the west Bokaro coalfield (Lower Gondwanas) of Bihar. *The Palaeobotanist,* 2: 9–17.

————, P. N. Srivastava, and Prem Singh. 1953. Microfossil analysis of some Lower Gondwana coal seams of west Bokaro, Bihar. *Bull. Nat. Institi. Sci. India,* 2: 111–127.

Teichert, Curt, and James M. Schopf. 1958. A Middle or Lower Devonian psilophyte flora from central Arizona and its paleogeographic significance. *Journ. Geol.,* 66: 208–217.

Termier, Henri, and Geneviéve Termier. 1950. La Flore Eifélienne de Dechra Aït Abdallah (Moroc central). *Bull. Soc. Geol. France,* ser. 5, 20: 197–224.

Thomas, H. Hamshaw. 1952. A *Glossopteris* with whorled leaves. *The Palaeobotanist,* 1: 435–438.

Walkom, A. B. 1915–1917. Mesozoic floras of Queensland. *Queensland Geol. Survey.* Pub. 252, pt. 1, pp. 1–50, 1915; Pub. 257, pt. 1 cont., pp. 1–66, 1917; Pub. 259, pt. 1 concl., pp. 1–48, 1917.

Walton, John, and Jessie A. R. Wilson. 1932. On the structure of *Vertebraria*. *Proc. Roy. Soc. Edinburgh,* 52: 200–207.

Wegener, Alfred. 1922. *The Origin of Continents and Oceans.* E. P. Dutton and Co., New York, 3rd ed. 212 pp.

17

an introduction to palynology

by Charles J. Felix

The term "palynology" was introduced by Hyde and Williams in 1944 to formally include all work with pollen and spores. It encompasses a wide variety of investigations in pure and applied botany, both modern and fossil. Palynology is a research tool in such diverse fields as paleobotany, systematic botany, plant geography, archaeology, geochronology, allergies, melittopalynology, and petroleum exploration. Because of this heterogeneity, comprehensive reference works are lacking and only a few segments are adequately covered by textbooks. However, palynology is unique in possessing a near complete bibliography, which has been published regularly since 1927.

Although palynology is commonly regarded as a new botanical science, it is more applicable to say that it is only now coming into its own. Actually palynological history is associated with the development of the microscope, and nearly 300 years ago both Nehemiah Grew and Marcello Malpighi observed and described pollen. There have been many botanists who have contributed to our knowledge of plant microfossils, and Wodehouse gives an excellent historical review, which surveys pollen work from its earliest days to the modern era. Carl Hugh Fischer is widely accepted as the founder of pollen morphology from the modern viewpoint, although only his late nineteenth century work dealt with pollen. Lennart von Post presented the first modern percentage pollen analyses in 1916 and brought realization of the true potentialities of the method. Erdtman presented von Post's methods outside of the Scandinavian countries in 1921. Erdtman's subsequent contributions have been numerous, and today he stands as a major figure in world palynology.

Fossil spore and pollen research has grown considerably in the last decade. The journal *Micropaleontology* gave names and addresses for 90 North American palynologists in its 1958 directory. *Palynologie Bibliographie* in 1959 listed the same data for 516 palynological personnel on a world-wide basis. There are certainly several thousand scientific workers with direct or indirect interests in palynology. Two international journals are devoted solely to palynology, and numerous publications regularly present plant microfossil articles. The first international palynological conference met at Stockholm in 1950, and in 1954 palynology was represented for the first time with its own section at the Eighth International Congress of Botany in Paris. In Europe and America palynological meetings with restricted membership now meet regularly. The first American national pollen conference met in 1953 and has continued informally, with the fifth national conference held in 1958.

It is perhaps the successful utilization of plant microfossils in stratigraphic work of petroleum geology that has given palynology its greatest impetus. However, plant spores have been a proven correlation tool in coal investigations for several decades both in North America and Europe. Nearly every segment of the plant kingdom has contributed the microscopic reproductive dissimules, termed pollen or spores, to palynology. This comprises a number of different sources, but only a small number of major groups produce most of the plant microfossils, and palynology is largely concerned with those of the vascular plants. Pollen is produced by members of the angiosperms and gymnosperms, the former commonly referred to as the flowering plants and the latter as non-flowering. Spores are contributed by a number of groups such as the ferns, fern allies, bryophytes, fungi, and algae. The ferns are by far the major source, with fern allies a minor contributor and the remainder to a lesser degree.

A frequently overlooked phase of palynology is the role of the pollen specialist in certain allergies. Many people are sensitive or susceptible to pollen of certain plants, and such an allergy is popularly called hayfever. To induce hayfever the offending pollen must be buoyant and thus easily transported; it must be toxic to allergic people. Most hayfever is caused by wind-transported pollen although all such plants do not cause hayfever. For illustration, the cat-tails and many of the conifers produce large amounts of pollen, but with the exception of a few conifer species, these do not cause hayfever. A few insect-pollinated plants, such as the goldenrod, have pollen toxic to some, but they require intimate handling to produce an allergy and are regarded as minor hayfever causes. Among the better known such plants responsible for wide-spread hayfever are the grasses and ragweeds. Allergy pollen is of such importance that most major cities now make atmospheric pollen counts during hayfever seasons by exposing slides covered with an adhesive, and many newspapers publish daily pollen counts for the interest of hayfever sufferers. Little is known of the role of fern spores in allergies, but it is suspected that they may be a contributing factor.

There should be no objection to including melittology in palynology, since knowledge of pollen is recognized as an essential part of honey investigations. The authenticity of honey sources can be verified by examination of the included pollen of honey. Pollen is indispensable as a food material for bees and all pollen does not contain the same food value. Protein content varies from 7 to 11% in pine, to 35% in date palm pollen, and average fat content ranges from 1% in birch to 17.5% in black walnut. Bee-collected pollen is known to be an excellent source of vitamins B and C, as well as some vitamins D and E. However, vitamins A and K are not known to be present in pollen. On occasions when there is a scarcity of pollen in nature, the beekeeper finds a knowledge of pollen to be valuable in preparation of pollen substitutes for bees.

Although the scope of palynology is botanical, the uniqueness of floras through the ages has made it necessary to approach pollen from a different aspect from spores. From the advent of pollen-bearing plants the problems of morphology, nomenclature, and techniques in pollen research have been different from those of spore studies. Cryptogam spores and gymnosperm pollen differ from angiosperm pollen both in chemical composition and cell wall morphology, and parallelism of structure is still ill-defined. Presumed angiosperm pollen has been recorded from the Jurassic of Scotland and Sweden; it became abundant throughout the world in the lower Cretaceous and dominant in the upper Cretaceous. Gymnospermous plants have an ancient lineage, with some of them extending back possibly into the Devonian. Some primitive gymnosperms have spores which show considerable organizational advance beyond the cryptogamic microspores, yet lack some significant features of modern pollen. For these some have proposed to use the term "prepollen"; however, the existence of pollen tubes in any Paleozoic plants has never been demonstrated. There are references to spores from the Cambrian, but the most authentic published assignments of vascular plant spores places their earliest known occurrence in the Silurian.

Although it is a general practice to use the terms "pollen" and "spore" as synonyms, literature abounds with discussions on the distinction between spores and pollen and there are divergent views on the homology of the two. In spore studies particularly, there is disagreement on usage of the terms "microspore" and "megaspore." In the botanical sense, the microspore is a reproductive body which germinates to form the male gametophyte, whereas the megaspore germinates to form a female gametophyte. A pollen grain can be said to be the germinated microspore of the seed plants. The megaspore of the seed plants gives rise to the embryo sac containing a reduced female gametophyte. In a strict sense the microspore is determined solely by the presence of the male gametophyte. In isosporous plants the spores develop directly into both male and female gametophytes. Since isospores of homosporous plants are very similar to microspores of heterosporous plants in size and form, discrimination between them is difficult in living plants and often impossible in fossils. It is true that microspores are frequently small, but all small spores are not microspores. In several instances the microspore is larger than the megaspore, especially in the gymnosperms where in the majority of instances the so-called microspore is the larger. Therefore the term "microspore" refers to fundamentally functional differences entirely aside from size distinctions, for it is possible to be unable to differentiate in the fossil state as to whether a spore is iso-, micro-, or megaspore. Among suggested terminology by palynologists is the term "miospore" for all fossil spores less than 200 μ in diameter. The term "polospore" has been introduced into palynology to include pollen and/or spores. Also suggested is 200 μ as the arbitrary size limit, with megaspores exceeding 200 μ, and those below this size comprising microspores, isospores, and prepollen. Such new terminology or artificial limits are of questionable value since botanically spores and pollen are well defined on functional characters. Additional terminology will not aid in recognition in the isolated fossil state unless functions can be inferentially established, and only detailed botanical research can do this.

POLLEN MORPHOLOGY

The development of the pollen tube in the angio- and gymnosperms provides the best means of differentiation between pollen and spores, but a general descriptive terminology for pollen is still lacking. Probably the most detailed system for describing pollen is that of Erdtman, but its great number of morphological characteristics renders it more acceptable for the specialist, and the beginner would do well to utilize a less detailed classification. In general, the pollen grain consists of three concentric layers. The inner layer, consisting of the living cell or protoplast, disappears quickly if pollination is not achieved. The middle layer is the *intine,* which is a cellulosic coat and easily destroyed. The third and outer layer is the *exine* and is a typically complex structure of extraordinary durableness. It is the exine which survives ages of decay processes and lengthy chemical treatment that is relatively unchanged; it may be said that its remarkable durability is the basis of palynology. Its resistance to destructive forces is due to its unusual chemical composition. The exine of pollen, as well as the walls of fungal spores, contains a highly polymerized, cyclic alcohol termed "sporopollenin" by Frey Wyssling. It is related to suberine and cutin but is more resistant than either. The quantity of sporopollenin differs specifically, and resistance to decay evidently changes in accordance with the quantity of the chemical. The exine is as variable morphologically as in chemical composition. It seems to be completely absent in aquatic species, and thus their remains are seldom preserved. The exine is very simple in some species, such as *Larix,* where it is a thin, homogeneous sheet.

Typically the exine is complex structurally, with at least two layers (Fig. 17-1). An inner layer, the *endexine,* is a continuous, homogeneous membrane. The outer layer, the *ektexine,* is composed of numerous small elements whose development and distribution produce great variability in structure. Two principal types

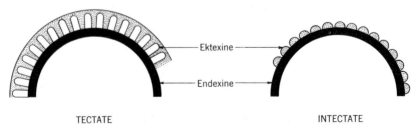

TECTATE   INTECTATE

Fig. 17-1. Principal exine types.

of wall structure are recognized because of ektexine variability. If ektexine elements are free and isolated or form an open pattern in such manner as not to place a continuous cover over the endexine, the type is called *intectate*. If the ektexine forms a continuous coat outside the endexine such as by a fusion of ektexinous elements, it is labeled as *tectate*. These two types are not absolutely distinct, and several transitional wall types exist. There are numerous exceptions to the ektexine division, as in some species (*Zostera*) both exine layers appear rudimentary or are absent, and in *Juncus* a very thin exine may correspond to the endexine.

The casual use of the term "pollen morphology" by palynologists obviously has a restricted meaning since it refers only to the morphology of the exine. Most investigations are concerned with fossil pollen that has lost intine and protoplasmic contents or recent pollen in similar conditions due to acetylation or other chemical means. The majority of diagnostic features are found in the exine, and presence of other layers obscure these features. Consequently, untreated grains are not suitable for morphological analysis of the exine.

Apertures

Most pollen grains possess openings or thin areas of the exine through which the pollen tube emerges at germination. Two general types of such *apertures* exist, and the arrangement and number of these on the grain produce a variety of pollen types. A few grains possess no visible aperture, and the exine is continuous over the entire surface (*Populus*). When apertures are present, the number varies from one to as high as 100. The terminology of apertures is still unsettled, but generally as purely morphologic definitions they are designated as *furrows* and *pores*. However, the nomenclature of Erdtman is more elaborate and creates some difficulty in application. Furrows are somewhat boat-shaped depressions in the exine, the ektexine being much reduced but with the endexine less affected. The openings are also the harmonegathi of Wodehouse and exist as flexible parts of the exine to accommodate changes in volume with varying water content. The typical exit is a rather isodiametric pore. In some instances it is closed by an operculum, and it is often surrounded by an annular area (*annulus*). Unlike pores, furrows do not completely penetrate the exine, and consequently if pores are lacking, the pollen tube must force its growth through the covering membrane of the exine at germination. Arrangement of apertures varies, with one-furrowed grains frequent in monocotyledonous (Fig. 17-3, 8) and gymnospermous plants. Three furrows and/or pores are common in dicotyledonous plants. They are often arranged equidistantly from each other and meridionally along the grain's equator. If more than three openings are present (Fig. 17-3, 6), they may be arranged as above or evenly distributed over the entire grain surface. The number and distributional pattern of apertures are generally easily observed. They provide valuable diagnostic characters, and several practical classification systems have been devised on aperture types. The numbers and disposition of apertures promise to be useful in classification, as major taxonomic units are generally uniform in this respect, and the use of aperture characters would afford a natural classification.

Size

Size is important since structural differences are sometimes inadequate for distinguishing species, and size becomes a reliable criterion. In *Picea,* for instance, such measurements have aided in species identification. Pollen grains of angiosperms range from about 5 to over 200 μ in diameter, although extreme sizes are rare and many such records are open to question. Most grains seem to fall in a range of 25 to 100 μ in living angiosperms. There are a few authentic instances of pollen about 5 μ in greatest diameter; the Boraginaceae, Piperaceae, Crypteroniaceae, and Cunoniaceae are families having species in the lowest size ranges. There are also recorded instances of grains in excess of 200 μ, with species in the Dipsacaceae, Nyctaginaceae, and Oenotheraceae having occasional grains of such extraordinary dimensions. Cranwell has observed that most of her monocotyledonous pollen studied lies between 15 and 80 μ, with very small grains being rare. The pollen of *Zostera,* an aquatic monocot, is among the most unusual in the plant world with regard to size. The grains resemble a membranous tube and exceed 2500 μ in length, although only 3 to 4 μ wide. There often appears to be a direct relation between grain size and number per anther. The very large pollen of *Mirabilis* occurs 32 grains per loculus, and the minute grains of *Borago* probably exceed 50,000. It also appears that, as a rule, the largest grains are produced by ephemeral flowers lasting only a day. Strict rules for size measurements are not standardized, but it should always be clear in descriptions which layers are included, whether over-all dimensions are given, or if processes are given separately.

GYMNOSPERMS

As in the angiosperms, the pollen tube usually serves as a fertilization agent in the gymnosperms. However, in certain members of the Cycadales it is simply a nutritive haustorium. A general description of gymnosperm pollen is impossible since there is great diversity among the orders. Among gymnosperms the pollen of the Coniferales is perhaps most familiar and most frequently encountered in sediments. The most conspicuous character is the presence of air sacs or bladders in genera of the Abietineae (Fig. 17-3, 1, 2, 3, 5) and Podocarpineae, although both tribes contain genera without bladders. When such appendages are present, the Abietineae have two, whereas the Podocarpineae may have one to six. The saccate grains in recent gymnosperms, as exemplified by the genus *Pinus,* are generally bilateral with proximal distal surfaces that are distinctly different (heteropolar). The aperture is distal between the bladders and is rather poorly defined. It is often referred to as a "sulcus," but is seldom more than a thin area merging into the adjacent exine.

SPORE MORPHOLOGY

The main types of spore configurations are bilateral and radial. Spores are seldom completely spherical due to close proximity in the tetrad. Depending upon tetrad arrangement, divergence from a sphere is toward an elongate shape (Fig. 17-2, 2), producing bilateral spores, or toward a tetrahedral shape with three contact faces and radial symmetry (Fig. 17-2, 3-16). That part of the spore nearest the tetrad's center and in contact with adjacent members is the "proximal surface" and the opposite free surface is the "distal surface." The safest and most universal distinction between the two fundamental spore types is the tetrad scar. It is usually well defined and readily discernible as "monolete" (Fig. 17-2, 2) or "trilete" (Fig. 17-2, 3-16). There are a few instances in extant and fossil spores where specimens are considered to be inaperturate (alete). Spore differentiation based on tetrad characters refers almost entirely to microspores or isospores. The megaspore is generally identified as such upon size features or an awareness of its botanical relationship. Even spore studies have their specialists and there is an increasing trend toward specializa-

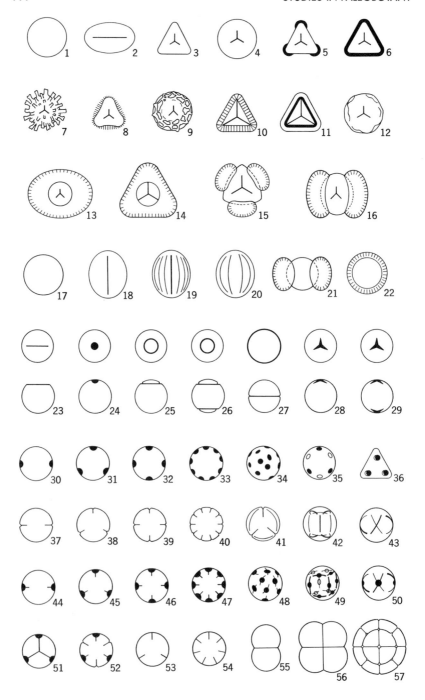

tion in either microspores or megaspores. This is in part because megaspores are not produced by all vascular plants, and true microspores are present in vastly greater numbers than megaspores. Microspores and homosporous spores are much more suitable for the prevalent mass treatment methods than are the larger megaspores, which are subject to damage in drilling.

The exine in spores is generally structureless. Some stratification may be visible, but the elaborate divisions seen in pollen walls is lacking in spores. The exine does have, however, a variety of sculpturing which aids in identification. Living pteridophytes have a considerable size range with extremes of 15 to 125 μ for microspores and isospores. Harris found 67% of the New Zealand fern spores to be in a medium size class of 25 to 50 μ.

POLLEN AND SPORE DISSEMINATION

The method of pollination is important to the palynologist since it provides a clue to the dispersal of pollen. This is valuable in interpreting fossil floral assemblages, especially in oil geology where environment plays a significant role in oil accumulation. To attain correct interpretations, the investigator must be cognizant of a number of facts: (1) The amount of pollen produced by specific plants. (2) The chief mechanism of transport. (3) The possible distance from its growth site that pollen is likely to be carried. In

Fig. 17-2. Schematic drawings of important spore and pollen classes. 1–12. Microspores.

1. *Aletes*. 2. *Laevigatosporites*. 3. Granulatisporites. 4. *Punctatisporites*. 5. *Triquitrites*. 6. *Densosporites*. 7. *Raistrickia*. 8. *Reinschospora*. 9 *Reticulatisporites*. 10. *Cirratriradites*. 11. *Lycospora*. 12. *Calamospora*.

13–16. Prepollens.

13. *Florinites*. 14. *Endosporites*. 15. *Alatisporites*. 16. *Illinites*.

17–22. Gymnospermous pollen classes.

17. Inaperturate (also in 23–29, 30–57). 18. Monocolpate (also in 23–29). 19–20. Polyplicate. 21–22. Vesiculate.

23–29. Monocotyledonous pollen.

23. Monocolpate. 24. Monoporate. 25–27. Operculate. 28–29. Trichotomocolpate.

30–57. Dicotyledonous pollen (polar and equatorial view).

30. Diporate. 31. Triporate. 32–33. Stephanoporate. 34–36. Periporate. 37. Dicolpate. 38. Tricolpate. 39–40. Stephanocolpate. 41–43. Pericolpate. 44. Dicolporate. 45. Tricolporate. 46–47. Stephanocolporate. 48–50. Pericolporate. 51. Syncolpate. 52. Heterocolpate. 53–54. Zonorate. 55. Dyad. 56. Tetrad. 57. Polyad. (From Kuyl, Muller and Waterbolk, 1955.)

general, pollination mechanisms accountable for grain dispersal may be grouped as: Anemophilous (wind); Entomophilous (insect); Hydrophilous (water); and Ornithophilous (bird). In addition there are reported instances of Chiropteriphily (bat) and Malacophily (snail and slug pollination).

There are fundamental structural differences in pollen grains which afford clues as to their mode of distribution. In general, hydrophilous pollen grains lack an exine, whereas anemophilous grains are small and smooth-walled. Greater variation is noted in entomophilous pollen where the grains are generally adhesive, and the ektexine bears prominent protuberances and pits of a wide variety of forms. Most of the gymnosperms are widely distributed by wind, and catkin-bearing angiosperms such as *Quercus, Corylus,* and *Betula* lend themselves to wind dispersal of pollen. Some extraordinary distances of wind transport are recorded, with *Pinus* pollen recovered in the Arctic some 60 miles from the nearest coniferous forest, and Erdtman has cited examples of *Picea* and *Pinus* pollen being wind transported for a distance of over 600 miles. One of the most precise figures is Faegri and Iversen's of 30 to 60 miles as the natural limits of pollen dispersal. In all probability the extreme distance carried is much overrated, and movement from a source area by wind of more than 100 miles is an extreme case. Nevertheless, the researcher must remain aware of long distance pollen movement, especially when the question of species migration arises. It will be difficult to ascertain whether a low percentage of a pollen species is due to a small local occurrence or to long distance transport. It must be realized that with increasing distance from the forest site that the importance of long distance species increases. The spores of ferns and mosses are disseminated by wind and water almost exclusively, but there are few data available on comparative distances of transport.

Anemophilous species contribute vast amounts of the pollen deposited, and the quantity of pollen produced by some plants is enormous. Such pollination is a very inefficient process since vast numbers of pollen grains are involved, although an infinitesimal proportion complete their role in the life cycle. It has been estimated that a single anther of *Canabis sativa* produces 70,000 grains. Pollen productivity is usually highest among wind-pollinated species. For instance in *Linum catharticum,* which is entomophilous, the number of grains for anther is only about 100. Certainly, the forest trees, which contribute so greatly to pollen spectra, produce huge quantities of pollen. Dissemination from a mature *Pinus, Picea,* or *Quercus* may amount to hundreds of millions of grains yearly. Because of differences in habitat and plant habit, few Filicales may be expected to have as wide spore dispersal as anemophilous tree pollen in post-Paleozoic deposits. Spore production is varied in extant ferns and often abundant, but it does not approximate the prodigious numbers of the pollen producers. Production ranges from thousands of spores in each Eusporangiate sporangium to the single spore of the Hydropteridea megasporangium. *Ophioglossum* with about 15,000 spores per sporangium is probably most prolific, and some mature *Dryopteris* plants have been estimated to produce a half million spores in a single season. The Filicales are certainly productive and ubiquitous enough to warrant attention in statistical investigations. There is apparently no relationship in the ferns between spore size and number per sporangium. The quillwort, *Isoetes,* is another prolific spore producer; it has been estimated that as many as one million microspores are borne in a sporangium.

POLLEN IDENTIFICATION

A basic prerequisite in palynology is correct identification of the microfossils. This is a considerable task in pollen work in view of the vast number of pollen producers, and the evaluation of palynological data must be cautious. Although it may be said that pollen grains in related genera are usually more or less of the same type, striking differences exist. In the Caryophyllaceae the grains for the most part have many pores (cribellate), but *Spergula, Pteranthus,* and a few other genera have tricolpate grains. There are also numerous instances of pollen grains of similar types occurring in plants not at all related, such as in the genera *Salix* (Salicaceae) and *Adoxa* (Adoxaceae). Most genera do show a marked consistency of pollen form, but there are exceptions where more than one pollen type will occur in certain genera. *Tulipa* of the Liliaceae, a genus with constant floral characteristics and seemingly natural, bears monocolpate pollen in some species and tricolpate pollen in others. It is unique in being the only monocot known to have the latter type of grain. The genus *Crocus* has pollen either nonaperturate or with several parallel, ringlike furrows. Some other genera with comparable differences are *Symplocos, Anemone,* and *Morina.* A step further is the production of more than one pollen type by a single species. The dimorphic and heterostylous flowers of *Primula veris* have a decided difference in pollen from long and short styled flowers, the grains of the latter being smaller in size along with differences in shape and in aperture numbers. Instances of pollen dimorphism are also known in the Plumbaginaceae and Rubiaceae. A classic example of heterostyly is *Lythrum salicaria* of the Lythraceae. Three kinds of flowers, based on style and stamen length, are present in the species. Not only is there a size difference in pollen from each but the longest stamens produce green pollen, the other two lengths having yellow pollen. It can be said that the generic identification of pollen is usually possible, but specific identification with any degree of certainty is less easily accomplished. However, there are many families in which characteristic features may be sufficiently constant for diagnosis. The Cyperaceae have tetrahedral, psilate grains with a prominent basal pore and three slitlike lateral pores as the common aperture type. The Betulaceae have characteristic thickened streaks of exine, known as arci, which form sweeping curves from pore to pore.

Compound grains due to failure of the spore tetrads to separate are found in families and genera of both the monocots and dicots. The Ericaceae are notable because the pollen remains permanently in tetrads, a peculiarity shared with the Empetraceae, Typhaceae, (Fig. 17-3, 11) and Juncaceae among others. The permanent cohesion of pollen into groups larger than tetrads (polyads) is conspicuous in the Mimosoideae (Fig. 17-3, 9) of the Leguminosae. Even greater variation is present among the Orchidaceae where isolated grains, tetrads, and polyads occur in different genera.

Most of the notable anomalies occur in the dicotyledonous plants. Monocotyledonous pollen is generally recognizable, but the narrow range of diagnostic characters often render generic and family differentiation difficult. There are two basic monocot grain types: (1) essentially spheroidal without a recognizable aperture and (2) with bilateral symmetry and characterized by a single furrow (Fig. 17-3, 8) or pore (Fig. 17-3, 13). There are a few variants such as the many-pored *Alisma* and *Hypoxis* with two apertures.

COLLECTION AND STUDY TECHNIQUES

Reference Collection

Despite the increase in number of publications in recent years, studies of Tertiary and Quaternary pollen suffer from a lack of

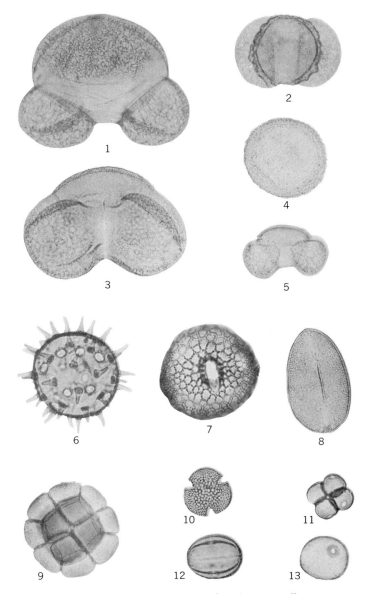

Fig. 17-3. Representative extant gymnosperm and angiosperm pollen.
1–5. Gymnosperm pollen.
   1. *Abies grandis.* 2. *Picea sitchensis.* 3. *Cedrus atlantica. Tsuga heterophylla.*
5. *Pinus strobus.*
6–13. Angiosperm pollen.
   6. *Hibiscus althea.* 7. *Ochroma limonensis.* 8. *Iris kaempferi.* 9. *Pithecolobium oblongum.* 10. *Fremontia californica.* 11. *Typha latifolia.* 12. *Cissus ampelopsis.*
13. *Hordeum vulgare.*

published illustrations for use in identifications. The necessity for determining botanic affinities of unknown entities has led to the establishment of reference collections of modern pollen and spores. This is one of the most valuable tools available, for regardless of the quality of illustrations, no drawing or photograph can show satisfactorily the minute diagnostic features that can be discerned in reference material. Illustrations at best can only convey suggestions as to the identification of pollen grains and spores and are not intended to replace pollen and spore preparations. Since the majority of Tertiary and Quaternary plants belong to extant genera or families, a carefully planned reference collection is indispensable. In enabling correct assignment of pollen and spores, it can help prevent creation of artificial genera for living ones. It is almost essential that the investigator make accurate botanical identifications if accurate interpretations are to be made of flora evolution or ancient climates, and the reference slides can provide the necessary comparative material.

There are over 200,000 species of living vascular plants. Therefore, the basic reference collection must be prepared with the specific problem in mind. Obviously a near complete collection would be a vast undertaking, and the collection of the Palynological Laboratory of the Swedish Natural Science Research Council in Solna, Sweden, is probably the largest in existence today. It comprises about 25,000 slides representing more than 20,000 species of plants. The most valuable collection is one in which specimens are referable to a specific sheet in an established herbarium. This would provide the most accurately identified material, and in the event of discrepancy, the exact specimen could be rechecked. A serious objection to this source of collecting is the danger of mutilating valuable taxonomic material, but conscientious collecting can alleviate this. It is quite possible that, as palynology grows and demands for reference specimens increase, herbaria sources for standards may be exhausted, and the palynological laboratories may conceivably have to build herbaria or contribute to existing ones.

By examination of floral material with a hand lens the capable botanist can usually determine whether pollen is present. Flower buds collected just before anthesis are the most reliable since little of the pollen will be shed. There are objections to this on the premise that the pollen may be immature, but the possibility is so remote as to be negligible. The fertility of gymnosperm cones and cryptogam sporangia can also be determined with a lens. Fern

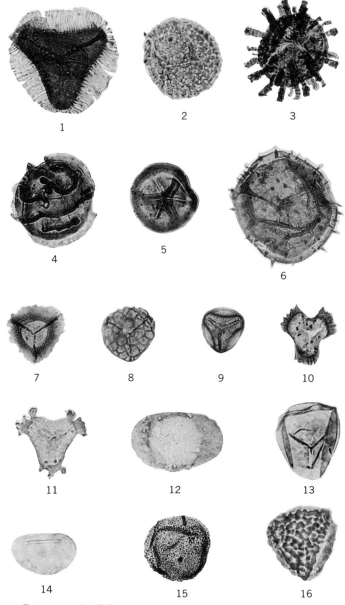

Fig. 17-4. Representative Paleozoic spore genera.
1. *Reinschospora.* 2. *Convolutispora.* 3. *Raistrickia.* 4. *Proprisporites.* 5. *Knoxisporites.* 6. *Grandispora.* 7. *Densosporites.* 8. *Dictyotriletes.* 9. *Rotaspora.* 10. *Tripartites.* 11. *Mooreisporites.* 12. *Schulzospora.* 13. *Calamospora.* 14. *Laevigatosporites.* 15. *Punctatisporites.* 16. *Callisporites.*

spores may create a somewhat different situation than pollen since several palynologists have experienced difficulty in obtaining satisfactory spore preparations from herbarium sheets. Evidently shedding is so efficient in the Filicales as to disperse most mature spores and, unlike pollen grains, fern spores appear to undergo a final developmental stage of external morphology prior to normal shedding and mature spores are often rare. Specimens collected are best stored in small vials or envelopes and all herbaria data should be recorded.

Various published preparation methods exist, but most are basically the acetylation procedure of Erdtman, frequently with modifications. It involves acetylation of the carbohydrate fraction of plant microfossils, leaving only the spore or pollen exine with its diagnostic features. This method involves treatment with nine parts acetic anhydride to one part concentrated sulfuric acid. The general practice is to divide the sample, submitting one fraction to a bleaching procedure and staining the unbleached fraction. This provides specimens useful for viewing wall sculpture when stained, and permits examination of internal wall structures in bleached specimens. The acetylation method has its chief disadvantage in that pollen grains are overexpanded. Numerous investigations have confirmed that recent and fossil grains expand to different degrees, and therefore acetylation is not reliable in statistical studies of size in recent pollen or in comparison of recent and fossil specimens. It is advisable to use untreated specimens when size determination is critical, even though permanent preparations may not be retained.

Proper mounting media are of inestimable importance, for many early preparations have deteriorated to uselessness due to inadequate mountants. Unfortunate, too, has been the practice of lactic acid mounts of pollen from herbarium sheets. This method results in such expansion of grains as to nullify their usefulness in morphological studies, and its clearing action is so rigorous that grains are virtually destroyed. Two main schools of thought exist on media. Absolute permanency in a medium is probably favored by most Paleozoic spore investigators. However, a viscous medium is considered desirable by many post-Paleozoic pollen workers. A viscous preservative permits orienting the grains for examination, and pollen is not generally subjected to the deformation by movement that occurs in moving the many winged and flanged forms of the Paleozoic. Factors to be considered in selecting a medium are: (1) refractive index; (2) viscosity; (3) permanency; and (4) crystalization.

Among better known palynological mounting media are Canada balsam, diaphane, and glycerin jelly. The first two require dehydration of the specimens before mounting. Glycerin is water-miscible and has its supporters among those desiring a viscous medium. A number of synthetic resins are marketed, but for proven durability balsam and diaphane remain the most reliable. A quick drying plastic medium used by many palynologists is Clearcol, and it has a good history of permanency. It is advisable to seal cover slips, especially glycerin mounts, to reduce the possibility of drying. Even though there are records of unsealed slides several decades old, the addition of a ringing cement is good insurance. If slides are permitted to dry in an inverted position, specimens settle on or near the cover glass and greatly facilitate microscopic examination.

Systematics

As with many phases of paleontology, nomenclatural problems are an important aspect of palynology. In Paleozoic investigations most spores have been treated as species of organ genera independent of any natural phylogenetic system. Their true botanic relationships can be determined only by associating them with fructifications of known affinity. Thus artificial systems with genera defined by arbitrary morphological features suffice quite well until true relationships of spores are discovered, and species can then be transferred to the appropriate natural genera. Some Paleozoic students have completely abandoned normal botanic practices and use systems of letters and numbers in classification. However, for eventual intelligent comparison of results and the incorporation of spores into a natural classification system, standard nomenclatural practices are most desirable.

In the Quaternary, spores and pollen are largely assignable to living genera, and modern nomenclature, often specific epithets, can be used. Tertiary palynological taxonomy has problems additional to those of the Paleozoic and Quaternary, and there is always the possibility that new artificial genera will be created without knowledge of extant genera. It is also possible that Tertiary pollen assigned to a living genus may instead belong to a related, extinct genus with near indistinguishable pollen. There is a wide disagreement among palynologists regarding taxonomy, and this has resulted in several nomenclatural systems. Some are compromises between the use of artificial systems of letters and symbols, and normal paleontologic taxonomic systems.

Among constructive developments have been the use of artificial organ genera where the microfossil's relationship is questionable, but showing the botanic relationship wherever it is known. Also there is a growing insistence that new taxonomic units be amply described, well illustrated, and that a holotype be designated.

Microscopy

Good optical equipment is most essential in palynology. Plant microfossils are not only so small as to require high magnifications, but the time that must be devoted to microscopy entails considerable eyestrain. A binocular microscope is recommended rather than a monocular type to ease eye fatigue and also because steroscopic relations of grain characters are more easily viewed. Working magnifications of about 100 to 1200 are generally considered most desirable. The more difficult specimens require about $1000 \times$ lens with a very high numerical aperture. Apochromatic and fluorite lenses have a wide preference among palynologists, and good synthetic fluospar objectives are now available. All of these lenses should be used in conjunction with wide-field compensating eyepieces; and very necessary for advanced pollen-analytic research is an immersion objective of high quality.

Electron optics is only beginning to be introduced into pollen morphology. Pollen is not the optimum subject for such study, but electron microscopy has already provided important information on the fine structure of exines. The high magnifications of the electron microscope promise to shed new light upon structural differences between wall layers that presently lie beyond the utmost powers of the light microscope. Phase-contrast microscopy also has its adherents, but it has wrought little material change in palynology investigations. Pollen is not well suited for phase-contrast observations, and little has been accomplished by these techniques beyond affording clearer resolution of some wall features.

Sampling Methods

The collection of the sample is one of the most important aspects of palynology, for the greatest care must be taken to prevent the introduction of foreign matter. Contamination is a constant source of error in palynology from the moment of collection of the specimen until the final slide preparation. Sources of impurity are so numerous that cleanliness and vigilance must be considered indispensable. The very atmosphere at the collecting site and in the laboratory are hazards. Improper collecting techniques, soiled laboratory equipment, and carelessness all contribute to the danger of contamination. Cleanliness in the palynological laboratory must be equal to that practiced in bacteriology if absolute confidence in final analysis is realized.

Both surface and subsurface material is used. The quantity varies with problems and investigators, but about 20 grams appears to be a reliable amount. In collecting surface samples, column samples from mine and outcrop exposures can be equally good, if uniform segments are taken. In coal sampling the seam is cleared of debris with hammer and divided into benches. Partings and bands often provide natural benches. A continuous sample, providing complete vertical representation, is made by sampling perpendicular to the bedding plane. Tools should be cleaned carefully after each collection, and the outcrop is collected from bottom to top to minimize contamination from sifting debris.

Subsurface samples are obtained most accurately from well cores, but the majority of wells drilled employ the rotary drilling technique in which a hollow drill stem terminated by a drill bit is rotated into the subsurface. A stream of drill mud, circulated down the hollow stem and back up the hole to a pit, facilitates drilling by lubricating the drill bit and by removing rock chips from the hole. The chips are collected at regular intervals and are the most common type of sample. They are also least satisfactory because of contamination due to caving from the hole walls. The lack of care taken by well crews in obtaining the cuttings is often a principal source of error. However, when properly collected, rotary cuttings have proved valuable in correlation studies. Despite these disadvantages, well cuttings must be utilized since most wells are drilled by this method, and pollen and spores are especially valued as most megafossils are severely damaged in drilling. Most dependable, but costly, are "standard cores" obtained by receiving a cored section in a hollow core barrel attached to the lower end of the drill stem. Other types of core are "wire line core" and "side wall core." Although expensive, core samples are most reliable in the establishment of standards.

Sample Preparation

Preparation techniques vary in palynology since fossils differ in size, chemical composition, and in lithologies, and preparation methods are about as numerous as rock types. Paleozoic spore studies have necessitated techniques different from those in pollen research. Particularly in coal, where spore preservation is generally excellent, treatment is quite specialized. A well-known preparation is the maceration method first described by Franz Schulze in 1855 when he treated coal with potassium chlorate and nitric acid to render microscopic structure clearer. Consisting of partial oxidation of coal and dispersal of the humic matter, the Schulze's method is still widely used today with various modifications. This solution oxidizes and partially dissolves the humic matrix. After being washed free of acid, the residue is treated with a basic solution such as potassium hydroxide or ammonium hydroxide until all humic material is thoroughly dispersed. No fixed procedure can be recommended for various coals. Weathered coal generally macerates more readily than fresh coal and the procedure best suited for each type of sample must be determined by trial. Hydrochloric acid is the most common reagent for dissolution of carbonates, whereas hydrofluoric acid eliminates silicates. After maceration of the sedimentary matrix, complete separation is often attained by differential flotation with heavy liquids, with the organic remains floating and the mineral fraction sinking. These procedures have many variations, depending upon the matrix. A new development in sample processing has been the use of the ultrasonic generator. Ultrasonics are useful in the disaggregation of some rock matrices. However, the principal utilization has been the removal of debris and foreign particles from the plant micro-fossils following the initial breakdown by chemical methods.

APPLICATION OF PALYNOLOGY

So vast is the scope of botany and the varied utilization of data that palynology has developed its specialized divisions in regard to time. Pleistocene palynology has encompassed extensive study of post-glacial vegetation, climatology, and paleoecology. Paleozoic research has concerned itself largely with the Pennsylvanian coals and their correlation. Mesozoic palynology has devoted most attention to the Cretaceous and early pollen floras, but the Jurassic spore assemblages are beginning to receive attention. The Tertiary includes such diverse flora and climatic changes that several specialized divisions have developed.

Every age therefore has its specifically different goals which are necessitated by the changing spectra throughout the geologic column due to plant evolution and the migration of vegetational units. The wind-transported entities are of interest to the geologist as they are not, as a rule, restricted to particular facies. In their dispersal by wind the pollen and spores are so well mixed as to settle in a depositional basin and leave a fairly representative picture of the regional vegetation. This picture may be distorted in several ways: (1) The microfossils are differentially resistant to decay not only in fossilization but also in withstanding chemical preparation treatment; (2) Local overrepresentation must be considered since an exceptionally abundant pollen producer in the proximity of the depositional area can distort the interpretation of a region's vegetational history; (3) Long distance transport is usually a minor factor, but it must be considered, as minor accessory pollen is often important and the possibility of its transport can affect a spectrum.

Plant microfossils are more likely to be found in unweathered sediments of reducing environments. These consist of marine, brackish and fresh-water shales, limestones, bituminous coals, lignite, and peat. A high organic content, usually characterized by a dark color, is often an indication of plant microfossils being present. Sandstones are generally barren since the coarse grain size permits entrance of oxygen and ground water to a destructive extent. Pollen and spores are independent of basin environments, which sets palynology apart from traditional paleontology. Most microfossils used for stratigraphic work (foraminifera, ostracodes, fusulinids) are products of their depositional basins and thus are controlled by ecologic conditions of the environment. Plant microfossils occur in marine, brackish, and continental sediments. The varied environments of plants, and the unique pollen transport mechanisms, serve to release pollen from the usual environmental restrictions of most organisms.

The defining of ancient shore lines bears special significance in oil accumulation, and palynology has proven a valuable aid to the geologist in determining strand lines. As the pollen rain is progressively less in the seaward direction, sedimentary environments with pollen assemblages are limited to near shore marine and lacustrine waters. In modern marine sediments pollen and spores are frequently associated with diatoms, microforaminifera, dinoflagellates, hystrichospherids, and various oceanic plankton. It is possible to determine distance and direction of ancient shore lines by the kinds and quantities of microfossils, as pollen and spores will decrease in density in a seaward direction, with a corresponding increase in marine forms.

It is in the Paleozoic that palynology has been utilized most effectively in correlation work. Especially in the Pennsylvanian, spore analyses of coal seams are an integral part of coal correlations, and coal beds may often be identified on the basis of relative abundance of certain spore genera. Figure 17-5 illustrates the remarkable uniformity of the microflora composition of the Indiana Coal V over a regional area. Such a seam traced laterally usually remains similar over a coal field, whereas an adjacent seam usually has a different histogram due to the different proportions between the constituent spore genera.

In Quaternary pollen analysis correlations are based on changes in relative abundance of carefully selected species or groups of pollen grains. These are mainly pollen grains of forest trees. Shifts in the relative abundance of these forms reflect important vegetational changes due to altered climatic conditions. Plant microfossils have proven of value in determining changes in climate associated with changes in the extent of the Pleistocene ice. Most plant genera of the Mesozoic and Paleozoic are extinct. Thus ecological evaluation of such assemblages is dependent on correlative evidence from associated fossils and the physical character of the deposit.

Although much of the work that has been done in palynology to date has been of a practical nature, in connection with stratigraphic problems, fossil spores and pollen give promise of adding greatly to our understanding of the past distribution of floras. In all probability floristic studies of the future will combine the evidence from microfossil and macrofossil remains, thus adding to the accuracy of our knowledge of the floras. Pollen will be particularly helpful in cases where they offer distinctive diagnostic characters and with plants that are not often preserved in macrofossil form. The Gnetales, a unique assemblage of gymnospermous plants, offers a case in point. Three living genera, the interrelationships of which are not clear, are included in the group: *Gnetum,* with 30 species, is found in moist tropical regions; *Welwitschia* is a monotypic genus found in desert areas of southwest Africa; *Ephedra* includes about 42 species that are widely distributed through arid tropical and temperate regions. None of the habitats is favorable for deposition.

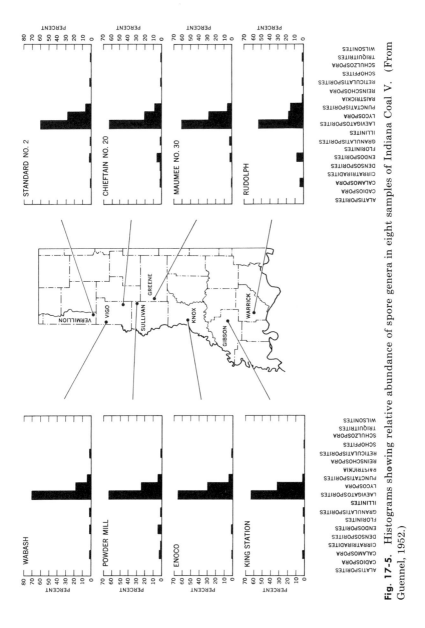

Fig. 17-5. Histograms showing relative abundance of spore genera in eight samples of Indiana Coal V. (From Guennel, 1952.)

459

The macrofossil record of the Gnetales is extremely scanty. However, the pollen grains are readily recognized as *Ephedra* and *Welwitschia* have distinctive longitudinal ridges; the former is polycolpate and *Welwitschia* monocolpate. *Gnetum* does not possess any visible germinal opening. Pollen attributed to *Ephedra* has been found at early Tertiary horizons in Tasmania, Victoria, and South Australia, in the Eocene Green River shales of Colorado and in the Paleocene of Hanover, Germany, and at Cretaceous horizons in Iraq, Venezuela, and Long Island in New York. *Welwitschia* pollen has been reported from Russia and Australia, as well as the United States. The most ancient records to date for both are from the Middle Permian of Oklahoma; L. R. Wilson has reported pollen (*Ephedripites* sp. and *Vittatina* sp.) which is believed to be that of early ancestral forms of *Ephedra* and *Welwitschia.*

REFERENCES

Arnold, Chester A. 1950. Megaspores from the Michigan Coal Basin. *Michigan Univ., Contr. Mus. Paleontology,* vol. 8, No. 5, pp. 59–111.

Bhardwaj, D. C. 1957*a*. The palynological investigations of the Saar coals. *Palaeontographica,* vol. 101, pt. B, pp. 73–125.

————. 1957*b*. The spore flora of Velener Schichten (Lower Westphalian D) in the Ruhr coal measures. *Palaeontographica,* vol. 102, pt. B, pp. 110–138.

Cookson, Isabel C. 1956. Pollen grains of the *Ephedra* type in Australian Tertiary deposits. *Nature,* vol. 177, pp. 47–48.

Couper, R. A. 1958. British Mesozoic microspores and pollen grains. A systematic and stratigraphic study. *Palaeontographica,* vol. 103, pt. B, pp. 75–179.

Cranwell, Lucy M. 1953. *New Zealand Pollen Studies. The monocotyledons. Bull. Auckland Institute and Museum,* No. 3, 91 pp.

Dijkstra, S. J. 1946. Eine monographische Bearbitung der karbonischer Megasporen. *Meded. Geol. Stichting.,* Ser. C-III-1. 101 pp.

————. 1956. Lower Carboniferous Megaspores. *Meded. Geol. Stichting.,* n. s. No. 10, pp. 5–18.

Erdtman, Gunnar. 1952. *Pollen Morphology and Plant Taxonomy (An Introduction to Palynology.* I). The Chronica Botanica Co., Waltham. 539 pp.

————. 1954. *An Introduction to Pollen Analysis.* The Chronica Botanica Co., Waltham. 239 pp.

————. 1957. *Pollen and Spore Morphology/Plant Taxonomy (An Introduction to Palynology.* II). The Ronald Press Co., New York. 151 pp.

Faegri, Knut. 1956. Recent trends in palynology. *Bot. Rev.,* vol. 22, No. 9, pp. 639–664.

————, and Iversen, J. 1950. *Textbook of Modern Pollen Analysis.* E. Munksgaard, Copenhagen. 168 pp.

Guennel, G. K. 1952. Fossil spores of the Alleghenian coals in Indiana. *Indiana Geol. Surv. Rpt. of Progress,* No. 4. 40 pp.

Harris, William F. 1955. A Manual of the spores of New Zealand Pteridophyta. *New Zealand Dept. Sci. and Indust. Research, Bull.,* No. 116, 186 pp.

Hyde, H. A., and Adams, K. F. 1958. *An Atlas of Airborne Pollen Grains.* MacMillan and Co. Ltd., London. 112 pp.

Ikuse, M. 1956. *Pollen Grains of Japan.* Hirokawa Publ. Co., Tokyo. 303 pp.

Kosanke, Robert M. 1950. Pennsylvanian spores of Illinois and their use in correlation. *Ill. Geol. Surv. Bull.,* No. 74. 128 pp.

Kuyl, O. S., Mueller, J., and Waterbolk, H. T. 1955. The application of palynology to oil geology with reference to western Venezuela. *Geologie en Mijnbouw,* No. 3, n. s., vol. 17, pp. 49–75.

Leopold, E. B., and Scott, R. A. 1958. Pollen and spores and their use in geology. *Smithsonian Institution Rpt. for 1957.* Pp. 303–323.

Morgan, J. L. 1955. Spores of McAlester coal. *Oklahoma Geol. Surv.,* Circ. 36. 52 pp.

Ogden, E. C. 1957. Survey of airborne pollen and fungus spores of New York State. *New York State Mus. and Sci. Ser. Bull.,* 356. 62 pp.

————, and Lewis, D. M. 1960. Airborne pollen and fungus spores of New York State. *New York State Mus. and Sci. Ser. Bull.,* 378. 104 pp.

Potonié, Robert. 1956. Synopsis der Gattungen der Sporae dispersae. *Beihefte z. Geol. Jahrbuch,* vol. 23. 103 pp.

————. 1958. Synopsis der Gattungen der Sporae dispersae. *Beihefte z. Geol. Jahrbuch,* vol. 31. 114 pp.

———— and Kremp, G. 1954. Die Gattungen der palaozoischen Sporae dispersae und ihre Stratigraphie. *Geol. Jahrbuch,* vol. 69, pp. 111–194.

Schopf, James M., Wilson, Leonard R., and Bentall, R. 1944. An annotated synopsis of Paleozoic fossil spores and the definition of generic groups. *Ill. Geol. Surv. Rpt. Inv.,* No. 91. 72 pp.

Selling, Olaf H. 1946. Studies in Hawaiian pollen statistics. I. *Bishop Mus. Spec. Publ.,* 37. 87 pp.

————. 1947. Studies in Hawaiian pollen statistics. II. *Bishop Mus. Spec. Publ.,* 38. 430 pp.

————. 1948. Studies in Hawaiian pollen statistics. III. *Bishop Mus. Spec. Publ.,* 39. 154 pp.

Traverse, Alfred. 1955. Pollen analysis of the Brandon lignite of Vermont. *Bur. Mines Rpt. Inv.,* 5151. 107 pp.

Van Der Hammen, Thomas. 1956. A palynological systematic nomenclature. *Boletin Geol.,* vol. 4, No. 2–3, Bogota, Colombia. Pp. 63–101.

Wenner, C. G. 1947. Pollen diagrams from Labrador. A contribution to the quaternary geology of Newfoundland-Labrador, with comparisons between North America and Europe. *Geografiska Annaler.* 241 pp.

Wilson, Leonard R. 1959. Geological History of the Gnetales. *Oklahoma Geol. Notes,* vol. 19, No. 2, pp. 35–40.

Wodehouse, Roger P. 1933. Tertiary Pollen II. The oil shales of the Eocene Green River formation. *Bull. Torrey Bot. Club,* vol. 60, pp. 479–524.

————. 1935. *Pollen Grains. Their Structure, Identification and Significance in Science and Medicine.* McGraw Hill Book Co., New York. 574 pp.

Periodicals Devoted to Palynology

Catalog of Fossil Spores and Pollen. College of Mineral Industries, The Pennsylvania State University.

Grana Palynologica: An International Journal of Palynology. Almqvist and Wiksell, Stockholm.

Palynologie Bibliographie. Service d'information Geologique du Bureau de Recherches Geologiques, Geophysiques et Minieres, Paris.

Pollen et Spores. Museum National d'Histoire Naturelle, Laboratoire de Palynologie, Paris.

18

techniques for
studying fossil plants;
some basic references

The increase in knowledge that may be made in any field
of science is in large part limited by the tools that are
available to work with; the tools need not always be elaborate but
they must be adequate. Great advances have been made in paleobotany in recent decades partly because new ways and means have
been found to extract information from fossil plants; certain older
methods have been improved and entirely new ones have been
developed.

This chapter is not, however, intended as a laboratory manual
nor as a survey of all the techniques that are in use at present.
Some of the basic methods are outlined, but the student is strongly
urged to consult the original accounts or better still visit laboratories in which various types of paleobotanical research are being
conducted. Even though space were available and the writer's
knowledge adequate to the task it would not be possible to compile a paleobotanical techniques manual that would lead one to
the best preparation methods for all types of preservation. The
latter are sufficiently variable so that a good deal of ingenuity
must frequently be brought into play to adapt standard techniques
to the material at hand. The advances of recent years should
leave no doubt that techniques will continue to improve and our
knowledge should expand correspondingly in the future. The
techniques used in certain highly specialized branches of paleobotany such as palynology and coal studies are not included here.

Since most fossil plant materials fall in the *compression* or *petrifaction* categories it seems most expedient to consider them
separately.

463

PETRIFACTIONS

Ground Thin Sections

This classical method of preparing petrifactions for study has been employed for about 150 years and is still useful. The major items of equipment needed are a diamond impregnated saw and one or more grinding laps. There are many different saws and laps available which vary considerably in price; the only feasible approach for a beginner who is setting up a laboratory is to visit several in which such equipment is in use and acquire some understanding of what seems most suitable. General purpose saws and laps are adequate for much paleobotanical work; however, if a great deal of thin section preparation is to be done special equipment is available. The preparation of a ground thin section is essentially as follows:

1. The specimen is cut to a convenient size; this will, of course, vary with the botanical nature of the material. For example, specimens of petrified coniferous or dicot wood about 1 cm square are usually adequate.

2. The surface to be studied is smoothed with #400 or #600 Carborundum on a grinding lap or a sheet of plate glass.

3. The smoothed surface is then affixed to a glass slide. This is most effectively accomplished on a hot plate by warming both specimen and slide, applying melted resin to a slide, then placing the specimen, smooth side down in the resin, pressing and rotating gently to eliminate any air bubbles; then allow the mounted specimen to cool until the resin hardens. This will take only a few minutes.

Synthetic resins prepared especially for the purpose should be used as they will harden to the proper viscosity. Paleobotanists often require larger slides than the small ones used in petrographic work; these can be cut, at very little expense, from ordinary double weight window glass and are safe and satisfactory if the edges are ground slightly. If one is doing much thin sectioning, it is well to have a variety of sizes on hand.

4. Fasten the specimen (not the slide) to the jaws of the saw and cut off as close to the glass as possible.

5. The slice of the petrifaction thus affixed to the slide is then ground on a revolving lap. It is usually safe to use #100 Carborundum until light begins to appear through the specimen; then

wash thoroughly and transfer to another lap or a piece of plate glass, and finish with #400 Carborundum or finer. This procedure is an art and requires some practice before one is able to prepare a uniformly thin section. A rather simple grinding instrument has been described by Croft (1950) which facilitates the preparation of uniformly thin sections. Petrifactions differ tremendously in hardness and correspondingly in the time required for grinding; in the last stages, with the fine abrasive, the slide should be washed and examined frequently until the desired thinness is reached.

6. When sufficiently thin the slide should be washed, the excess resin cut away, and the remainder washed off with a soft cloth and the resin solvent. A cover slip may then be mounted over the section using a liquid synthetic resin mounting medium.

There are several disadvantages to this method, which may be partially overcome with certain minor innovations. The finished section is usually very fragile, but can be much less so by the addition of this step: following (2) the smoothed surface may be etched very lightly and a film of peel solution poured on and allowed to dry (see Peel Method for details). The method continues without change, the only difference being that there is now a very thin cellulose film between the specimen and the slide. Two advantages accrue: the specimen does not tend to chip away as readily in the late stages of grinding, and the film serves as a backing for the section so that it can be removed from the slide if necessary by soaking in a dish of the resin solvent.

With extremely soft or porous petrifactions it may be helpful to allow the specimen to stand for a few minutes in melted resin before affixing to the slide.

If material is scarce or for other reasons it is desirable to cut sections closer together and thinner, this may be accomplished with a fine copper wire; the wire is drawn back and forth over the specimen using a sludge of Carborundum to do the cutting. The chief disadvantages of the ground section technique is that it is wasteful of material and at best the resultant sections are several millimeters apart.

The Peel Method

This is applicable to well-preserved petrifactions, that is, ones in which a considerable amount of organic material is still present. It has several advantages over the ground section method: it is rapid; essentially serial sections may be prepared; the size of sections is limited only by the cutting and grinding equipment, and the resultant sections ("peels") are durable and easily stored.

A. *The (original) liquid technique*

1. The surface to be sectioned is smoothed with #400 Carborundum. Prior cutting and rough grinding are as given above.

2. The smooth surface is etched in dilute acid. In most of the petrifactions one encounters, the minerals are silica or corbonates. With carbonates HCl is used, while HFl is employed with silica. *Great care must be taken with the use of hydrofluoric acid* and the beginning student should consult a person who is experienced in its use.

A convenient procedure, with carbonate petrifactions, is to sprinkle a few small pieces of siliceous gravel (⅛ to ¼ inch in diameter) over the bottom of a shallow glass tray and flood with HCl; the specimen is then placed with the smooth side down in the acid. A minimum quantity of acid should be used so that the sides of the specimen are attacked as little as possible; the gravel simply prevents the surface to be etched from contacting the bottom of the tray.

In the author's laboratory we maintain a constant solution of acid (about 2 or 3%) and regulate the etching by time, which will vary from one to five minutes. The etching dissolves away the mineral matter leaving the cell walls standing in relief; two or three trial peels should be made to determine the best thickness of the finished section.

3. The specimen is washed in a gentle stream of water and allowed to dry. The etched surface is very fragile and should not be touched.

4. Place the specimen in a tray of sand or gravel so that the etched surface is level. A small spirit level with a wire loop attached as a handle is satisfactory and, if allowed to contact the surface gently, will cause no damage. If many peels are to be made from one surface of a specimen it may save a little time to embed it in plaster so that the desired surface is level when the specimen is placed on a flat surface, thus eliminating leveling each time.

5. The peel solution is then poured over the dry, etched surface. A considerable variety of formulas have been used, the one given by Darrah being highly satisfactory:

Parlodion, 28 grams	xylol, 10 cc
butyl acetate, 250 cc	castor oil, 3 cc
amyl alcohol, 30 cc	ether, 3 cc

The solution must cover the surface evenly and a little practice is required to obtain a uniform peel. It may be desirable to add a few drops of butyl acetate first to prevent entrapping small air bubbles; if the resultant peel is too thick, less solution should be used and gently spread with a small piece of paper or the solution may be diluted with butyl acetate. It is then allowed to dry for 12 to 20 hours in a reasonably dust free place.

6. The peel is removed by starting an edge with a scalpel or razor and then carefully pulling it off. The result is an actual *thin section* of whatever plant materials were exposed and not merely an impression.

The peels may be studied directly with reflected light or mounted under a cover slip in the usual way if higher magnifications are needed. When mounting in balsam or a synthetic resin the rough side should be moistened with a drop or two of solvent to prevent entrapping air bubbles.

7. Before the next peel is made the surface should be smoothed very lightly with the fine abrasive; otherwise the specimen will etch differentially, ultimately forming a rough surface.

Each worker will devise his own modifications of the technique. One innovation that has frequently proven useful may be worth noting: it is possible to prepare sections in two planes, such as the transverse and longitudinal faces of a stem, and mount them as a single unit for study. The two surfaces are smoothed and etched as usual; one surface is then "poured" and allowed to set; the block is then turned and the second surface treated; after hardening, the peels from the two surfaces may be removed carefully and will remain attached at the adjoining edge.

B. *The cellulose sheet technique*

Recently the method as outlined above has been modified using cellulose acetate sheets instead of a solution, as follows:

1. Steps 1 through 4 as above.

2. The etched surface is flooded with acetone and a sheet of cellulose acetate 0.003 inch thick is quickly and gently rolled down onto the surface. If the film is bowed slightly, and started at one edge, air bubbles will be eliminated.

3. The film may be removed in about ½ hour.

This method is highly satisfactory and has two distinct advantages over the use of a solution: absolute uniformity of the resultant peels is assured and a dozen or more may be made in a day.

The peel method has largely supplanted the older grinding technique although the latter is by no means obsolete. It often produces better thin sections with certain materials, particularly where the preservation is of poorer quality. A paleobotanist working with petrified plants must become adept at the ground thin section method.

In his *Methods of Palaeobotany* Lacey has recorded a dramatic instance of the advantage of the peel thin section method, although one that we may hope will never be repeated. In 1940 certain slides of the articulate cone *Calamostachys,* preserved in the British Museum of Natural History, were partially destroyed by a bomb blast. Upon searching through the debris W. N. Croft was able to retrieve seven of the original eleven sections and, although the slides and cover glasses were shattered, the peel sections, which remained intact, were remounted and were essentially as good as ever.

Microtome Techniques for Sectioning Petrifactions

To the best of my knowledge very little has been done with this approach to handling petrifactions; it is mentioned as a technique that is possibly worth more attention than it has received. It is applicable with fossil woods in which much of the cell wall structure is intact:

1. Prepare cubes about 1 cm square.
2. These are allowed to stand in hydrofluoric acid (the usual commercial 52% should be diluted with equal parts of water for safety in handling) for a few days. The specimen can be tested with a needle to determine when the mineral has been largely dissolved away.
3. Wash in running water overnight.
4. Embed in celloidin using the usual technique for modern plant materials. Standard technique books such as Johansen, 1940, or Sass, 1940, should be consulted for details of this embedding procedure.

Treatment of Opaque Petrifactions

Some petrified plant materials are opaque, notably where sulfides of iron are the mineral agent, and the grinding or peeling methods are of little use. A good deal of information may be extracted from such fossils by studying polished surfaces under reflected light. Examples of what may be obtained with the use of modern techniques of this sort are found in Leclercq's study of *Rhacophyton* and Beck's investigation of *Tetraxylopteris* (see Chapter 3). In both cases these were compression fossils but fragmentary portions were found here and there with vascular tissue preserved. In the former study a plastic embedding medium was used and in the latter a resin; these are outlined briefly:

Balsam coating method (Beck, 1955)

A thin slice of the petrifaction is placed in a solution of three parts toluene and one part Canada balsam or a synthetic resin and boiled gently on a hot plate until most of the toluene has evaporated. The boiling should be carried far enough so that the resin, when cooled, will be firm but not excessively brittle. The section is then removed, excess resin scraped off, and the two surfaces ground with #600 Carborundum followed by chromium oxide; grinding should be kept to a minimum since there is little actual penetration of the petrifaction. The section is then washed, dipped in xylol or toluene, and mounted under a cover slip; both surfaces may then be examined under reflected light.

Plastic embedding method (Leclercq and Discry, 1950; Leclercq and Noël, 1953)

Since the use of plastic as an embedding medium is now widely applied in biology, it does not seem necessary to give a detailed description of this method.

As carried out in Professor Leclercq's laboratory the technique involves the following steps:

Small porcelain trays (as used in paraffin embedding) or boxes lined with aluminum foil may be used as a container. A small portion of the plastic is poured in and allowed to set; the specimen is then placed in the container and covered with the plastic which is allowed to harden overnight. It is then cut according to the surface desired and polished with a fine abrasive.

COMPRESSIONS

Well-preserved compression fossils may reveal an astonishing amount of information, and specimens of a type that were once examined only superficially or even discarded can now be made to yield critical data. The fragment of a Carboniferous fertile frond (*Senftenbergia*), shown in Figs. 1-2A–C, is representative. The study of compression fossils can for the most part be conducted in three rather different ways, depending on the nature of the plant materials: by direct *excavation, maceration,* or *transfer.*

Excavation Technique

When the plant remains ramify rather freely through the rock matrix and consequently are not confined to a narrow bedding plane, there is no method presently available for removing the fossil intact from the matrix (as in the transfer methods described below). A great deal may be learned, however, by judicious excavation with mircrovibratory machines. If such equipment is not available or if one is working with minute structures as is often the case, a small hammer and steel needles may be employed. The specimen may be held on a small sand bag and the excavation carried out under a binocular dissecting microscope. This is slow, tedious work but may be repaid with highly productive results that can be obtained in no other way; the structure of the minute leaves and sporangiophores of *Calamophyton* (see Chapter 10) were worked out using this method.

Transfer Techniques

When the compression is preserved in an essentially flat bedding plane and is too delicate (or lacking a tough cuticle) to allow removal by maceration (see below) one of the transfer methods may be used.

The Ashby cellulose film transfer method (Lang, 1926)

1. The plant compression and immediately adjoining rock surface is coated with the peel solution described above; two coats are usually necessary to produce a tough film.

2. When the film is dry, the rock matrix is cut away from the back side as much as possible without damaging the compression.

3. The specimen is then placed in a suitable container and covered with dilute hydrofluoric acid (about 25%). The acid will attack the rock matrix but will not affect the cellulose film or plant compression.

The time required to completely dissolve away the rock matrix will vary with the size of the specimen and the nature of the minerals. The dissolving action may be speeded by changing the acid each day and by carefully scraping away the disintegrating rock. The last of it may be removed under water with a small, soft artist's brush.

This procedure is, as noted, particularly helpful with fertile frond fragments. The sporangia are made more accessible for study and small portions of the transferred compression may be treated in a small glass dish with a few spoonfuls of nitric acid containing about 5% potassium chlorate in solution. This is a powerful oxidizing agent and will soon disintegrate all except epidermal cuticles and spores. It is necessary to watch the maceration process and stop it at the desired point by diluting the solution with water.

The Balsam transfer method (Walton, 1923)

1. The compression specimen, or a representative portion of it, is trimmed to a convenient size and as much as possible of the rock matrix in back is cut away.

2. The specimen is then pressed, compression side down, into a mass of melted balsam on a glass slide.

3. The entire preparation is then heavily coated with wax by dipping it into melted paraffin; when hard, the wax is scraped away from the back of the rock (the side opposite the compression).

4. The preparation is then immersed in a bath of hydrofluoric acid. The paraffin serves to protect the glass slide and allow the acid to act only on the rock. The process may be aided and the final remnants of rock cleaned away as indicated above.

Of these two transfer techniques the balsam one is perhaps more satisfactory with highly fragile material.

The Abbott transfer method

A modification of the film transfer has been developed by Abbott (1950, 1951) which allows the removal of large compressions. The compression is given a rather heavy coat of fingernail polish followed by sheet acetate or a nail polish-varnish mixture. The film is then pulled off bringing the compression with it. Compressions up to 4 by 6 inches may be removed in this way. For research work such large preparations are usually not necessary, but the method produces excellent display specimens and is especially valuable for preserving a compression that is found in a rock that is not durable.

Bulk macerations

When compression fragments are abundant in a shale or are tough enough so that they may be removed without the use of a transfer medium such as cellulose or balsam, a bulk maceration may be employed.

Harris' (1926) method

1. The shale specimen is immersed for several days in a solution of strong nitric acid containing about 5% potassium hydroxide. It is suggested that the beginner start with small specimens of perhaps 1 or 2 square inches and only as much of this rather violent oxidizing agent as is needed to cover it.

2. The acid is then removed with numerous changes of water.

3. The specimen is next placed in dilute sodium hydroxide; the matrix breaks down into a fine mud and the plant materials, which are now reduced for the most part to cuticularized parts, may be screened off.

4. The cuticular fragments are finally cleaned with 25% hydrofluoric acid.

Hydrofluoric acid method

Although it is perhaps a matter of personal preference it seems to me that it is sometimes advantageous to remove the plant materials intact before subjecting them to the oxidation process.

1. The fossiliferous rock sample is placed in about 25% hydrofluoric acid; several hours to a day or two may be required to break down the mineral matrix.

2. The sludge containing the plant fragments should be thoroughly washed with water to remove the acid.

3. The larger plant fragments may then be removed and the smaller ones separated by screening. They may then be treated individually with nitric acid and potassium chlorate followed by sodium hydroxide as outlined above. The time required for this treatment varies from an hour or less to several days.

REFERENCES

Abbott, Maxine L. 1950. A paleobotanical transfer method. *Journ. Paleont.*, 24: 619–621.

Abbott, Ralph E., and M. L. Abbott. 1952. A simple paleobotanical transfer technique. *Ohio. Journ. Sci.*, 52: 258–260.

Beck, Charles B. 1955. A technique for obtaining polished surfaces of sections of pyritized plant fossils. *Bull. Torrey Bot. Club.,* 82: 286–291.

Croft, W. N. 1950. A parallel grinding instrument for the investigation of fossils by serial sections. *Journ. Paleont.,* 24: 693–698.

Darrah, William C. 1936. The peel method in paleobotany. *Harvard Univ. Bot. Mus. Leaflets,* 4: 69–83.

Graham, Roy. 1933. Preparation of paleobotanical sections by the peel method. *Stain Technology,* 8: 65–68.

Harris, Thomas M. 1926. Notes on a new method for the investigation of fossil plants. *New Phyt.,* 25: 58–60.

Johansen, Donald A. 1940. *Plant Microtechnique.* McGraw-Hill Book Co., New York. 523 pp.

Joy, K. W., A. J. Willis, and W. S. Lacey. 1956. A rapid cellulose peel technique in palaeobotany. *Ann. Bot.,* n. s., 20: 635–637.

Lacey, William S. 1953. Methods in palaeobotany. *The North Western Naturalist,* Arbroath, Wales, 24: 234–249.

Lang, W. H. 1926. A cellulose-film transfer method in the study of fossil plants. *Ann. Bot.,* 40: 710–711.

Leclercq, Suzanne, and M. Discry. 1950. De l'utilization du plastique en paléontologie végétale. *Ann. Soc. Geol. Belgique,* 73: 151–155.

—————, and R. Noël. 1953. Plastic—a suitable embedding substance for petrographic study of coal and fossil plants. *Phytomorphology,* 3: 222–223.

Sass, John E. 1940. *Elements of Botanical Microtechnique.* McGraw-Hill Book Co., New York. 222 pp.

Selling, Olof H. 1944. Studies on Calamitean cone compressions by means of serial sections. *Svenska Bot. Tidskrift,* 38: 295–330.

Walton, John. 1923. On a new method of investigating fossil plant impressions or incrustations. *Ann. Bot.,* 37: 379–390.

—————. 1928. A method of preparing sections of fossil plants contained in coal balls or in other types of petrifaction. *Nature,* 122: 571.

—————. 1930. Improvements in the peel method. Nature 125: 413.

—————. 1952. Notes on the preparation and permanence of peel sections. *Compte Rendu, 3ieme Congr. Strat. Geol. Carbonifere, Heerlen,* 1951. Pp. 651–653.

Some Basic References

The problem of "keeping up with the literature" is a potential source of despair in almost any major branch of science. As to paleobotany it is hardly possible for an individual to be thoroughly familiar with all that is being done in the various branches, but an acquaintance with the standard reference works and literature reports allows one to keep up with what seem to be the more important works within his own area of interest.

In the foregoing chapters some of the basic as well as more recent references are listed and the bibliographies given in them will in turn lead the reader to a more detailed survey. Certain other useful works are cited below.

Standard Textbooks

Arnold, Chester A. 1947. *An Introduction to Paleobotany.* McGraw-Hill Book Co., New York. 433 pp.

Darrah, William C. 1960. *Principles of Paleobotany.* 2nd ed. Ronald Press Co., New York. 256 pp.

Emberger, Louis. 1954. *Les plantes fossiles dan leurs rapports avec les végétaux vivants.* Masson et Cie., Paris. 492 pp.

Hirmer, Max. 1927. *Handbuch der Palaobotanik.* R. Oldenbourg, Munich. 708 pp. This is an encyclopaedic treatment of the cryptogamic groups, copiously illustrated and with extensive bibliographies. It is volume I of what was apparently intended as a two-volume work.

Jongmans, Willem J. 1949. *Het Wisselend Aspect van het Bos in de Oudere Geol- ogische Formaties.* Issued as a part of "Hout in alle tijden," edited by W. Boerhave Beekman. Deventer, Netherlands. 164 pp. This is essentially a compilation of individual plant as well as general landscape restorations and it stands alone in this way as a valuable and remarkble volume.

Kräusel, Richard. 1950. *Versunkene Floren.* Waldemar Kramer, Frankfurt a. M. 152 pp.

Krystofovich, A. N. 1957. *Paleobotany.* Leningrad. 650 pp.

Mägdefrau, Karl. 1956. *Paläobiologie der Pflanzen.* 3rd ed. Gustav Fischer. Jena. 443 pp.

Seward, Albert C. *Fossil Plants.* 4 vols. Cambridge Univ. press. In addition to making many original contributions Sir Albert Seward was the greatest compiler that paleobotany has had. Although this work is out of date in many aspects, it contains a wealth of information and is still indispensable. The four volumes ap- peared as follows: I. 1898; II. 1910; III. 1917; IV. 1919.

Scott, Dukinfield H. *Studies in Fossil Botany.* vol. I. 1920; vol. II. 1923.

Walton, John. 1953. *An Introduction of the Study of Fossil Plants.* 2nd ed. A & C, Black Ltd., London. 201 pp.

Journals Devoted Exclusively to Paleobotany

The Palaeobotanist. Birbal Sahni Institute of Palaeobotany, Lucknow, India.
Palaeontographica (Abt. B). Stuttgart, Germany.

Catalogues and Literature Reports

Andrews, Henry N., Jr. 1955. Index of generic names of fossil plants, 1820–1950. *U. S. Geol. Survey Bull.,* 1013. 262 pp. This is taken from the United States Geological Survey's Compendium Index of Paleobotany, a card file that maintains a citation of the original source of publication of all species of fossil plants except the diatoms, spores, and pollen.

Knowlton, Frank H. 1919. A catalogue of the Mesozoic and Cenozoic plants of North America. *U. S. Geol. Survey Bull.,* 696. 815 pp.

Kremp, G. O. W., et al. 1957–. *Catalogue of Fossil Spores and Pollen.* Palynologi- cal laboratories, The Pennsylvania State University. Several volumes have appeared each year, starting in 1957.

LaMotte, Robert S. 1944. Supplement to Catalogue of Mesozoic and Cenozoic plants of North America 1919-1937. *U. S. Geol. Survey Bull.,* 924, 330 pp.

————. 1952. Catalogue of the Cenozoic plants of North America through 1950. *Geol. Soc. Amer. Mem.,* 51, 381 pp.

World Report on Palaeobotany. Edited by E. Boureau and collaborators. Prior to 1956 several regional literature reports were issued more or less regularly. These included Reports for Europe, Britain, India, and the United States. In 1956 the first World Report appeared as an official organ of the International Paleobotanical organization. The second number appeared in 1958.

Studies in the History of Palaeobotany

Edwards, Wilfred N. 1931. Guide to an exhibition illustrating the early history of paleontology. *Brit. Mus. Nat. Hist.,* 68 pp.

Gordon, William T. 1934. Plant life and philosophy of geology. *Pan-Amer. Geologist,* 62: 161–186, 329–346.

Ward, Lester F. 1884. Sketch of paleobotany. *U. S. Geol. Survey. Fifth Ann. Rept.,* pp. 363–452. Washington.

INDEX